Noteables™

Interactive Study Notebook with FOLDABLES™

Pre-Algebra

Contributing Author
Dinah Zike

FOLDABLES™

Consultant
Douglas Fisher, PhD
Director of Professional Development
San Diego State University
San Diego, CA

New York, New York Columbus, Ohio Chicago, Illinois Peoria, Illinois Woodland Hills, California

The McGraw-Hill Companies

Send all inquiries to:
The McGraw-Hill Companies
8787 Orion Place
Columbus, OH 43240-4027

ISBN: 0-07-868217-7

Pre-Algebra (Student Edition)
Noteables™: Interactive Study Notebook with Foldables™

6 7 8 9 10 047 09 08 07 06

Contents

Contents

Organizing Your Foldables

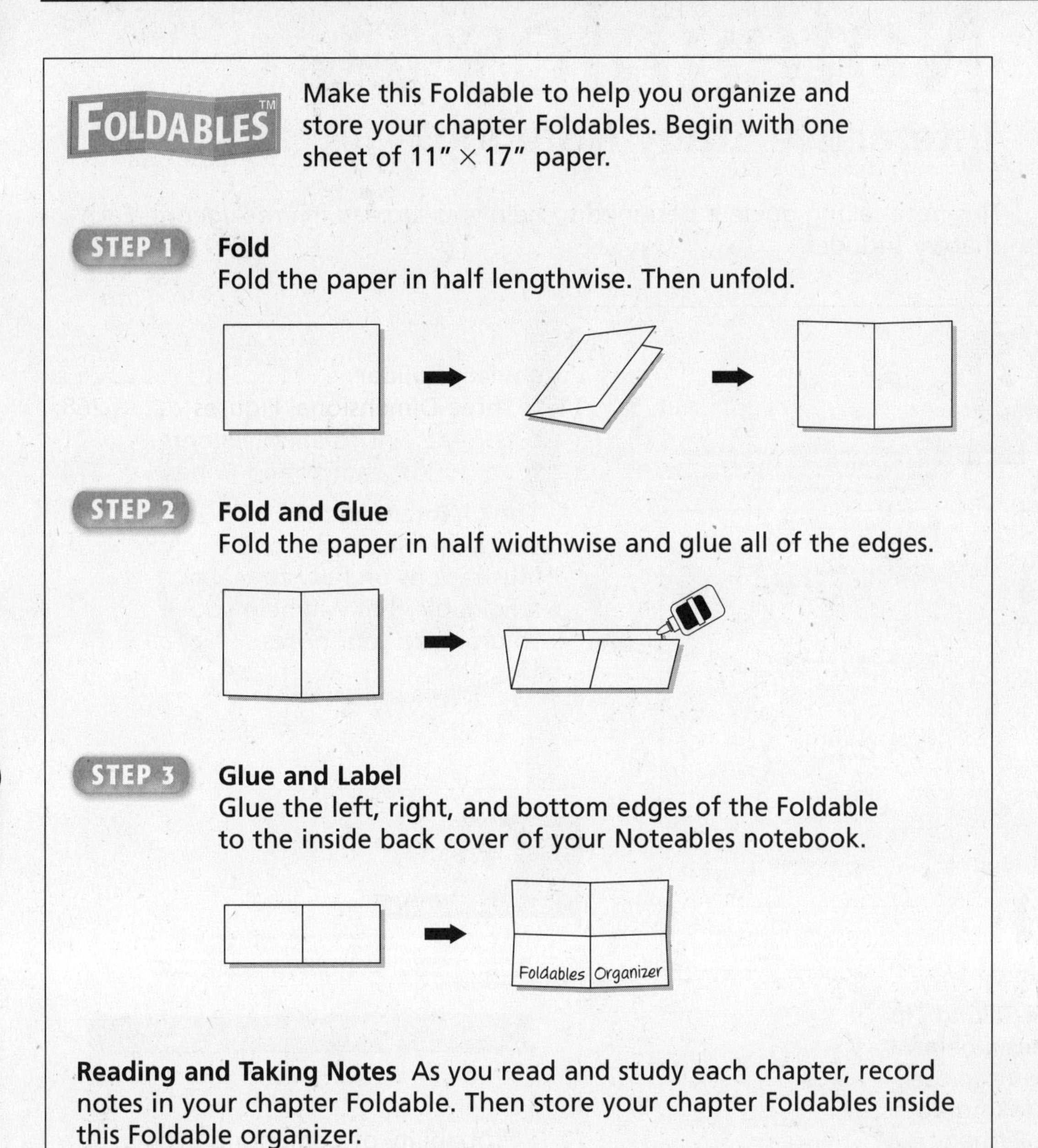

Make this Foldable to help you organize and store your chapter Foldables. Begin with one sheet of 11″ × 17″ paper.

STEP 1 **Fold**
Fold the paper in half lengthwise. Then unfold.

STEP 2 **Fold and Glue**
Fold the paper in half widthwise and glue all of the edges.

STEP 3 **Glue and Label**
Glue the left, right, and bottom edges of the Foldable to the inside back cover of your Noteables notebook.

Reading and Taking Notes As you read and study each chapter, record notes in your chapter Foldable. Then store your chapter Foldables inside this Foldable organizer.

Using Your Noteables™ with Foldables Interactive Study Notebook

This note-taking guide is designed to help you succeed in *Pre-Algebra*. Each chapter includes:

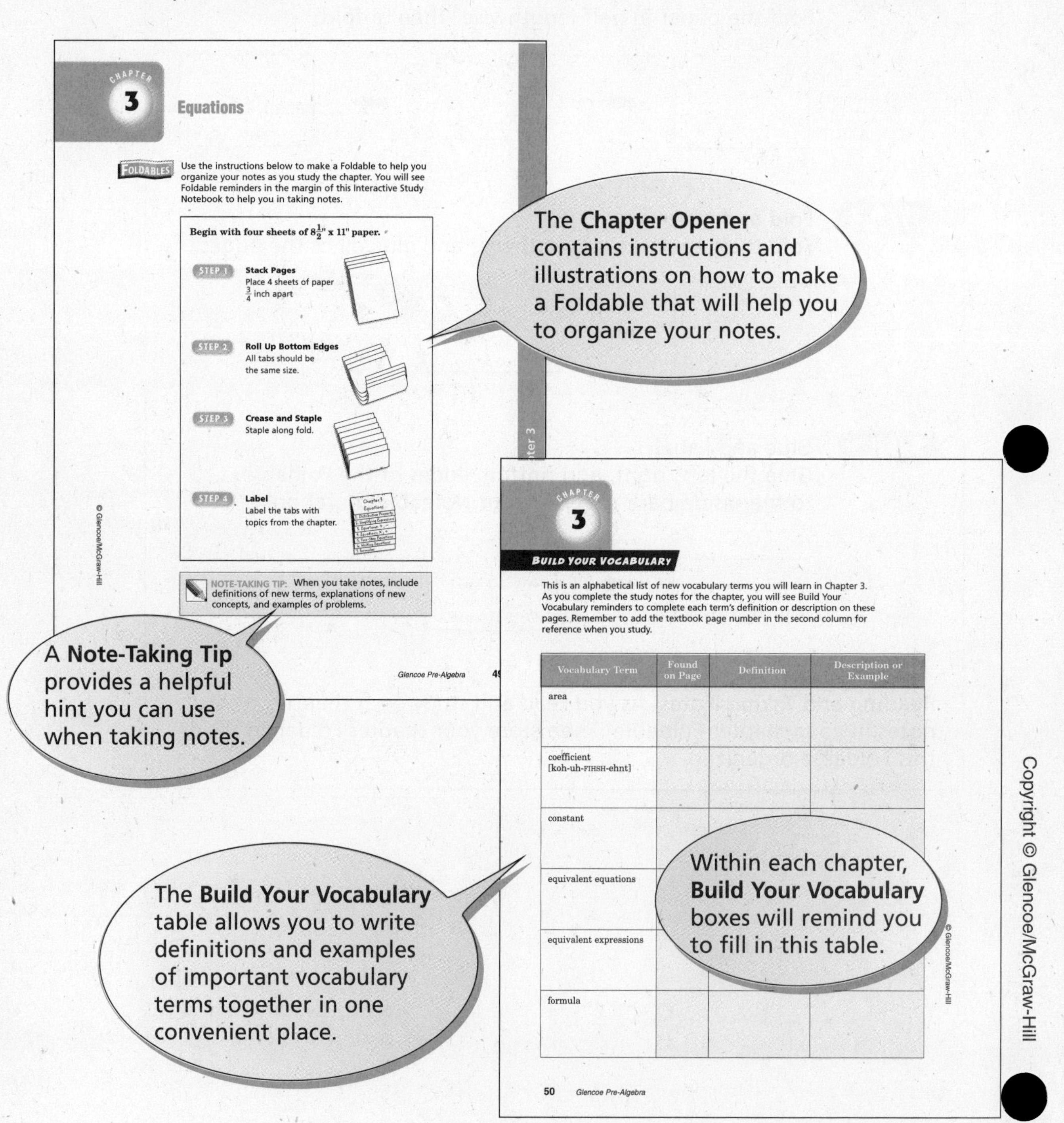

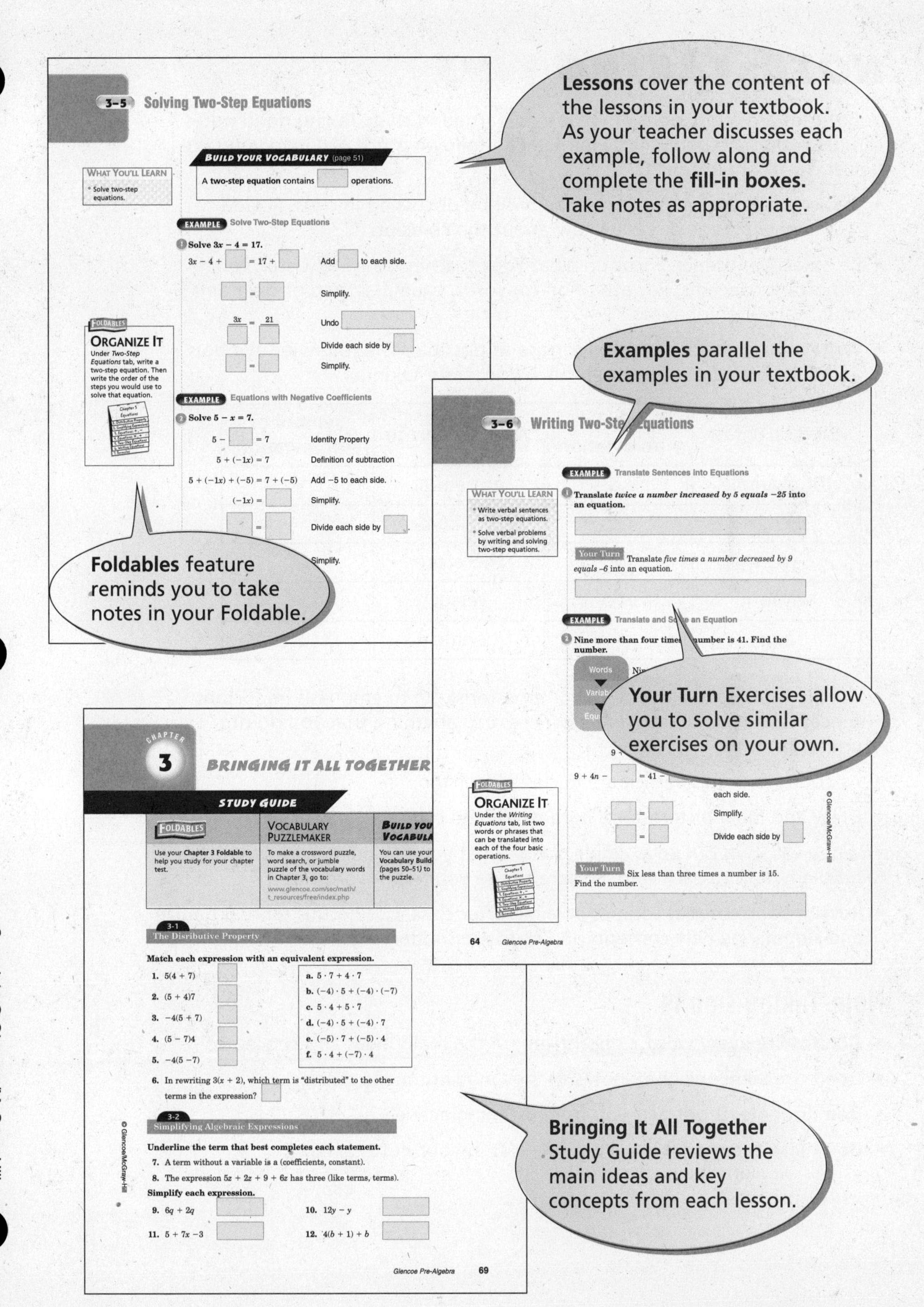

3-5 Solving Two-Step Equations
Lessons cover the content of the lessons in your textbook. As your teacher discusses each example, follow along and complete the fill-in boxes. Take notes as appropriate.
Examples parallel the examples in your textbook.
3-6 Writing Two-Step Equations
Foldables feature reminds you to take notes in your Foldable.
Your Turn Exercises allow you to solve similar exercises on your own.
CHAPTER 3 BRINGING IT ALL TOGETHER
STUDY GUIDE
Bringing It All Together Study Guide reviews the main ideas and key concepts from each lesson.

NOTE-TAKING TIPS

Your notes are a reminder of what you learned in class. Taking good notes can help you succeed in mathematics. The following tips will help you take better classroom notes.

- Before class, ask what your teacher will be discussing in class. Review mentally what you already know about the concept.
- Be an active listener. Focus on what your teacher is saying. Listen for important concepts. Pay attention to words, examples, and/or diagrams your teacher emphasizes.
- Write your notes as clear and concise as possible. The following symbols and abbreviations may be helpful in your note-taking.

Word or Phrase	Symbol or Abbreviation	Word or Phrase	Symbol or Abbreviation
for example	e.g.	not equal	$\neq$
such as	i.e.	approximately	$\approx$
with	w/	therefore	$\therefore$
without	w/o	versus	vs
and	+	angle	$\angle$

- Use a symbol such as a star (★) or an asterisk (*) to emphasis important concepts. Place a question mark (?) next to anything that you do not understand.
- Ask questions and participate in class discussion.
- Draw and label pictures or diagrams to help clarify a concept.
- When working out an example, write what you are doing to solve the problem next to each step. Be sure to use your own words.
- Review your notes as soon as possible after class. During this time, organize and summarize new concepts and clarify misunderstandings.

Note-Taking Don'ts

- **Don't** write every word. Concentrate on the main ideas and concepts.
- **Don't** use someone else's notes as they may not make sense.
- **Don't** doodle. It distracts you from listening actively.
- **Don't** lose focus or you will become lost in your note-taking.

The Tools of Algebra

Use the instructions below to make a Foldable to help you organize your notes as you study the chapter. You will see Foldable reminders in the margin of this Interactive Study Notebook to help you in taking notes.

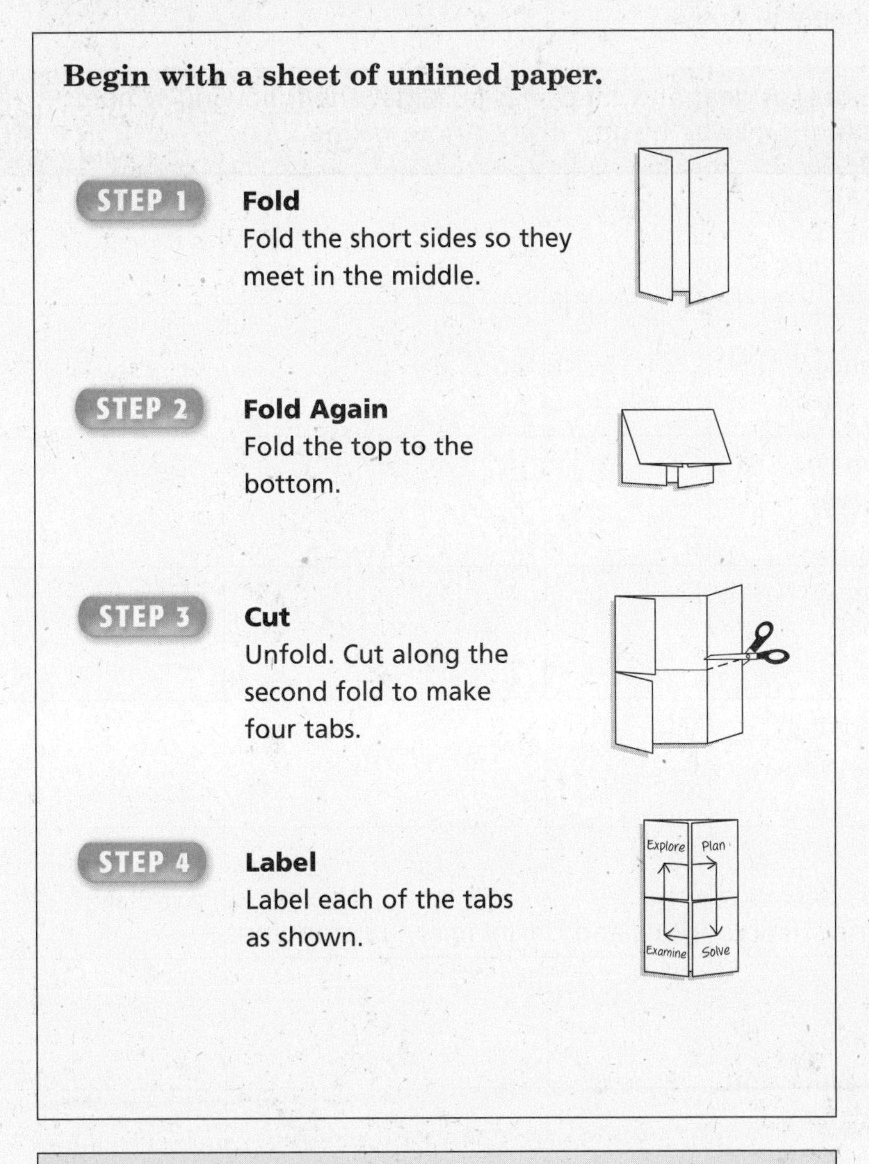

Begin with a sheet of unlined paper.

STEP 1 **Fold**
Fold the short sides so they meet in the middle.

STEP 2 **Fold Again**
Fold the top to the bottom.

STEP 3 **Cut**
Unfold. Cut along the second fold to make four tabs.

STEP 4 **Label**
Label each of the tabs as shown.

NOTE-TAKING TIP: When you take notes, be sure to describe steps in detail. Include examples of questions you might ask yourself during problem solving.

CHAPTER 1

Build Your Vocabulary

This is an alphabetical list of new vocabulary terms you will learn in Chapter 1. As you complete the study notes for the chapter, you will see Build Your Vocabulary reminders to complete each term's definition or description on these pages. Remember to add the textbook page numbering in the second column for reference when you study.

Vocabulary Term	Found on Page	Definition	Description or Example
algebraic expression [al-juh-BRAY-ik]			
conjecture [cuhn-JEHK-shoor]			
coordinate plane or coordinate system			
counterexample			
deductive reasoning			
domain			
equation			
evaluate			
inductive reasoning [in DUHK-tihv]			

Vocabulary Term	Found on Page	Definition	Description or Example
numerical expression			
open sentence			
order of operations			
ordered pair			
properties			
range			
relation			
scatter plot			
simplify			
solution			
variable			

1–1 Using a Problem-Solving Plan

WHAT YOU'LL LEARN

- Use a four-step plan to solve problems.
- Choose an appropriate method of computation.

EXAMPLE Use the Four-Step Problem-Solving Plan

1 PIZZA **The price of a large cheese pizza at Paul's Pizza is \$9.25. You receive a \$0.50 discount for each additional pizza ordered, up to 10. So, one pizza costs \$9.25, two pizzas cost \$8.75 each, three pizzas cost \$8.25 each and so on. If you need 8 pizzas, what is the cost per pizza?**

EXPLORE The problem gives the cost for the first pizza and the discount for each additional pizza ordered. Find the cost per pizza for 8 pizzas.

PLAN Look for a pattern in the costs. Extend the pattern to find the cost per pizza for 8 pizzas.

SOLVE First, find the pattern.

1 pizza: ______

2 pizzas: ______ – ______ or ______

3 pizzas: ______ – ______ or ______

Now, extend the pattern.

4 pizzas: ______ – ______ or ______

5 pizzas: ______ – ______ or ______

6 pizzas: ______ – ______ or ______

7 pizzas: ______ – ______ or ______

8 pizzas: ______ – ______ or ______

The cost per pizza for 8 pizzas is ______.

EXAMINE It costs \$9.25 for one pizza with a discount of \$0.50 for each additional pizza ordered. For an order of 8 pizzas, the cost per pizza would be

\$9.25 – (7 × ______) or

\$9.25 – ______ = ______.

REMEMBER IT

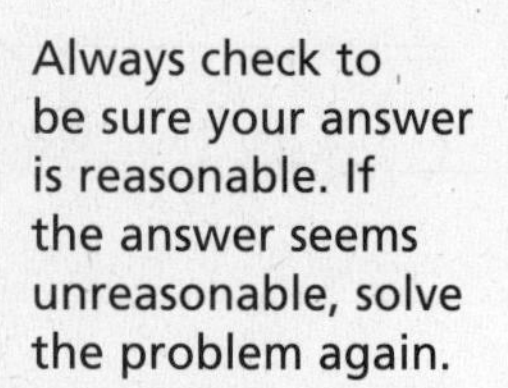

Always check to be sure your answer is reasonable. If the answer seems unreasonable, solve the problem again.

FOLDABLES™

ORGANIZE IT

Write this Your Turn Exercise under the Examine tab of the Foldable. Then under the remaining tabs, record how you will plan, solve, and examine to reach a solution.

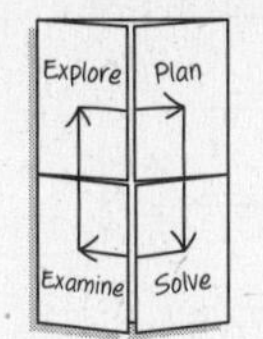

Your Turn The cost of renting movies at Mike's Marvelous Movie House is advertised as $5 for the first movie and $3.50 for each additional movie. Find the cost of renting 6 movies.

BUILD YOUR VOCABULARY (page 2)

A **conjecture** is an  guess.

When you make a conjecture based on a pattern of examples or past events, you are using **inductive reasoning**.

EXAMPLE Use Inductive Reasoning

2 **a. Find the next term in 1, 4, 16, 64, 256, . . .**

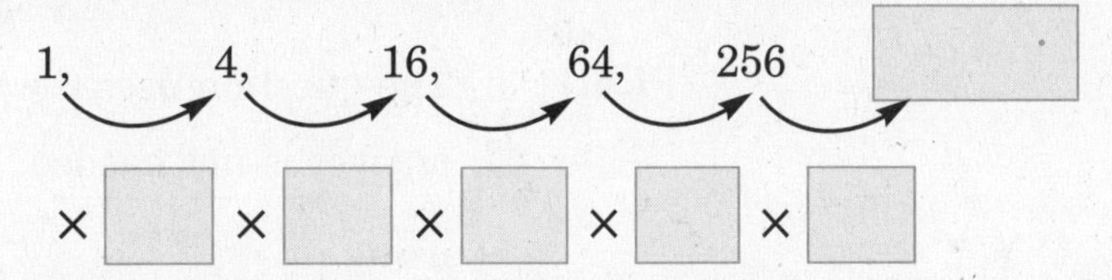

b. Draw the next figure in the pattern.

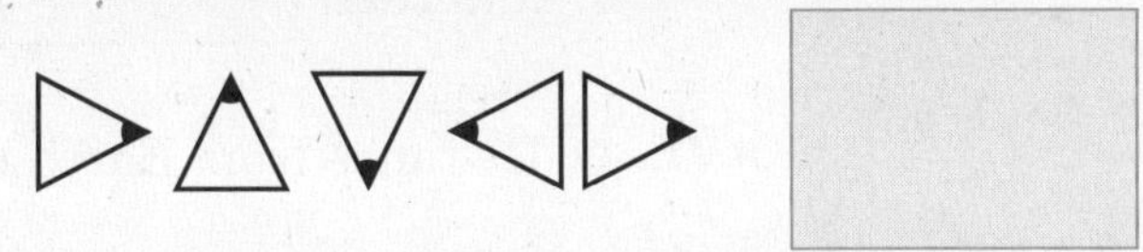

The shaded point on the triangle moves in the pattern: right, top, bottom, left, right, etc. If the pattern continues, the shaded point will be at the ______ of the next figure.

Your Turn

a. Find the next term in 48, 43, 38, 33, 28, . . .

b. Draw the next figure in the pattern.

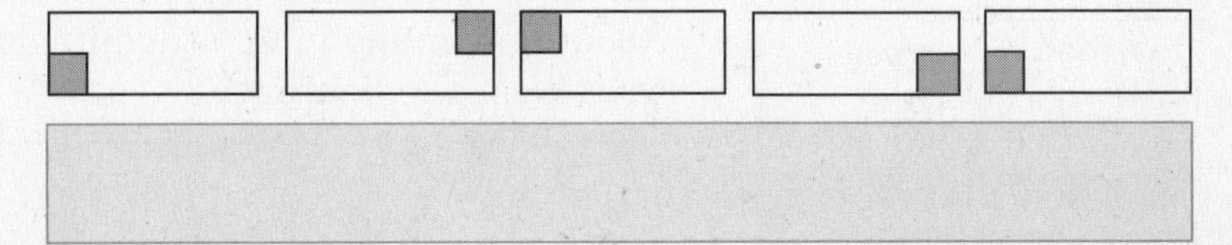

EXAMPLE Choose the Method of Computation

3 PLANETS **The chart shows the distance of selected planets from the Sun. About how much farther is it from Earth to the Sun than from Mercury to the Sun?**

Planet	Distance from Sun (millions of miles)
Mercury	36.00
Venus	67.24
Earth	92.90
Mars	141.71

EXPLORE You know the distance from Earth to the ______ and the distance from ______ to the Sun. You need to find about how much farther it is from ______ to the ______.

PLAN The question uses the word *about*, so an exact answer is not needed. We can solve the problem using ______. Estimate each distance and then ______.

SOLVE Distance from Earth to the Sun: 92.9 → ______

Distance from Mercury to the Sun: 36.0 → ______

So, Earth is about ______ – ______ or ______ million miles farther from the Sun than Mercury is.

EXAMINE Since ______ + ______ = 93, the answer makes sense.

Your Turn East Elementary School has 792 students enrolled. West Elementary School has 518 students enrolled. About how many more students does East Elementary have than West Elementary?

______

HOMEWORK ASSIGNMENT

Page(s):

Exercises:

1–2 Numbers and Expressions

WHAT YOU'LL LEARN

- Use the order of operations to evaluate expressions.
- Translate verbal phrases into numerical expressions.

BUILD YOUR VOCABULARY (pages 2–3)

Numerical expressions contain a combination of numbers and operations such as addition, subtraction, multiplication, and division.

When you **evaluate** an expression, you find its numerical value.

To avoid confusion when evaluating expressions, mathematicians have agreed upon an **order of operations**.

EXAMPLE Evaluate Expressions

1 Find the value of each expression.

a. $24 \div 8 \times 3$

$24 \div 8 \times 3 = \_\_\_ \times 3$ Divide 24 by 8.

$= 9$ Multiply 3 and .

b. $5(4 + 6) - 7 \cdot 7$

$5(4 + 6) - 7 \cdot 7 = 5(\_\_\_) - 7 \cdot 7$ Evaluate (4 + 6).

$= \_\_\_ - 7 \cdot 7$ Multiply 5 and ___.

$= \_\_\_ - \_\_\_$ Multiply 7 and 7.

$= \_\_\_$ Subtract.

c. $3[(18 - 6) + 2(4)]$

$3[(18 - 6) + 2(4)] = 3[\_\_\_ + 2(4)]$ Evaluate (18 − 6).

$= 3(\_\_\_ + \_\_\_)$ Multiply 2 and 4.

$= 3(\_\_\_)$ Add.

$= \_\_\_$ Multiply.

REVIEW IT

Explain how to simplify expressions inside grouping symbols. (*Prerequisite Skill*)

REMEMBER IT 

Grouping symbols include parentheses, brackets, and fraction bars.

d. $\dfrac{(49 + 31)}{(19 - 14)}$

$\dfrac{(49 + 31)}{(19 - 14)}$

$= (49 + 31)$ ______ $(19 - 14)$ Rewrite as a division expression.

$=$ ______ $\div$ ______ Evaluate each expression.

$=$ ______ Divide ______ by ______.

Your Turn **Find the value of each expression.**

a. $63 \div 7 + 2$ 

b. $3(12 - 10) + 14 \div 2$

c. $4[(3 + 8) - 2(4)]$

d. $\dfrac{(21 - 3)}{4(2) + 1}$

EXAMPLE Translate Phrases into Expressions

2 **Write a numerical expression for each verbal phrase.**

a. the quotient of eighteen and six

Phrase the quotient of eighteen and six

Key Word 

Expression

b. the sum of nine and five

Phrase the sum of nine and five

Key Word 

Expression 

Your Turn **Write a numerical expression for each verbal phrase.**

a. the product of three and five

b. the difference of seventeen and six

EXAMPLE Use an Expression to Solve a Problem

3 EARNINGS **Madison earns an allowance of $5 per week. She also earns $4 per hour baby-sitting, and usually baby-sits 6 hours each week. Write and evaluate an expression for the total amount of money she earns in one week.**

First, write an expression.

$5 allowance per week | plus | $4 per hour spent baby-sitting

_____ + _____

Then evaluate the expression.

$5 + 4 \times 6 =$ _____ Multiply.

$=$ _____ Add.

Madison earns _____ in one week.

Your Turn The Good Price Grocery Store advertises a special on 2-liter bottles of soft drinks. The first bottle purchased is $1.50 and each bottle after that is $1.20. Write and evaluate an expression for the total cost when 8 bottles are purchased.

HOMEWORK ASSIGNMENT

Page(s):

Exercises:

1–3 Variables and Expressions

What You'll Learn

- Evaluate expressions containing variables.
- Translate verbal phrases into algebraic expressions.

Build Your Vocabulary (pages 2–3)

A **variable** is a ______ for any ______.

An **algebraic expression** contains sums and/or products of ______ and ______.

EXAMPLE Evaluate Expressions

1 Evaluate $x - y + 6$ if $x = 27$ and $y = 12$.

$x - y + 6 =$ ______ Replace x with ______ and y with ______.

$=$ ______ Subtract ______ from ______.

$=$ ______ Add ______ and ______.

Your Turn Evaluate $12 + a - b$ if $a = 7$ and $b = 11$.

______

EXAMPLE Evaluate Expressions

Key Concept

Substitution Property of Equality If two quantities are equal, then one quantity can be replaced by the other.

2 Evaluate each expression if $x = 3$, $y = 4$, and $z = 7$.

a. $6y - 4x$

$6y - 4x = 6(\ \ ) - 4(\ \ )$ $y =$ ______, $x =$ ______.

$=$ ______ Multiply.

$=$ ______ Subtract.

WRITE IT

What is the name of the property that allows you to replace a variable with a number?

____

____

____

____

b. $\dfrac{(z - x)}{y}$

$\dfrac{(z - x)}{y} = ($ [ ] $) \div$ [ ] — Rewrite as a division expression.

$= ($ [ ] $-$ [ ] $) \div$ [ ] — Replace z with 7, x with 3, and y with 4.

$=$ [ ] $\div 4$ — Subtract.

$=$ [ ] — Divide.

c. $5z + (x + 4y) - 15$

$5z + (x + 4y) - 15$

$= 5$ [ ] $+ ($ [ ] $+ 4 \cdot$ [ ] $) - 15$ — Replace z with 7, x with 3, and y with 4.

$= 5$ [ ] $+ ($ [ ] $+$ [ ] $) - 15$ — Multiply [ ] and [ ].

$= 5$ [ ] $+$ [ ] $- 15$ — Add [ ] and [ ].

$=$ [ ] $+$ [ ] $- 15$ — Multiply [ ] and [ ].

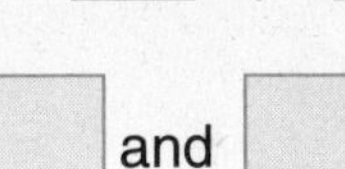

$=$ [ ] $- 15$ — Add [ ] and [ ].

$=$ [ ] — Subtract.

Your Turn **Evaluate each expression if $m = 9$, $n = 4$, and $p = 6$.**

a. $5p - 3m$

b. $\dfrac{mn}{p}$

c. $p + (8n - 3m)$

EXAMPLE Translate Verbal Phrases into Expressions

REVIEW IT

List eight words or phrases that suggest addition or subtraction. (*Prerequisite Skill*)

____

____

____

____

____

____

____

____

3 **Translate each phrase into an algebraic expression.**

a. 35 more than the number of tickets sold

Words	35 more than the number of tickets sold
Variable	Let t represent the number of tickets sold
Expression	35 more than the number of tickets sold

The expression is ________.

b. the difference of six times a number and ten

Words	the difference of six times a number and ten
Variable	Let n represent the number.
Expression	the difference of six times a number and ten

The expression is ________.

Your Turn **Translate each phrase into an algebraic expression.**

a. eight less than the number of cookies baked

b. the sum of twelve and five times a number

EXAMPLE Use an Expression to Solve a Problem

4 THEATER **East Middle School sold tickets for a school play. The price of an adult ticket was \$3 and the price of a student ticket was \$1.**

a. Write an expression that can be used to find the total amount of money collected.

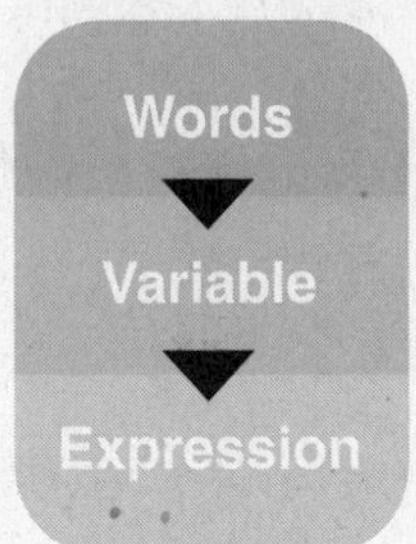

\$3 for an adult ticket and \$1 for a student ticket

Let a = number of adult tickets and s = number of students tickets.

\$3 for an adult ticket and \$1 for a student ticket.

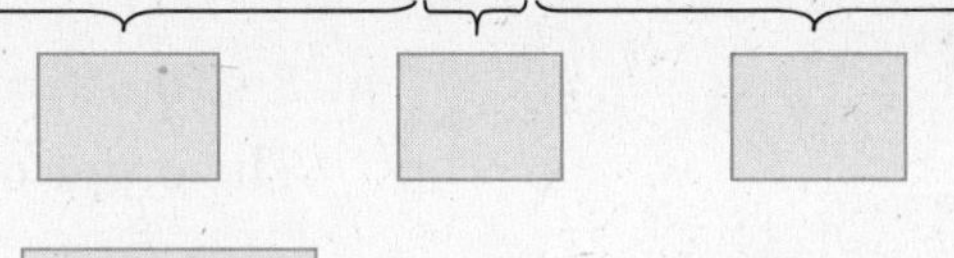

The expression is ________.

b. Suppose 70 adult tickets and 85 student tickets were sold. How much money was collected?

$3a + 1s = 3(\quad) + 1(\quad)$ $a = 70$, $s = 85$.

= ________ Multiply.

= ________ Add.

The amount of money collected was ________.

Your Turn **The Read It Bookstore is advertising a sale. The price of hardback books is \$9.50 and the price of paperback books is \$4.50.**

a. Write an expression that can be used to find the total amount of money spent at the bookstore.

b. Suppose Emily buys 5 hardback books and 4 paperback books. Find the total amount she spent at the book sale.

HOMEWORK ASSIGNMENT

Page(s):

Exercises:

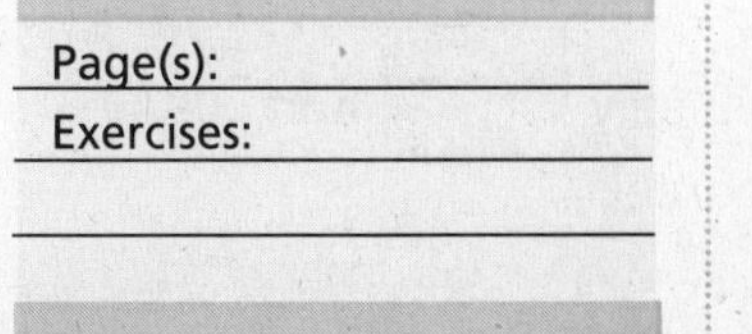

1–4 Properties

What You'll Learn

- Identify and use properties of addition and multiplication.
- Use properties of addition and multiplication to simplify algebraic expressions.

Key Concepts

Commutative Properties of Addition and Multiplication The order in which numbers are added or multiplied does not change the sum or product.

Associative Properties of Addition and Multiplication The way in which numbers are grouped when added or multiplied does not change the sum or product.

Additive Identity When 0 is added to any number, the sum is the number.

Multiplicative Identity When any number is multiplied by 1, the product is the number.

Multiplicative Property of Zero When any number is multiplied by 0, the product is 0.

Build Your Vocabulary (page 3)

In algebra, **properties** are statements that are true for any numbers.

EXAMPLE Identify Properties

1 Name the property shown by each statement.

a. $3 \cdot 10 \cdot 2 = 3 \cdot 2 \cdot 10$

The order of the numbers changed. This is the

____

b. $17 \cdot 1 = 17$

The number was multiplied by one.

____

Your Turn **Name the property shown by each statement.**

a. $(4 \cdot 6) \cdot 2 = 4 \cdot (6 \cdot 2)$ ______________________

b. $3 \cdot 0 = 0$ ______________________

EXAMPLE Mental Math

2 Find $(18 \cdot 20) \cdot 5$ mentally.

$(18 \cdot 20) \cdot 5 = 18 \cdot (\_\_\_\_)$ Associative Property of Multiplication

$= 18 \cdot \_\_\_\_$ Multiply ____ and ____ mentally.

$= \_\_\_\_$ Multiply 18 and ____ mentally.

Your Turn Find $4 \cdot 8 \cdot 25$ mentally.

BUILD YOUR VOCABULARY (page 3)

To **simplify** algebraic expressions means to write them in a ______.

The process of using facts, properties, or rules to ______ or reach valid ______ is called **deductive reasoning**.

EXAMPLE Simplify Algebraic Expressions

4 **a. Simplify $5 \cdot (3 \cdot r)$.**

$5 \cdot (3 \cdot r) =$ ______ Associative Property of Multiplication

$=$ ______ Substitution Property of Equality

b. Simplify $12 + (x + 18)$.

$12 + (x + 18)$

$= 12 +$ ______ Commutative Property of Addition

$=$ ______ $+ x$ Associative Property of Addition

$=$ ______ Substitution Property of Equality

Your Turn **Simplify each expression.**

a. $7 + (12 + m)$

b. $(6 \cdot a) \cdot 4$

HOMEWORK ASSIGNMENT

Page(s):

Exercises:

1–5 Variables and Equations

WHAT YOU'LL LEARN

- Identify and solve open sentences.
- Translate verbal sentences into equations.

BUILD YOUR VOCABULARY (pages 2–3)

A mathematical ________ that contains an ________ is called an **equation**.

An ________ that contains a ________ is an **open sentence**.

A ________ for the variable that makes an ________ true is called a **solution**.

EXAMPLE Solve an Equation

1 Find the solution of $44 + p = 53$. Is it 11, 9, or 7?

Replace p with each value.

Value for p	$44 + p = 53$	True or False?
11	$44 + \square \stackrel{?}{=} 53$	
9	$44 + \square \stackrel{?}{=} 53$	
7	$44 + \square \stackrel{?}{=} 53$	

Your Turn Find the solution of $24 - a = 9$. Is it 11, 13, or 15?

EXAMPLE Solve Simple Equations Mentally

2 a. Solve $7x = 56$ mentally.

$7 \cdot \square = 56$ Think: What number times 7 is 56?

$x = \square$

b. Solve $x - 15 = 40$ mentally.

$\square - 15 = 40$ Think: What number minus 15 is 40?

$x = \square$

Your Turn **Solve each equation mentally.**

a. $\frac{m}{3} = 7$ ______ **b.** $6 + h = 19$ ______

EXAMPLE Identify Properties of Equality

KEY CONCEPTS

Symmetric Property of Equality If one quantity equals a second quantity, then the second quantity also equals the first.

Transitive Property of Equality If one quantity equals a second quantity, and the second quantity equals a third quantity, then the first equals the third.

3 Name the property of equality shown by each statement.

a. If $3x + 1 = 10$, then $10 = 3x + 1$.

If ______, then ______. ______ Property of Equality.

b. If $z + 6 = 8$ and $8 = 2 + 6$, then $z + 6 = 2 + 6$.

If ______ and ______,

then ______. ______ Property of Equality.

Your Turn **Name the property of equality shown by each statement.**

a. If $x - 9 = 8$ and $8 = 17 - 9$, then $x - 9 = 17 - 9$.

______

b. If $12 + 3x = 24$, then $24 = 12 + 3x$.

______

EXAMPLE Sentences Into Equations

4 The quotient of a number and four is nine. Find the number.

Words	The quotient of a number and four is nine.
Variable	Let n = ______. Define the variable.
Expression	The quotient of a number and four is nine.

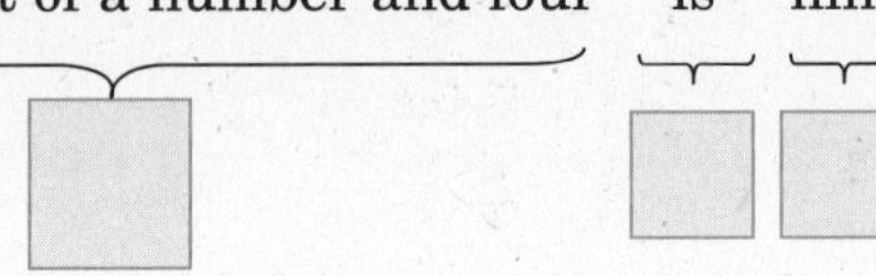

 Write the equation.

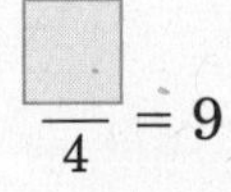 $\frac{\square}{4} = 9$ Think: What number divided by 4 is 9?

$n =$ ______

Your Turn The sum of a number and seven is twelve. Find the number.

HOMEWORK ASSIGNMENT

Page(s):

Exercises:

1–6 Ordered Pairs and Relations

WHAT YOU'LL LEARN

- Use ordered pairs to locate points.
- Use tables and graphs to represent relations.

BUILD YOUR VOCABULARY (pages 2–3)

The **coordinate system** is formed by the intersection of two number lines that meet at right angles at their zero points.

The ______ is also called the **coordinate plane**.

An **ordered pair** of numbers is used to locate any ______ on a coordinate plane.

The ______ number of an ______ is called the ***x*-coordinate**.

EXAMPLE Graph Ordered Pairs

1 **Graph each ordered pair on a coordinate system.**

a. (3, 4)

Step 1 Start at the ______.

Step 2 Since the x-coordinate is 3, move ______ units to the ______.

Step 3 Since the y-coordinate is 4, move ______ units ______. Draw a dot.

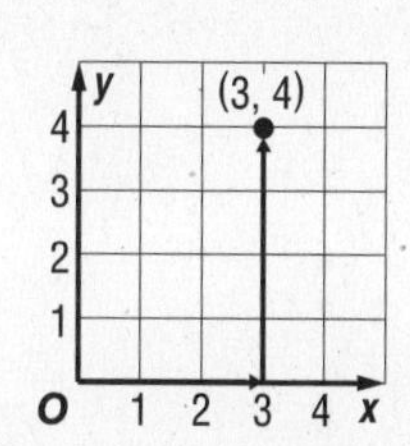

b. (0, 2)

Step 1 Start at the origin.

Step 2 Since the ______ is ______, you will not need to move to the right.

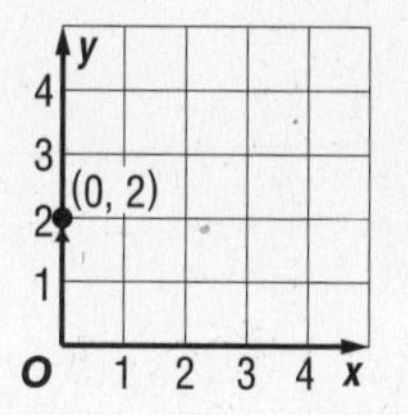

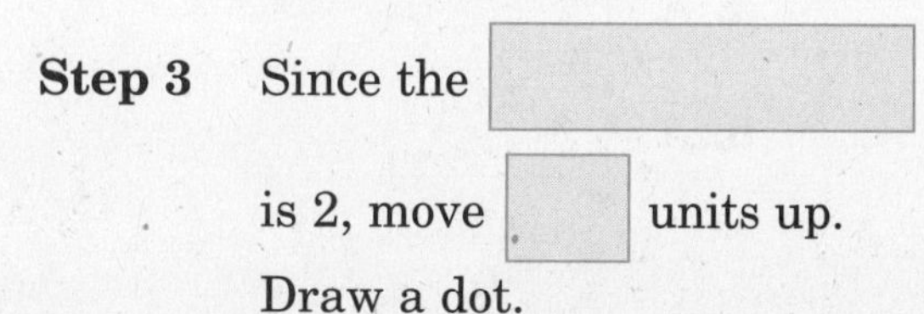

Step 3 Since the ______ is 2, move ______ units up. Draw a dot.

WRITE IT

Where is the graph of (5, 0) located?

____

____

____

____

____

Your Turn **Graph each ordered pair on a coordinate system.**

a. (2, 5)

b. (4, 0)

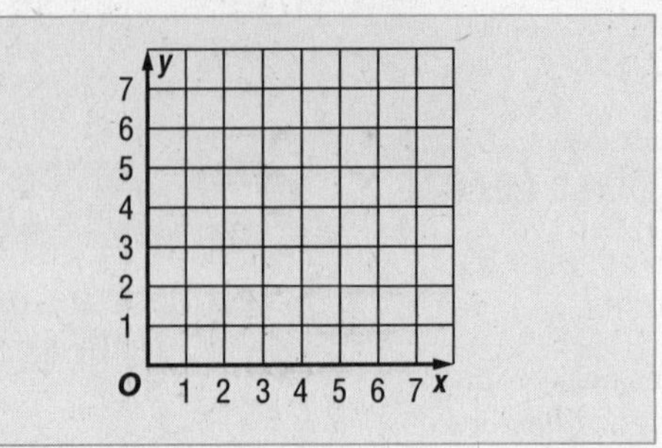

EXAMPLE Identify Ordered Pairs

2 **Write the ordered pair that names each point.**

a. Point *G*

Start at the origin. Move right on the x-axis to find the x-coordinate of point G, which is ____. Move up the y-axis to find the y-coordinate, which is ____.

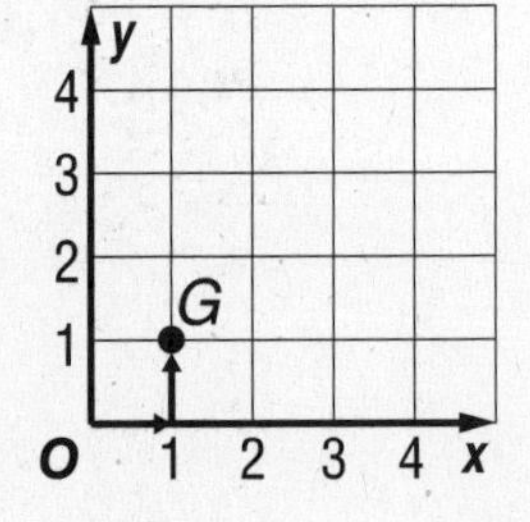

The ordered pair for point G is ____.

b. Point *H*

Start at the origin. Move right on the x-axis to find the x-coordinate of point H, which is ____. Since the y-coordinate is ____, you will not need to move up.

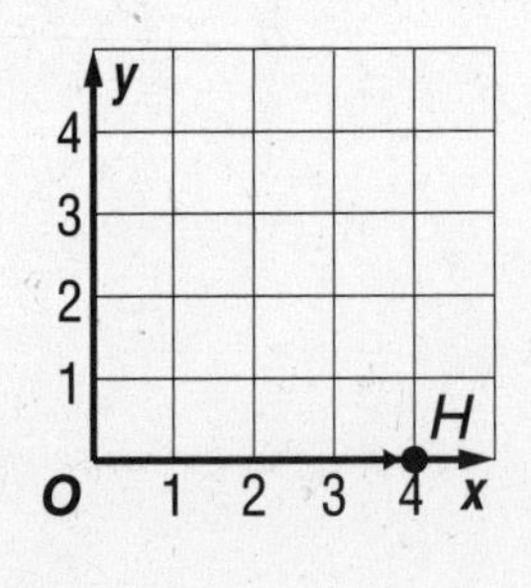

The ordered pair for point H is ____.

Your Turn **Write the ordered pair that names each point.**

a. A ____

b. B ____

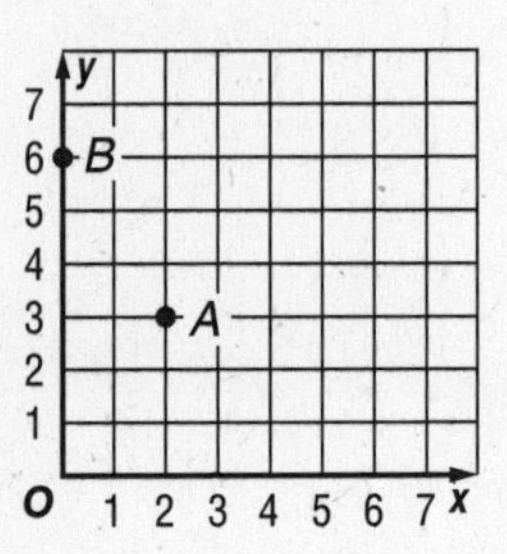

BUILD YOUR VOCABULARY (pages 2–3)

A set of ________ such as {(1, 2), (2, 4), (3, 0), (4, 5)} is a **relation**.

The **domain** of a relation is the set of ________ coordinates.

The **range** of a relation is the set of ________ coordinates.

EXAMPLE Relations as Tables and Graphs

3 Express the relation {(1, 4), (2, 2), (3, 0), (0, 2)} as a table and as a graph. Then determine the domain and range.

x	y
	4
2	
3	
	2

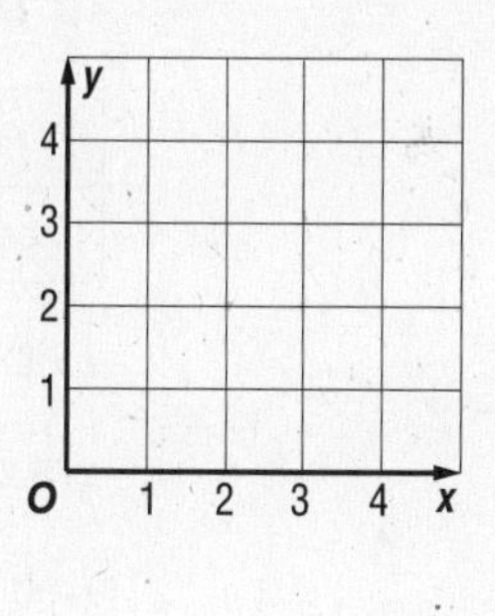

The domain is { ________ }. The range is {  }.

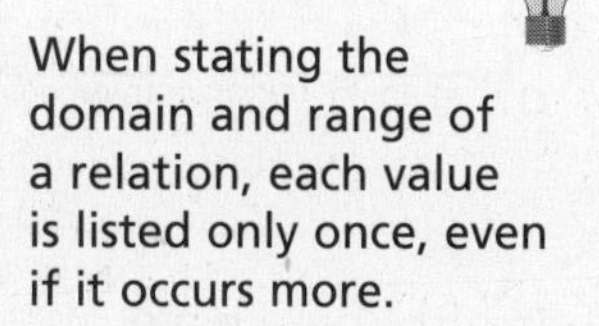

When stating the domain and range of a relation, each value is listed only once, even if it occurs more.

Your Turn Express the relation {(4, 1), (3, 2), (0, 1), (2, 3)} as a table and as a graph. Then determine the domain and range.

x	y

HOMEWORK ASSIGNMENT

Page(s):

Exercises:

1–7 Scatter Plots

What You'll Learn

- Construct scatter plots.
- Interpret scatter plots.

BUILD YOUR VOCABULARY (pages 2–3)

A **scatter plot** is a ______ that shows the ______ between two sets of data. The two sets of data are graphed as ______ on a coordinate system.

EXAMPLE Construct a Scatter Plot

1 RETAIL SALES **The table shows the average cost of a loaf of bread from 1920–2000. Make a scatter plot of the data.**

Year	1920	1930	1940	1950	1960	1970	1980	1990	2000
Cost (¢)	12	9	8	14	20	24	52	72	99

Let the horizontal axis, or x-axis, represent the ______.

Let the vertical axis, or y-axis, represent the ______.

Then graph ordered pairs ______.

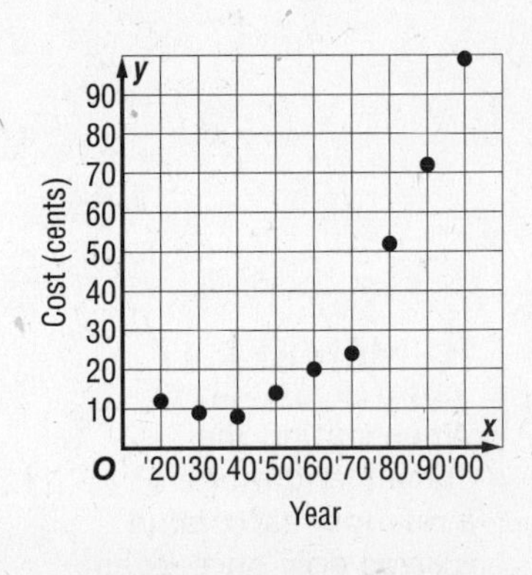

Your Turn The table shows the number of babies born at Central Hospital during the past eight months. Make a scatter plot of the data.

Month	Jan.	Feb.	Mar.	Apr.	May	June	July	Aug.
Babies	12	21	17	9	15	26	18	11

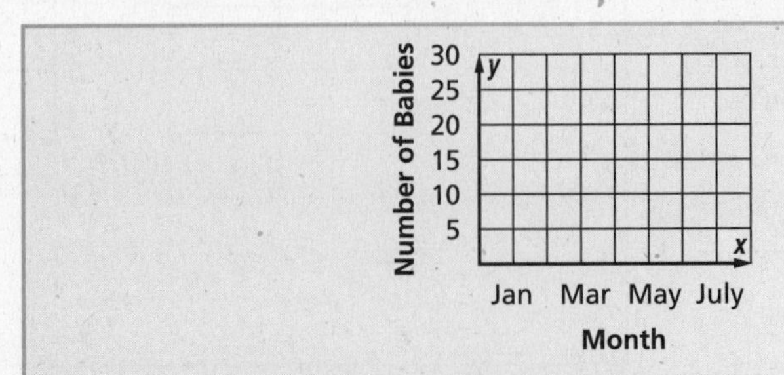

EXAMPLE Interpret Scatter Plots

2 **Determine whether a scatter plot of the data for the following might show a *positive*, *negative*, or *no* relationship. Explain your answer.**

a. height of basketball player and number of rebounds

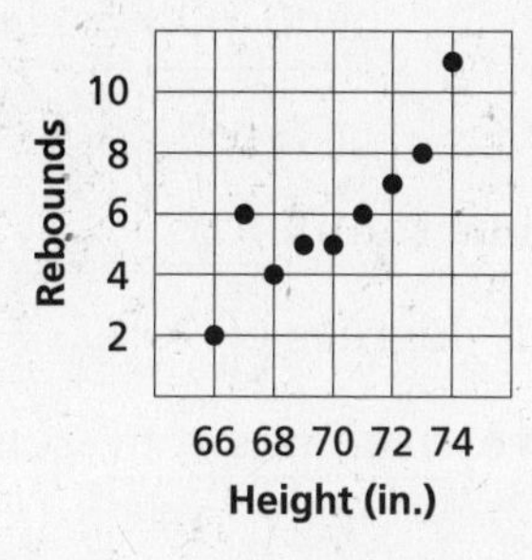

As the height ______, the number of rebounds ______.

______ relationship

b. shoe size and test scores

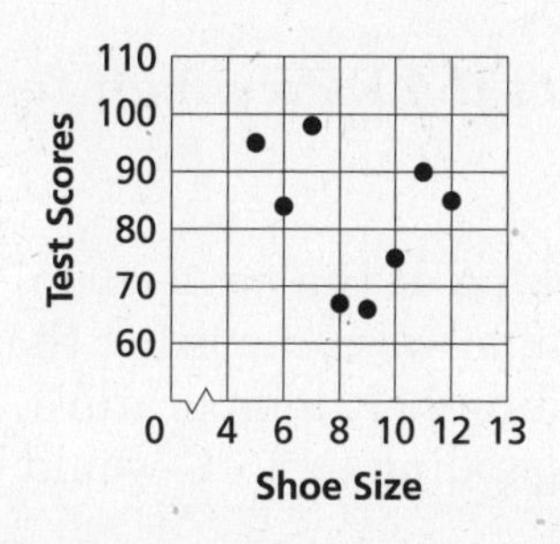

As shoe size ______, test scores fluctuate.

______ relationship

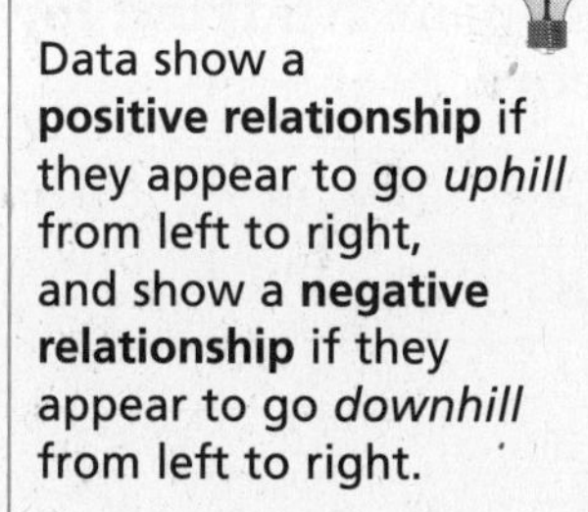

REMEMBER IT

Data show a **positive relationship** if they appear to go *uphill* from left to right, and show a **negative relationship** if they appear to go *downhill* from left to right.

Your Turn **Determine whether a scatter plot of the data for the following might show a *positive*, *negative*, or no relationship. Explain your answer.**

a. outside temperature and heating bill

b. eye color and test score

EXAMPLE Use Scatter Plots to Make Predictions

3 a. **TEMPERATURE The table shows temperatures in degrees Celsius and the corresponding temperatures in degrees Fahrenheit. Make a scatter plot of the data.**

°F	32	41	50	59	68	77	86
°C	0	5	10	15	20	25	30

Let the horizontal axis represent degrees ________.

Let the vertical axis represent degrees ________.

Graph the data.

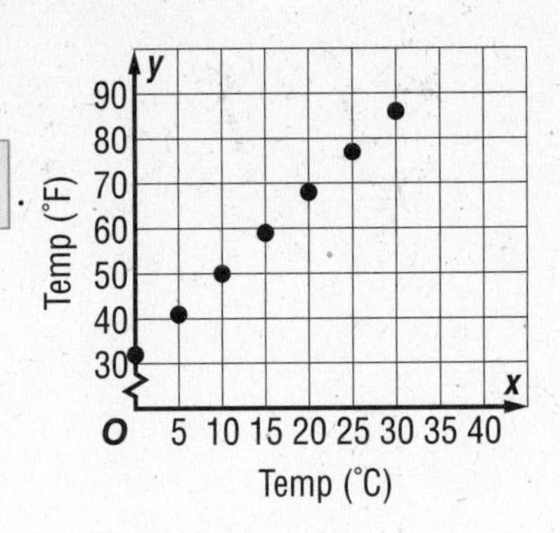

b. **Predict the Fahrenheit temperature for 35°C.**

By looking at the pattern on the graph, we can predict that the Fahrenheit temperature corresponding to 35°C would be about ________.

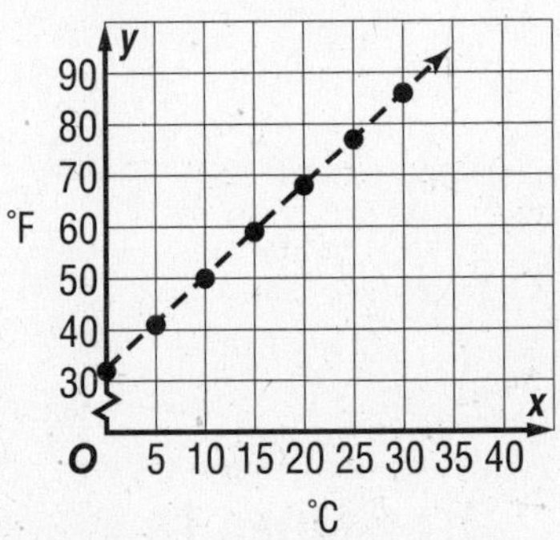

Your Turn **The table shows hours spent studying for a test and the corresponding test score.**

Hours	3	2	5	1	4	2	6
Score	72	75	90	68	85	70	92

a. Make a scatter plot of the data.

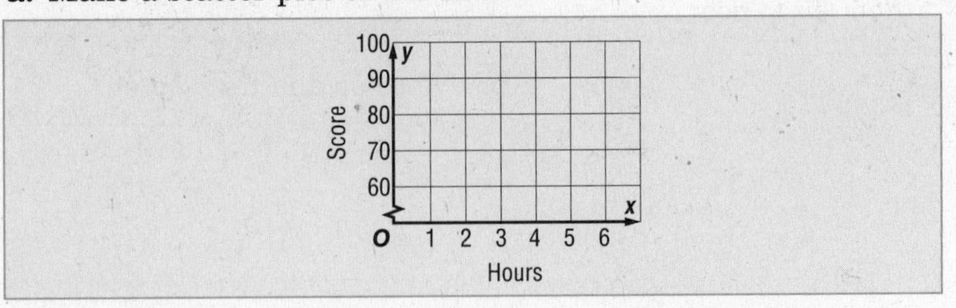

b. Predict the test score for a student who spends 7 hours studying.

HOMEWORK ASSIGNMENT

Page(s):

Exercises:

CHAPTER 1

BRINGING IT ALL TOGETHER

STUDY GUIDE

FOLDABLES	VOCABULARY PUZZLEMAKER	BUILD YOUR VOCABULARY
Use your **Chapter 1 Foldable** to help you study for your chapter test.	To make a crossword puzzle, word search, or jumble puzzle of the vocabulary words in Chapter 1, go to: www.glencoe.com/sec/math/t_resources/free/index.php	You can use your completed **Vocabulary Builder** (pages 2–3) to help you solve the puzzle.

1-1 Using a Problem-Solving Plan

Underline the correct term or phrase to fill the blank in each sentence.

1. A ________________ is an educated guess. (reason, strategy, conjecture)

2. When you make a conjecture based on a pattern of examples or past events, you are using ________________________. (inductive reasoning, reasonableness, problem-solving)

3. What is the next term: 3, 6, 12, 24 . . .? Explain.

4. Complete this sentence. In the ________ step of the four-step problem-solving plan, you check the reasonableness of your answer.

1-2 Numbers and Expressions

State whether each sentence is true or false. If false, replace the underlined word to make a true sentence.

5. Numerical expressions contain a combination of numbers and <u>operations</u>. ________

6. When you evaluate an expression, you find its <u>numerical</u> value. ________

1-3

Variables and Expressions

State whether each sentence is true or false. If false, replace the underlined word to make a true sentence.

7. A variable is a placeholder for any <u>operator</u>.

8. Any <u>letter</u> can be used as a variable.

9. Name three things that make an algebraic expression.

1-4

Properties

Match each statement with the property it shows.

10. $8 \cdot 2 = 2 \cdot 8$

11. $(3 + 2) + 7 = 3 + (2 + 7)$

12. $3x + 0 = 3x$

13. $6(st) = 6s(t)$

14. $10 + 2 = 2 + 10$

a. Additive Identity Property
b. Associative Property of Addition
c. Commutative Property of Addition
d. Associative Property of Multiplication
e. Commutative Property of Multiplication

1-5

Variables and Equations

Underline the correct term or phrase to fill the blank in each sentence.

15. A mathematical sentence that contains an equals sign (=) is called an ________________. (equation, expression, operation)

16. A value for the variable that makes an equation ______________ is called a solution. (reasonable, true, false)

17. Consider $x - 4 = 6$. Find a value for x that makes the sentence true and another value that makes it false.

1-6 Ordered Pairs and Relations

For Exercises 18–20, use the relation {(2, 1), (4, 7), (3, 2), (5, 4)}.

18. Express the relation as a table.

x	y

19. Express the relation as a graph.

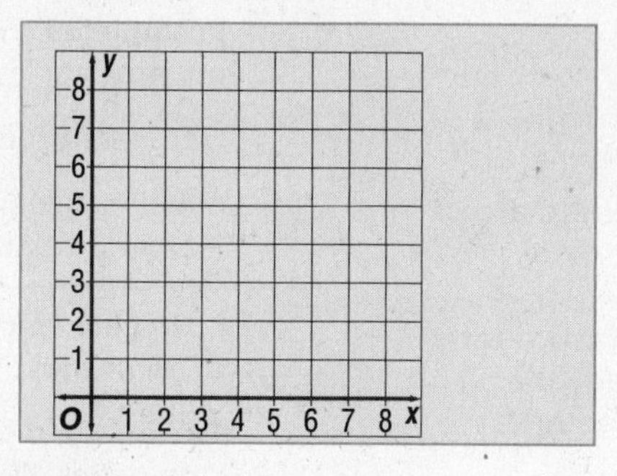

20. Determine the domain and range of the relation.

1-7 Scatter Plots

Underline the correct term or phrase to complete each sentence about the relationship shown by a scatter plot.

21. For a positive relationship, as x increases, y ________________ (increases, decreases, stays constant).

22. For a negative relationship, as x increases, y ________________ (increases, decreases, stays constant).

The scatter plot compares the weights and heights of the players on a high school football team.

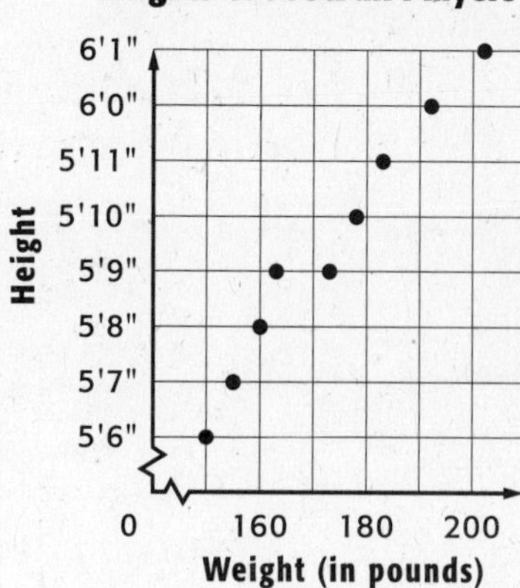

23. What type of relationship exists, if any?

24. Based on the scatter plot, predict the weight of a 5'5" player who decided to join the team.

Checklist

ARE YOU READY FOR THE CHAPTER TEST?

Math Online

Visit **pre-alg.com** to access your textbook, more examples, self-check quizzes, and practice tests to help you study the concepts in Chapter 1.

Check the one that applies. Suggestions to help you study are given with each item.

☐ **I completed the review of all or most lessons without using my notes or asking for help.**

- You are probably ready for the Chapter Test.
- You may want to take the Chapter 1 Practice Test on page 51 of your textbook as a final check.

☐ **I used my Foldable or Study Notebook to complete the review of all or most lessons.**

- You should complete the Chapter 1 Study Guide and Review on pages 47–50 of your textbook.
- If you are unsure of any concepts or skills, refer back to the specific lesson(s).
- You may also want to take the Chapter 1 Practice Test on page 51.

☐ **I asked for help from someone else to complete the review of all or most lessons.**

- You should review the examples and concepts in your Study Notebook and Chapter 1 Foldable.
- Then complete the Chapter 1 Study Guide and Review on pages 47–50 of your textbook.
- If you are unsure of any concepts or skills, refer back to the specific lesson(s).
- You may also want to take the Chapter 1 Practice Test on page 51.

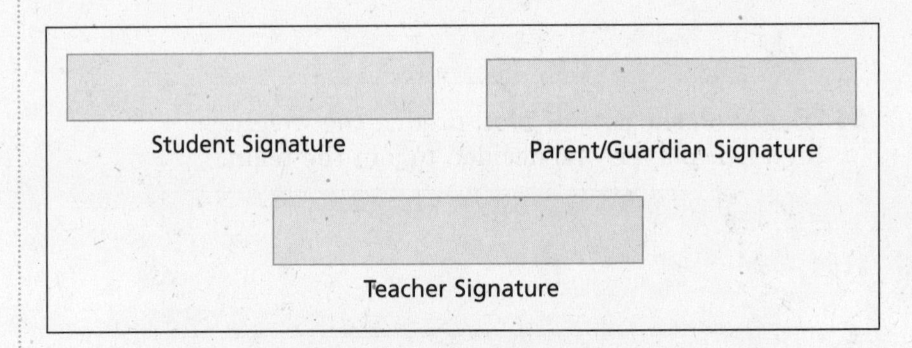

Integers

Use the instructions below to make a Foldable to help you organize your notes as you study the chapter. You will see Foldable reminders in the margin of this Interactive Study Notebook to help you in taking notes.

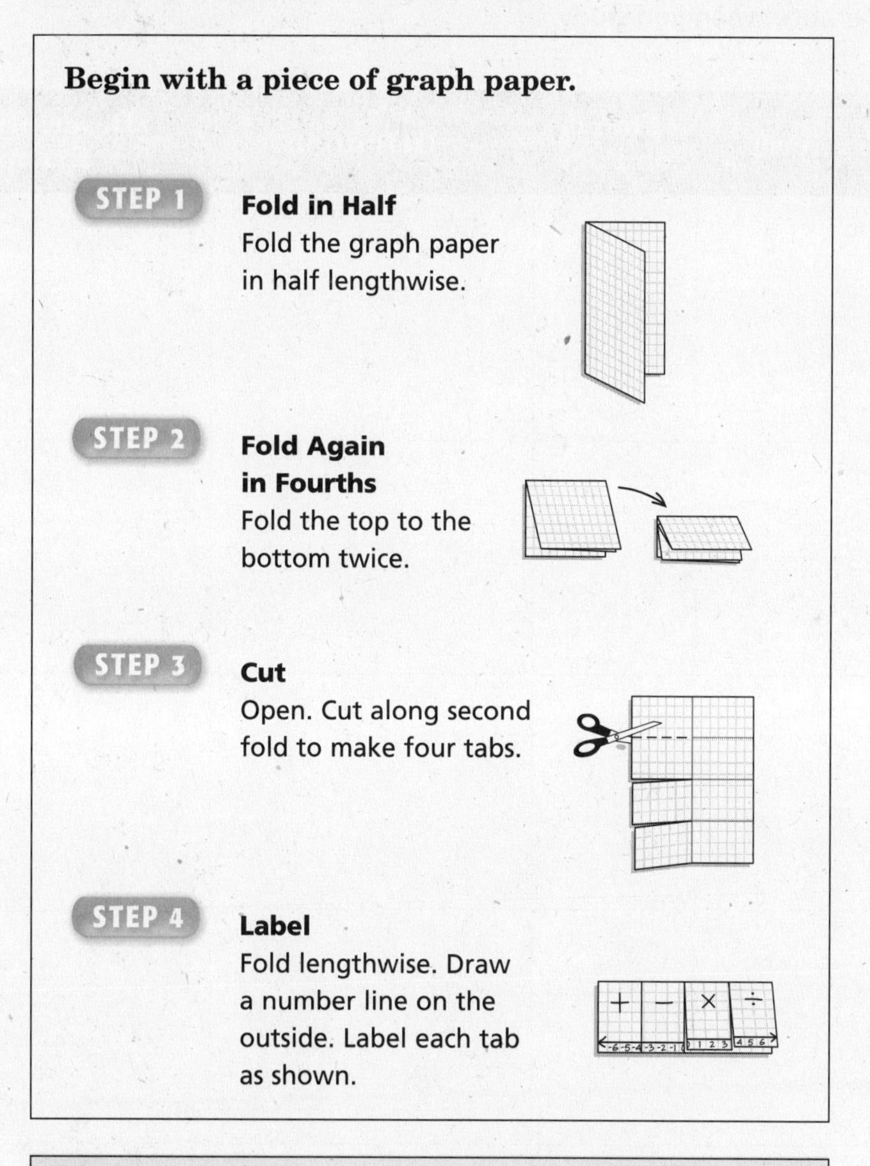

NOTE-TAKING TIP: When searching for the main idea of a lesson, ask yourself, "What is this paragraph or lesson telling me?"

CHAPTER 2

BUILD YOUR VOCABULARY

This is an alphabetical list of new vocabulary terms you will learn in Chapter 2. As you complete the study notes for the chapter, you will see Build Your Vocabulary reminders to complete each term's definition or description on these pages. Remember to add the textbook page number in the second column for reference when you study.

Vocabulary Term	Found on Page	Definition	Description or Example
absolute value			
additive inverse			
average			
coordinate			
inequality			

Vocabulary Term	Found on Page	Definition	Description or Example
integer			
mean			
negative number			
opposites			
quadrants			

2–1 Integers and Absolute Value

WHAT YOU'LL LEARN

- Compare and order integers.
- Find the absolute value of an expression.

BUILD YOUR VOCABULARY (pages 30–31)

A **negative number** is a number less than zero.

Negative numbers like −8, positive numbers like +6, and ______ are members of the set of **integers**.

The ______ that corresponds to a ______ is called the **coordinate** of that point.

Any mathematical sentence containing ______ or ______ is called an **inequality**.

WRITE IT

List 5 words or phrases that indicate positive or negative numbers.

______

______

______

______

______

EXAMPLE Write Integers for Real-World Situations

1 Write an integer for 32 feet under ground.

The integer is ______.

Your Turn Write an integer for a loss of 12 yards.

______

EXAMPLE Compare Two Integers

2 Replace the • with <, >, or = in −2 • 3 to make a true sentence.

−2 is less than 3 since it lies to the left of 3 on a number line.

So write −2 ______ 3.

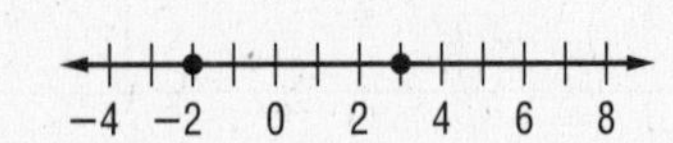

Your Turn Replace the • with <, >, or = in 6 • −7 to make a true sentence.

______

EXAMPLE Order Integers

3 WEATHER **The high temperatures for the first seven days of January were $-8°$, $10°$, $2°$, $-3°$, $-11°$, $0°$ and $1°$. Order the temperatures from least to greatest.**

Graph each integer on a number line.

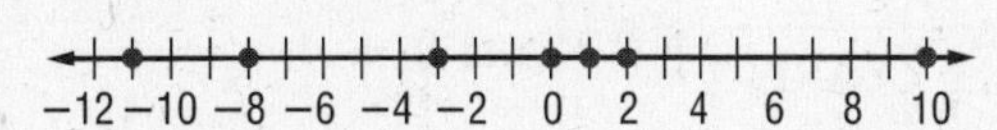

The order from least to greatest is

______________________.

Your Turn The yards gained during the first six plays of the football game were 5, −3, 12, −9, 6 and −1. Order the yards from least to greatest.

____

BUILD YOUR VOCABULARY (page 30)

Two numbers have the same **absolute value** if they are on ________ sides of zero, and are the same ________ from zero.

KEY CONCEPT

Absolute Value The absolute value of a number is the distance the number is from zero on the number line. The absolute value of a number is always greater than or equal to zero.

EXAMPLE Expressions with Absolute Value

4 **a. Evaluate $|5|$.**

$|5| =$ ____

b. Evaluate $|-8| + |-1|$.

$|-8| + |-1| =$ ________ $|-8| =$ ____

$|-1| =$ ____

$=$ ____ Simplify.

Your Turn **Evaluate each expression.**

a. $|-9|$ ________

b. $|-3| + |2|$ ________

HOMEWORK ASSIGNMENT

Page(s): ________

Exercises: ________

2-2 Adding Integers

WHAT YOU'LL LEARN

- Add two integers.
- Add more than two integers.

EXAMPLE Add Integers on a Number Line

1 **Find 3 + 4.**

Start at ______.

Move ______ units to the ______.

From there, move ______ more units to the ______.

$3 + 4 =$ ______

Your Turn Find $-2 + -5$.

KEY CONCEPT

Adding Integers with the Same Sign To add integers with the same sign, add their absolute values. Give the result the same sign as the integers.

EXAMPLE Add Integers with the Same Sign

2 **Find $-5 + (-4)$.**

$-5 + (-4) =$ ______ Add ______ and ______.

Both numbers are ______

so the sum is ______.

Your Turn Find $-3 + -8$.

FOLDABLES

ORGANIZE IT

Under the "+" tab, write a sum of integers with different signs, and explain how to add them on a number line.

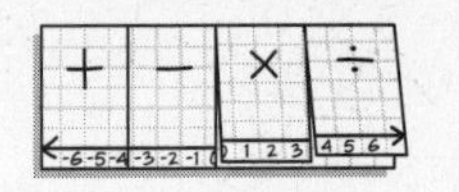

EXAMPLE Add Integers on a Number Line

3 **a. Find $7 + (-11)$.**

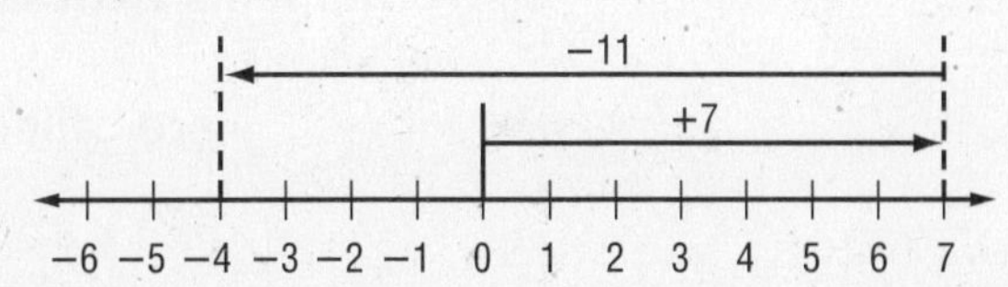

Start at ______.

Move ______ units to the ______.

From there, move ______ units to the ______.

$7 + (-11) =$ ______

Your Turn **Find each sum.**

a. $-5 + 8$

b. $3 + (-6)$

EXAMPLE Add Integers with Different Signs

4 **a. Find $-9 + 10$.**

$-9 + 10 =$ ______

To find $-9 + 10$, subtract ______ from ______. The sum is positive because $|10| > |-9|$.

b. Find $8 + (-15)$.

$8 + (-15) = -7$

To find $8 + (-15)$, subtract ______ from ______.

The sum is negative because $|-15| > |8|$.

KEY CONCEPT

Adding Integers with Different Signs To add integers with different signs, subtract their absolute values. Give the result the same sign as the integer with the greater absolute value.

Your Turn **Find each sum.**

a. $4 + (-7)$

b. $-6 + 11$

BUILD YOUR VOCABULARY (pages 30–31)

Two numbers with same absolute value but different ______ are called **opposites**.

An integer and its ______ are called **additive inverses**.

EXAMPLE Add Three or More Integers

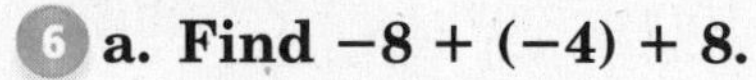

6 **a. Find $-8 + (-4) + 8$.**

$-8 + (-4) + 8 = -8 +$ ______	Commutative Property
$=$ ______ $+ -4$	Additive Inverse Property
$=$ ______	Identity Property of Addition

KEY CONCEPT

Additive Inverse Property The sum of any number and its additive inverse is zero.

b. Find $6 + (-3) + (-9) + 2$.

$6 + (-3) + (-9) + 2$

$= 6 +$ ______	Commutative Property
$= [6 + 2] +$ ______	Associative Property
$= 8 +$ ______ or ______	Simplify.

Your Turn **Find each sum.**

a. $3 + (-9) + (-3)$

b. $-2 + 11 + (-4) + 5$

HOMEWORK ASSIGNMENT

Page(s):

Exercises:

2–3 Subtracting Integers

WHAT YOU'LL LEARN

- Subtract integers.
- Evaluate expressions containing variables.

KEY CONCEPT

Subtracting Integers To subtract an integer, add its additive inverse.

EXAMPLE Subtract a Positive Integer

1 a. Find 9 − 14.

$9 - 14 = 9 + \_\_\_\_$ To subtract 14, add ____.

$= \_\_\_\_$ Simplify.

b. Find −10 − 8.

$-10 - 8 = -10 + \_\_\_\_$ To subtract 8, add ____.

$= \_\_\_\_$ Simplify.

Your Turn **Find each difference.**

a. $6 - 8$

b. $-9 - 13$

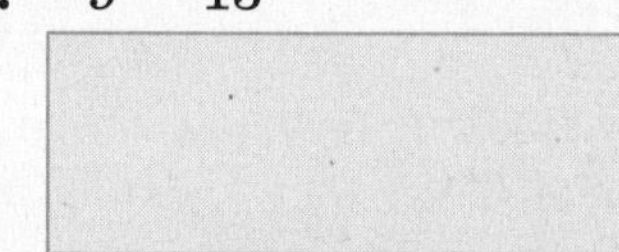

EXAMPLE Subtract a Negative Integer

FOLDABLES

ORGANIZE IT

Write two examples of subtracting a negative number from a positive number under the "–" tab.

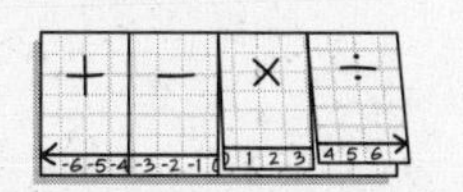

2 a. Find 15 − (−4).

$15 - (-4) = 15 + \_\_\_\_$ To subtract −4, add ____.

$= \_\_\_\_$ Simplify.

b. Find −10 − (−7).

$-10 - (-7) = -10 + \_\_\_\_$ To subtract −7, add ____.

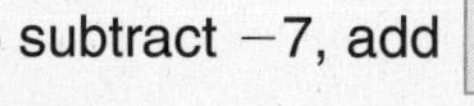

$= \_\_\_\_$ Simplify.

Your Turn **Find each difference.**

a. $8 - (-2)$

b. $-12 - (-5)$

EXAMPLE Evaluate Algebraic Expressions

3 **a. Evaluate $m - (-2)$ if $m = 4$.**

$m - (-2) = \_\_\_ - (-2)$ Replace m with ___.

$= \_\_\_$ To subtract -2, add ___.

$= \_\_\_$ Add ___ and ___.

b. Evaluate $x - y$ if $x = -14$ and $y = -2$.

$x - y = \_\_\_ - (\_\_\_)$ Replace x with ___ and y with ___.

$= \_\_\_$ To subtract -2, add ___.

$= \_\_\_$ Add ___ and ___.

Your Turn

a. Evaluate $p - (-6)$ if $p = -4$.

b. Evaluate $m - n$ if $m = -9$ and $n = -3$.

HOMEWORK ASSIGNMENT

Page(s):

Exercises:

2–4 Multiplying Integers

WHAT YOU'LL LEARN

- Multiply integers.
- Simplify algebraic expressions.

EXAMPLE Multiply Integers with Different Signs

1 Find 8(−12).

$8(-12) =$ ______ The factors have different signs.

The product is ______.

KEY CONCEPT

Multiplying Integers
The product of two integers with different signs is negative.

The product of two integers with the same sign is positive.

EXAMPLE Multiply Integers with the Same Sign

2 Find −4(−16).

$-4(-16) =$ ______ The two factors have the same sign.

The product is ______.

EXAMPLE Multiply More Than Two Integers

3 Find −7(11)(−2).

$-7(11)(-2) = [$ ______ ______ $](-2)$ Associative Property

$=$ ______ (-2) $(-7)(11) =$ ______

$=$ ______ Multiply ______ and (−2).

Your Turn **Find each product.**

a. $-4(12)$

b. $-3(-8)$

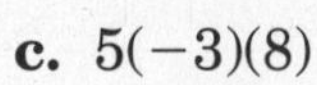

c. $5(-3)(8)$

FOLDABLES™

ORGANIZE IT

In your own words, describe how to multiply integers under the "×" tab. Give examples of all cases.

EXAMPLE Simplify and Evaluate Algebraic Expressions

3 a. Simplify $8a(-5b)$.

$8a(-5b) = (8)(a)(-5)(b)$

$= (8 \cdot -5)(ab)$ — Commutative Property of Multiplication

$= \square$ — $8 \cdot -5 = \square$, $a \cdot b = \square$

b. Evaluate $-3xy$ if $x = 4$ and $y = 9$.

$-3xy = -3\square$ — $x = -4$ and $y = 9$.

$= \square(9)$ — Associative Property of Multiplication

$= \square(9)$ — The product of $\square$ and $\square$ is positive.

$= \square$ — The product of $\square$ and $\square$ is positive.

WRITE IT

What is the name of the property that allows you to regroup the numbers and the variables being multiplied?

____

____

____

____

Your Turn

a. Simplify $5m(-7n)$.

b. Evaluate $-9ab$ if $a = -3$ and $b = -6$.

HOMEWORK ASSIGNMENT

Page(s):

Exercises:

2-5 Dividing Integers

WHAT YOU'LL LEARN

- Divide integers.
- Find the average of a set of data.

KEY CONCEPTS

Dividing Integers with the Same Sign The quotient of two integers with the same sign is positive.

Dividing Integers with Different Signs The quotient of two integers with different signs is negative.

EXAMPLE Divide Integers with the Same Sign

1 a. Find $-28 \div (-4)$.

$-28 \div (-4) = $ ______ The dividend and the divisor have the same sign. The quotient is ______.

b. Find $\frac{96}{8}$.

$\frac{96}{8} = 96 \div 8$ The dividend and the divisor have the same sign.

$= $ ______ The quotient is ______.

Your Turn **Find each quotient.**

a. $35 \div 7$

b. $\frac{-64}{-4}$

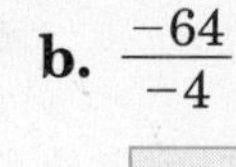

EXAMPLE Divide Integers with Different Signs

2 a. Find $54 \div (-3)$.

$54 \div (-3) = $ ______ The signs are different. The quotient is ______.

b. Find $\frac{-42}{6}$.

$\frac{-42}{6} = -42 \div 6$ The signs are different. The quotient is ______.

$= $ ______ Simplify.

Your Turn **Find each quotient.**

a. $72 \div (-8)$

b. $\frac{-36}{4}$

EXAMPLE Evaluate Algebraic Expressions

3 Evaluate $6x \div y$ if $x = -4$ and $y = -8$.

$6x \div y = 6$ ☐ $\div$ ☐ — $x = -4$ and $y = -8$

$=$ ☐ $\div (-8)$ — $6(-4) =$ ☐

$=$ ☐ — The quotient is ☐.

Your Turn Evaluate $-4m \div n$ if $m = -9$ and $n = -3$.

BUILD YOUR VOCABULARY (pages 30–31)

To find the **average**, or **mean**, of a set of numbers, find the ☐ of the numbers and then ☐ by the number in the set.

FOLDABLES

ORGANIZE IT

Describe how to find the average of a set of numbers in your own words under the "÷" tab.

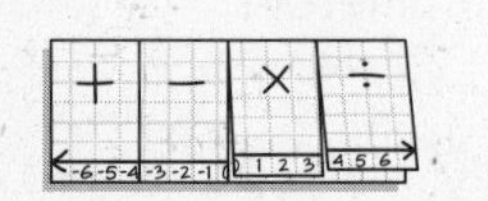

EXAMPLE Find the Mean

4 a. Sam had quiz scores of 89, 98, 96, 97, and 95. Find the average (mean) of his scores.

$\frac{89 + 98 + 96 + 97 + 95}{\square}$ — Find the sum of the scores. Then divide by the number of scores.

$=$ ☐ or ☐ — Simplify.

Your Turn Kyle had test scores of 89, 82, 85, 93, and 96. Find the average (mean) of his test scores.

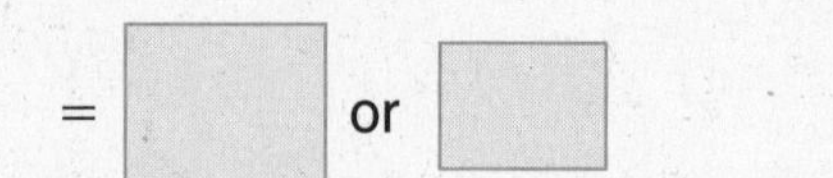

HOMEWORK ASSIGNMENT

Page(s):

Exercises:

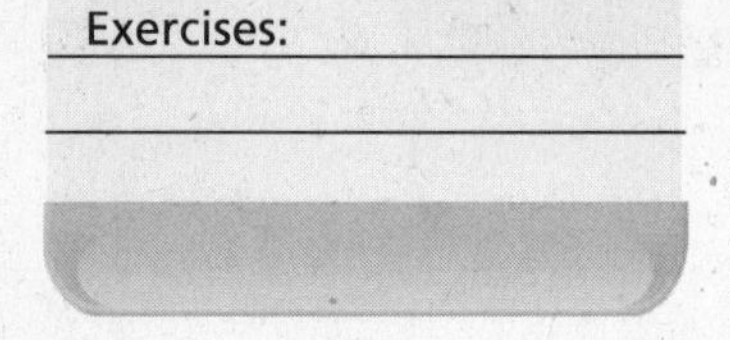

2–6 The Coordinate System

WHAT YOU'LL LEARN

- Graph points on a coordinate plane.
- Graph algebraic relationships.

EXAMPLE Write Ordered Pairs

1 **Write the ordered pair that names each point.**

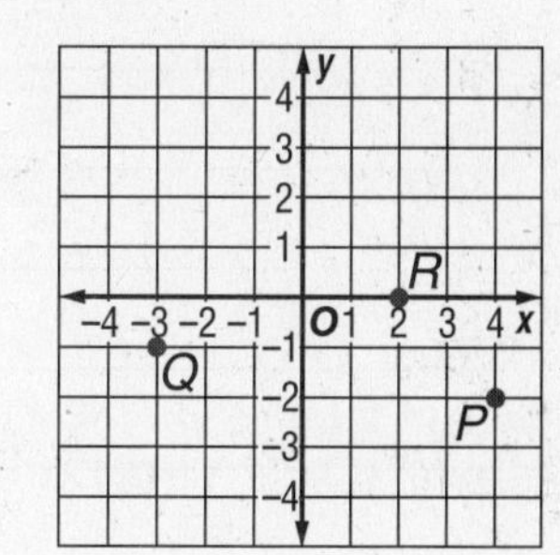

a. ***P***

The x-coordinate is ______.

The y-coordinate is ______.

The ordered pair is ______.

b. ***Q***

The x-coordinate is ______.

The y-coordinate is .

The ordered pair is .

c. ***R***

The x-coordinate is .

The point lies on the x-axis. The y-coordinate is 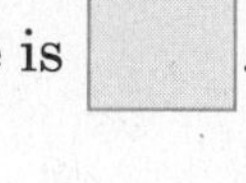.

The ordered pair is ______.

REMEMBER IT

The coordinates in an ordered pair (x, y) are listed in alphabetical order.

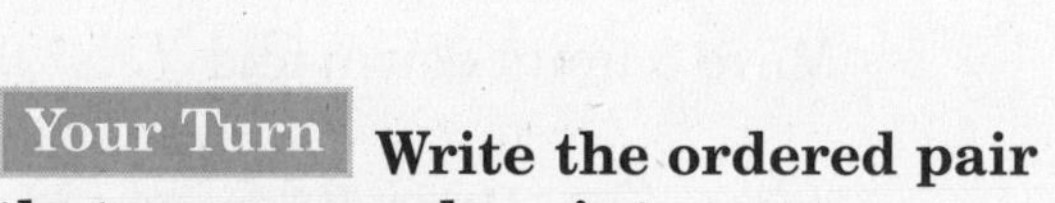

Your Turn **Write the ordered pair that names each point.**

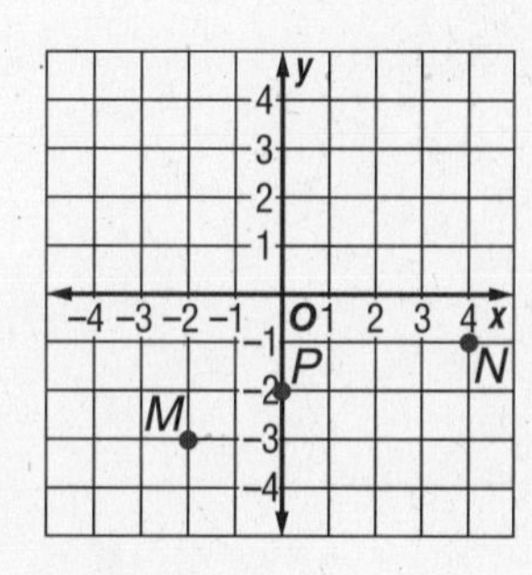

a. M 

b. N ______

c. P ______

BUILD YOUR VOCABULARY (pages 30–31)

The x-axis and the y-axis separate the coordinate plane into ______ regions, called **quadrants**.

EXAMPLE Graph Points and Name Quadrant

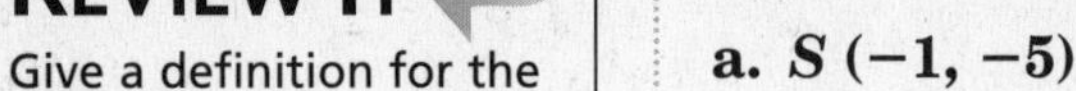

REVIEW IT

Give a definition for the origin of a coordinate system.
(Lesson 1-5)

____

____

____

____

____

____

____

____

2 Graph and label each point on a coordinate plane. Then name the quadrant in which each point lies.

a. $S(-1, -5)$

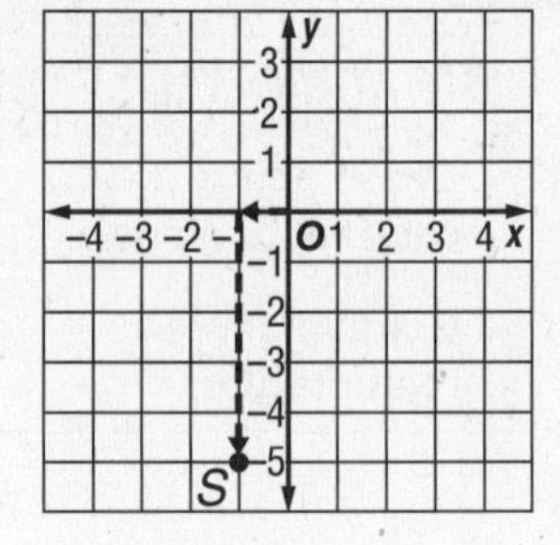

Start at the origin.

Move ______ unit ______.

Then move ______ units ______

and draw a dot. Quadrant ______.

b. $U(-2, 3)$

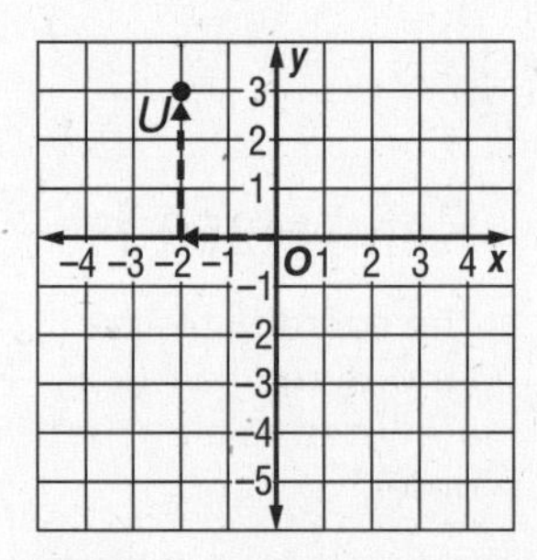

Start at the origin.

Move ______ units ______.

Then move ______ units ______

and draw a dot. Quadrant ______.

c. $T(0, -3)$

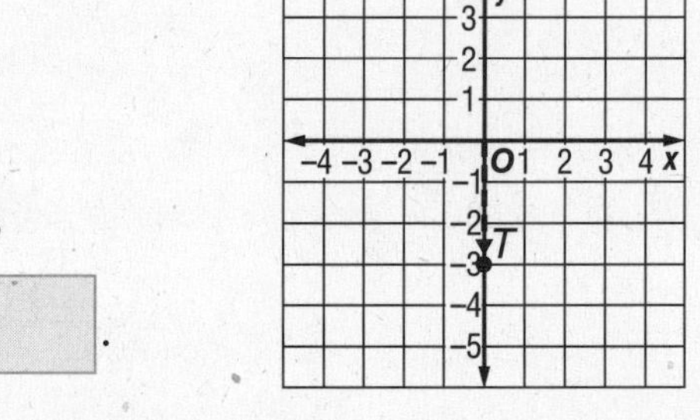

Start at the origin.

Since the x-coordinate is 0,

the point lies on the ______.

Move 3 units down, and

draw a dot. Point T is not in any quadrant.

Your Turn **Graph and label each point on a coordinate plane. Then name the quadrant in which each point lies.**

a. $A(3, -4)$ **b.** $B(-2, 1)$ **c.** $C(-4, 0)$

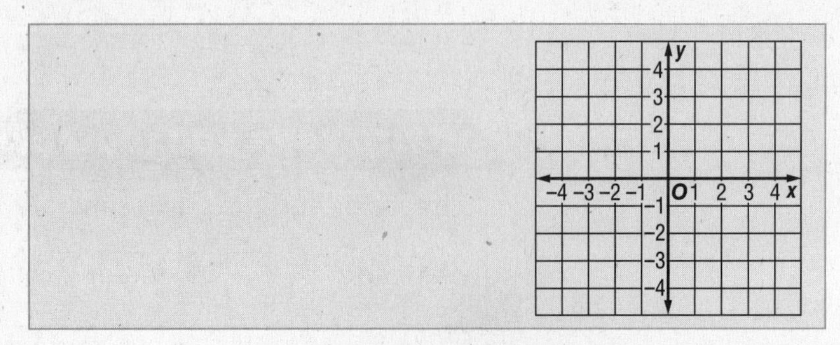

EXAMPLE Graph an Algebraic Relationship

3 **The difference between two integers is 4. If x represents the first integer and y is subtracted from it, make a table of possible values for x or y. Then graph the ordered pairs and describe the graph.**

First, make a table. Choose values for x and y that have a difference of 4.

$x - y = 4$		
x	y	(x, y)
2		
1		
0		
−1		
−2		

Then graph the ordered pairs on a coordinate plane.

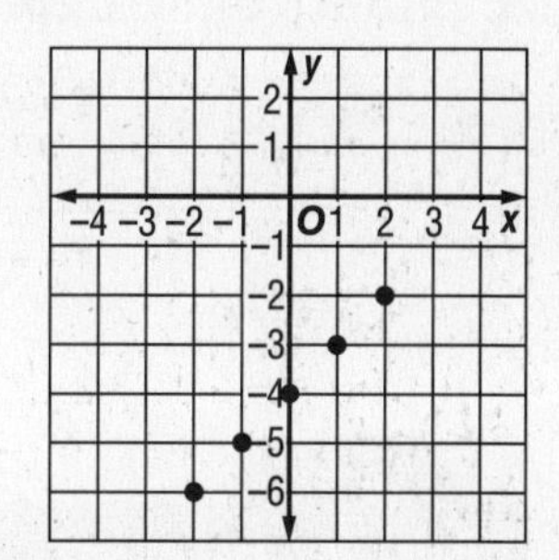

The points on the graph are in a line that slants downward to the left. The lines crosses the y-axis at −4.

Your Turn The sum of two integers is 3. If x is the first integer and y is added to it, make a table of possible values for x and y. Graph the ordered pairs and describe the graph.

$x + y = 3$		
x	y	(x, y)

HOMEWORK ASSIGNMENT

Page(s):

Exercises:

CHAPTER 2

BRINGING IT ALL TOGETHER

STUDY GUIDE

FOLDABLES™	Vocabulary PuzzleMaker	Build your Vocabulary
Use your **Chapter 2 Foldable** to help you study for your chapter test.	To make a crossword puzzle, word search, or jumble puzzle of the vocabulary words in Chapter 2, go to: www.glencoe.com/sec/math/t_resources/free/index.php	You can use your completed **Vocabulary Builder** (pages 30–31) to help you solve the puzzle.

2-1 Integers and Absolute Value

1. Order the integers {21, −1, 9, 7, 0 −4, −11} from least to greatest.

Evaluate each expression if $r = 3$, $s = -2$, and $t = -7$.

2. $|t| - 6$

3. $12 - |s - 5|$

4. $|s + t| - r$

5. $|rt - 1| \div s$

2-2 Adding Integers

Find each sum.

6. $-52 + 9$

7. $7 + (-31) + 4$

8. $(-8) + 22 + (-15) + 5$

9. $6 + (-10) + (-12) + 4$

2-3 Subtracting Integers

Find each difference.

10. $-17 - 26$

11. $35 - (-14)$

12. $42 - 19$

13. $11 - (-18)$

Evaluate each expression if $p = -6$, $q = 9$, and $r = -2$.

14. $q - 16$

15. $r - 4$

16. $p - q - r$

17. $q - r - p$

2-4 Multiplying Integers

Find each product.

18. $-4(-16)$

19. $3(-4)(-11)(2)$

Simplify each expression.

20. $5b \cdot (-7c)$

21. $2p(-7q)(-3)$

2-5 Dividing Integers

Find each quotient.

22. $72 \div -9$

23. $-28 \div 4$

24. $\frac{-49}{-7}$

25. $\frac{-144}{18}$

26. Find the average (mean) of 9, −6, 11, 7, 2, and −5.

2-6 The Coordinate System

Name the ordered pair for each point graphed on the coordinate plane.

27. *A*

28. *B*

29. *C*

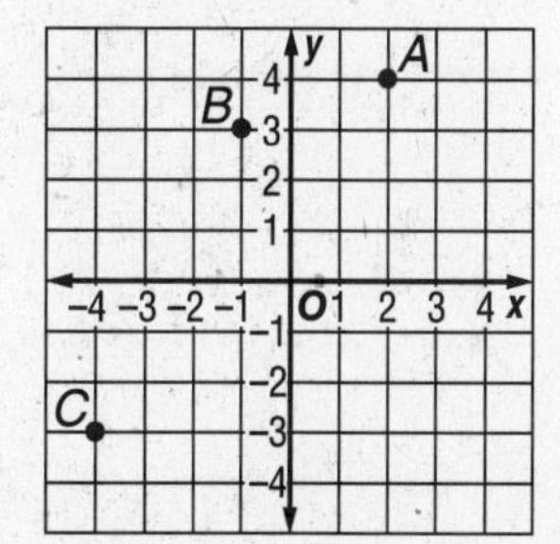

ARE YOU READY FOR THE CHAPTER TEST?

Math Online

Visit **pre-alg.com** to access your textbook, more examples, self-check quizzes, and practice tests to help you study the concepts in Chapter 2.

Check the one that applies. Suggestions to help you study are given with each item.

☐ **I completed the review of all or most lessons without using my notes or asking for help.**

- You are probably ready for the Chapter Test.
- You may want to take the Chapter 2 Practice Test on page 93 of your textbook as a final check.

☐ **I used my Foldable or Study Notebook to complete the review of all or most lessons.**

- You should complete the Chapter 2 Study Guide and Review on pages 90–92 of your textbook.
- If you are unsure of any concepts or skills, refer back to the specific lesson(s).
- You may also want to take the Chapter 2 Practice Test on page 93.

☐ **I asked for help from someone else to complete the review of all or most lessons.**

- You should review the examples and concepts in your Study Notebook and Chapter 2 Foldable.
- Then complete the Chapter 2 Study Guide and Review on pages 90–92 of your textbook.
- If you are unsure of any concepts or skills, refer back to the specific lesson(s).
- You may also want to take the Chapter 2 Practice Test on page 93.

Student Signature

Parent/Guardian Signature

Teacher Signature

Equations

Use the instructions below to make a Foldable to help you organize your notes as you study the chapter. You will see Foldable reminders in the margin of this Interactive Study Notebook to help you in taking notes.

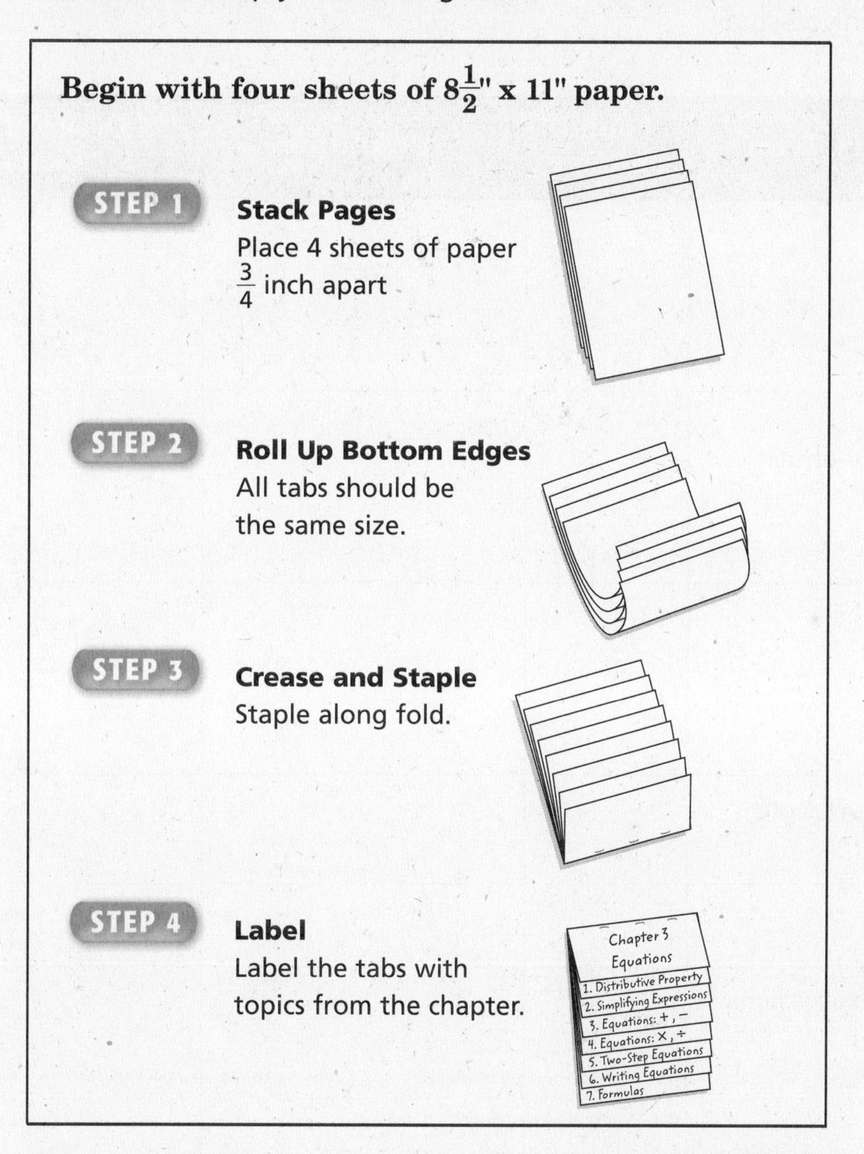

NOTE-TAKING TIP: When you take notes, include definitions of new terms, explanations of new concepts, and examples of problems.

BUILD YOUR VOCABULARY

This is an alphabetical list of new vocabulary terms you will learn in Chapter 3. As you complete the study notes for the chapter, you will see Build Your Vocabulary reminders to complete each term's definition or description on these pages. Remember to add the textbook page number in the second column for reference when you study.

Vocabulary Term	Found on Page	Definition	Description or Example
area			
coefficient [koh-uh-FIHSH-ehnt]			
constant			
equivalent equations			
equivalent expressions			
formula			

Vocabulary Term	Found on Page	Definition	Description or Example
inverse operations			
like terms			
perimeter			
simplest form			
simplifying an expression			
term			
two-step equation			

3–1 The Distributive Property

What You'll Learn

- Use the Distributive Property to write equivalent numerical expressions.
- Use the Distributive Property to write equivalent algebraic expressions.

Key Concept

Distributive Property To multiply a number by a sum, multiply each number inside the parentheses by the number outside the parentheses.

Include the Distributive Property in your Foldable.

Build Your Vocabulary (page 50)

The ______ $3(4 + 2)$ and $3 \cdot 4 + 3 \cdot 2$ are **equivalent expressions** because they have the same ______, 18.

EXAMPLE Use the Distributive Property

1 Use the Distributive Property to write each expression as an equivalent expression. Then evaluate the expression.

a. $4(5 + 8)$

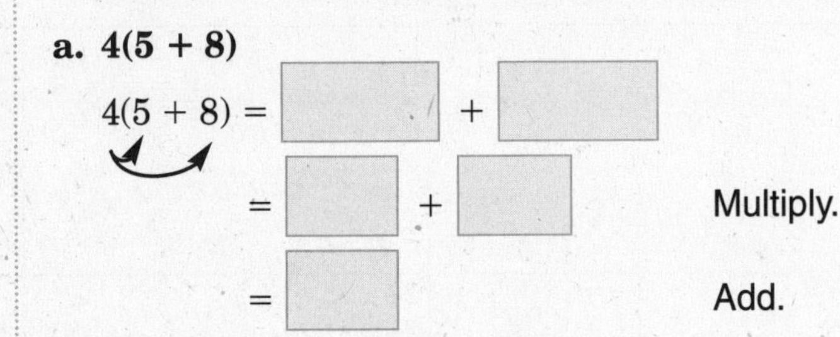

b. $(6 + 9)2$

$(6 + 9)2$ = ______ + ______

= ______ + ______ Multiply.

= ______ Add.

Your Turn Use the Distributive Property to write each expression as an equivalent expression. Then evaluate the expression.

a. $3(9 + 2)$

b. $(7 + 3)5$

REVIEW IT

Write a definition of algebraic expression in your own words. *(Lesson 1-3)*.

____

____

____

____

____

____

EXAMPLE Simplify Algebraic Expressions

2 Use the Distributive Property to write $2(x + 4)$ as an equivalent algebraic expression.

$2(x + 4) =$ ______ $+$ ______ ______ Property

$=$ ______ Simplify.

Your Turn Use the Distributive Property to write $4(m + 7)$ as an equivalent algebraic expression.

____

EXAMPLE Simplify Expressions with Subtraction

3 Use the Distributive Property to write $4(x - 2)$ as an equivalent algebraic expression.

$4(x - 2)$

$= 4[$ ______ $]$ Rewrite $x - 2$ as ______.

$=$ ______ $+$ ______ ______ Property

$=$ ______ $+$ ______ Simplify.

$=$ ______ Definition of subtraction

Your Turn Use the Distributive Property to write $2(a - 9)$ as an equivalent algebraic expression.

____

HOMEWORK ASSIGNMENT

Page(s):

Exercises:

3–2 Simplifying Algebraic Expressions

WHAT YOU'LL LEARN

- Use the Distributive Property to simplify algebraic expressions.

BUILD YOUR VOCABULARY (pages 50–51)

When plus or minus signs separate an algebraic expression into parts, each part is a **term**.

The ________ part of a term that contains a variable is called the **coefficient** of the ________.

Like terms are terms that contain the same ________, such as $2n$ and $5n$ or $6xy$ and $4xy$.

A term without a variable is called a **constant**.

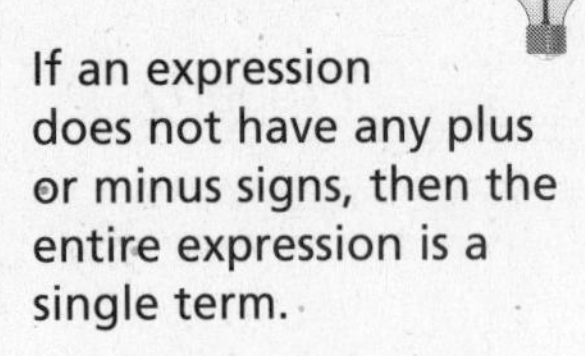

REMEMBER IT

If an expression does not have any plus or minus signs, then the entire expression is a single term.

EXAMPLE Identify Parts of Expressions

1 **Identify the terms, like terms, coefficients, and constants in the expression $4x - x + 2y - 3$.**

Rewrite $4x - x + 2y - 3$ as $4x + (-x) + 2y + (-3)$.

The terms are ____, ____, ____, and ____.

The like terms are ____ and ____.

The coefficients are ____, ____, and ____.

The constant is ____.

Your Turn Identify the terms, like terms, coefficients, and constants in the expression $5x + 3y - 2y + 6$.

BUILD YOUR VOCABULARY (pages 50–51)

An algebraic expression is in **simplest form** if it has no ______ and no parentheses.

When you use the Distributive Property to ______ like terms, you are **simplifying the expression**.

FOLDABLES — ORGANIZE IT

Under the *Simplifying Expressions* tab explain how you know when an expression can be simplified. Write an expression that can be simplified and one that cannot.

Chapter 3 Equations
1. Distributive Property
2. Simplifying Expressions
3. Equations: +, −
4. Equations: ×, ÷
5. Two-Step Equations
6. Writing Equations
7. Formulas

EXAMPLE Simplify Algebraic Expressions

2 **Simplify each expression.**

a. $5x + 4x$

$5x + 4x =$ ______ Distributive Property

$=$ ______ Simplify.

b. $8n + 4 + 4n$

$8n + 4 + 4n$

$= 8n +$ ______ Commutative Property

$=$ ______ Distributive Property

$=$  Simplify.

c. $6x + 4 - 5x - 7$

$6x + 4 - 5x - 7$

$= 6x + 4 +$ ______ $+$ ______ Definition of subtraction

$= 6x +$ ______ $+ 4 +$ ______ Commutative Property

$=$ ______ $+ 4 + (-7)$ Distributive Property

$=$ ______ Simplify.

WRITE IT

What does it mean for two expressions to be equivalent?

______

______

______

______

Your Turn **Simplify each expression.**

a. $7a + 2a$

b. $5x + 3 + 7x$

c. $3m + 9 - m - 6$

EXAMPLE Translate Verbal Phrases into Expressions

3 WORK **You and a friend worked in the school store last week. You worked 4 hours more than your friend. Write an expression in simplest form that represents the total number of hours you both worked.**

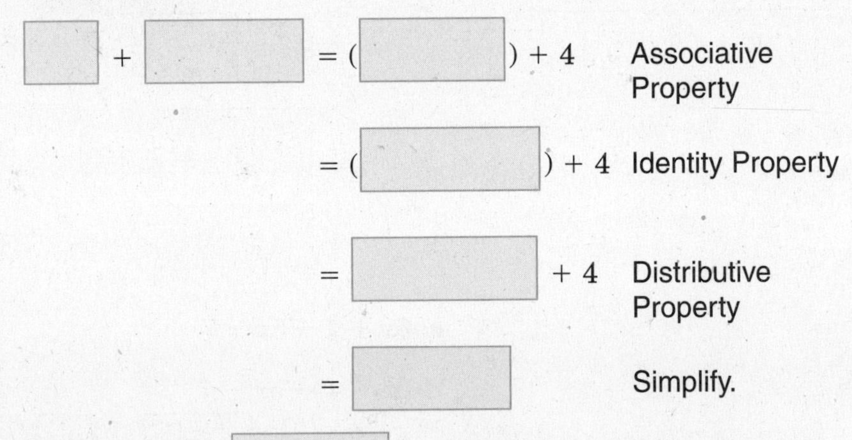

The expression ______ represents the total number of hours you both worked.

Your Turn You and a friend went to the library. Your friend borrowed three more books than you did. Write an expression in simplest form that represents the total number of books you both borrowed.

HOMEWORK ASSIGNMENT

Page(s):

Exercises:

3–3 Solving Equations by Adding or Subtracting

WHAT YOU'LL LEARN

- Solve equations by using the Subtraction Property of Equality.
- Solve equations by using the Addition Property of Equality.

BUILD YOUR VOCABULARY (pages 50–51)

Inverse operations "undo" each other.

The equations $x + 4 = 7$ and $x = 3$ are **equivalent equations** because they have the same ______, 3.

EXAMPLE Solve Equations by Subtracting

1 **Solve $x + 4 = -3$.**

$x + 4 = -3$

$x + 4 -$ ___ $= -3 -$ ___ Subtract ___ from each side.

___ = ___ $4 - 4 =$ ___ and

$-3 - 4 =$ ___

___ = ___ Identity Property

KEY CONCEPT

Subtraction Property of Equality If you subtract the same number from each side of an equation, the two sides remain equal.

Your Turn Solve $y + 7 = 3$.

______

EXAMPLE Graph the Solutions of an Equation

2 **Graph the solution of $x + 8 = 7$ on a number line.**

$x + 8 = 7$ Write the equation.

$x + 8 -$ ___ $= 7 -$ ___ Subtract ___ from each side.

___ = ___ Simplify.

Graph the solution on a number line.

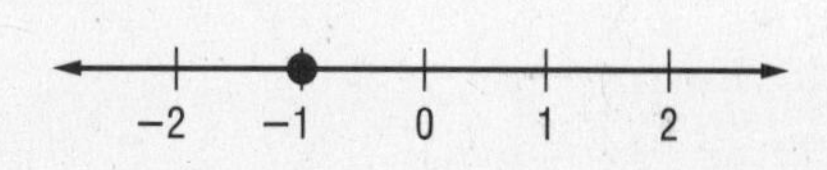

Your Turn Graph the solution of $x + 5 = 9$ on a number line.

KEY CONCEPT

Addition Property of Equality If you add the same number to each side of an equation, the two sides remain equal.

FOLDABLES Under the *Equations: +, – tab,* write one equation that can be solved by subtracting and one that can be solved by adding.

EXAMPLE Solve Equations by Adding

3 **Solve $y - 3 = -14$.**

$y - 3 = -14$

$y + \square = -14$ Rewrite $y - 3$ as $\square$.

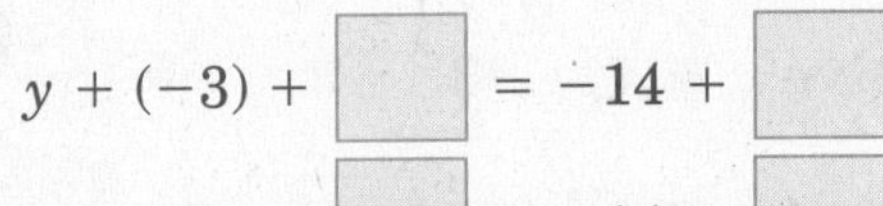

$y + (-3) + \square = -14 + \square$ Add $\square$ to each side.

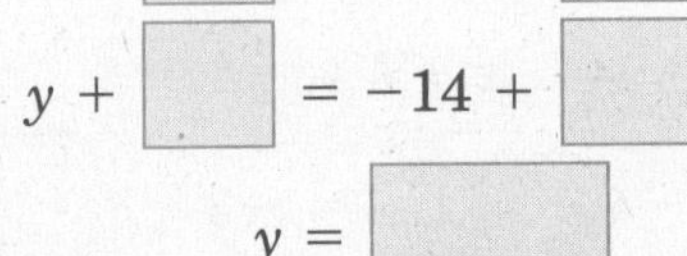

$y + \square = -14 + \square$ Additive Inverse Property

$y = \square$ Identity Property

Your Turn Solve $x - 2 = -9$.

EXAMPLE Use an Equation to Solve a Problem

4 ENTERTAINMENT **Movie A earned \$225 million at the box office. That is \$38 million less than Movie B earned. Write and solve an equation to find the amount Movie B earned.**

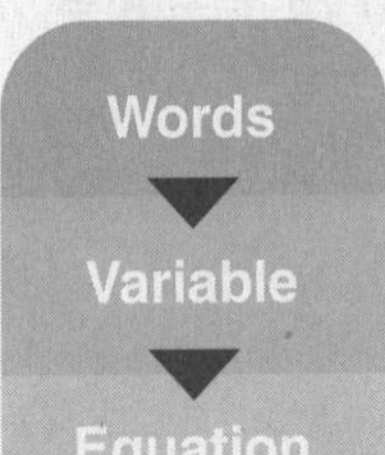

Movie A earned \$38 million less than Movie B earned.

Let b = amount Movie B earned.

Movie A earned \$38 million less than Movie B.

$\square \quad \square \quad \square$

Solve the equation.

Write the equation.

225 ____ $= b - 38$ ____ Add ____ to each side.

____ $=$ ____ Simplify.

Movie B earned ____ at the box office.

Your Turn Board A measures 22 feet. That is 9 feet more than the measure of board B. Write and solve an equation to find the measure of board B.

WRITE IT

Describe how you would check the solution to a problem.

EXAMPLE Solve Equations

5 **What value of x makes $x - 1 = 8$ a true statement?**

To find the value of x, solve the equation.

$x - 1 = 8$ Write the equation.

$x - 1 +$ ____ $= 8 +$ ____ Add ____ to each side.

____ Simplify.

Your Turn What value of x makes $x - 3 = -5$ a true statement?

HOMEWORK ASSIGNMENT

Page(s):

Exercises:

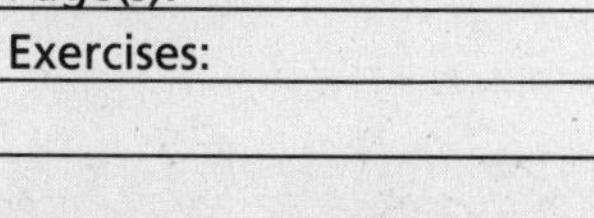

3–4 Solve Equations by Multiplying or Dividing

EXAMPLE Solve Equations by Dividing

WHAT YOU'LL LEARN

- Solve equations by using the Division Property of Equality.
- Solve equations by using the Multiplication Property of Equality.

KEY CONCEPTS

Division Property of Equality When you divide each side of an equation by the same nonzero number, the two sides remain equal.

Multiplication Property of Equality When you multiply each side of an equation by the same number, the two sides remain equal.

1 Solve $7x = -56$. Graph the solution on a number line.

$7x = -56$ Write the equation.

$\frac{7x}{\square} = \frac{-56}{\square}$ Divide each side by $\square$.

$\square = \square$ $7 \div \square = \square$, $-56 \div \square = \square$

$\square = \square$ Identity Property; $1x = \square$

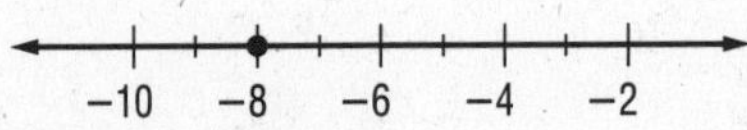

Your Turn Solve $4x = -12$. Graph the solution on a number line.

EXAMPLE Use an Equation to Solve a Problem

2 Esteban spent $112 on boxes of baseball cards If he paid $14 per box, how many boxes of cards did Esteban buy?

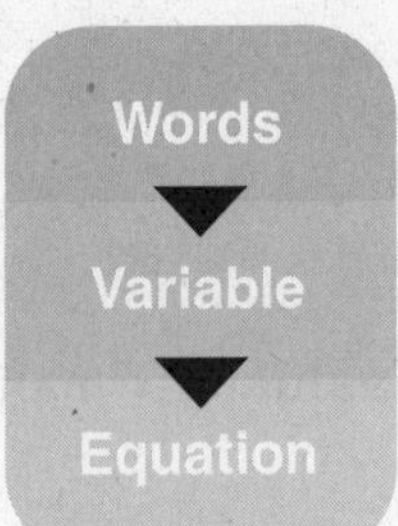

$14 times the number of boxes equals the total.

Let x represent the number of boxes.

cost per box	times	number of boxes	equals	total
$\square$	$\cdot$	$\square$	$=$	$\square$

Solve the equation.

☐ = ☐ Write the equation.

☐ = ☐ Divide each side by ☐.

Your Turn Drew spent \$18 on toy cars. If the cars cost \$2 each, how many cars did Drew buy?

EXAMPLE Solve Equations by Multiplying

FOLDABLES

ORGANIZE IT

Under the *Equations:* ×, ÷ *tab,* write one equation that can be solved by multiplying and one that can be solved by dividing.

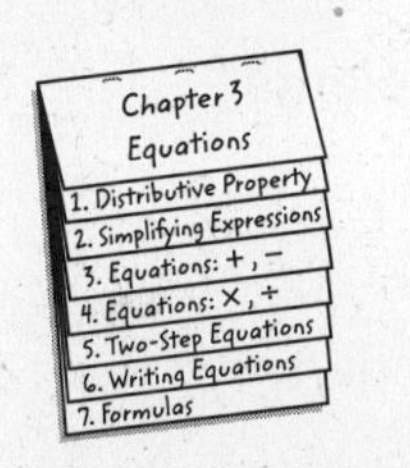

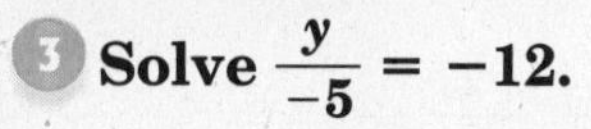

3 Solve $\frac{y}{-5} = -12$.

$\frac{y}{-5} = -12$ Write the equation.

$\frac{y}{-5}$ ☐ $= -12$ ☐ Multiply each side by ☐ to undo the division.

☐ = ☐ Simplify.

Your Turn Solve $\frac{m}{4} = -9$.

HOMEWORK ASSIGNMENT

Page(s):

Exercises:

3–5 Solving Two-Step Equations

WHAT YOU'LL LEARN

- Solve two-step equations.

BUILD YOUR VOCABULARY (page 51)

A **two-step equation** contains ______ operations.

EXAMPLE Solve Two-Step Equations

1 Solve $3x - 4 = 17$.

$3x - 4 +$ ______ $= 17 +$ ______ Add ______ to each side.

______ $=$ ______ Simplify.

$\dfrac{3x}{\_\_\_\_} = \dfrac{21}{\_\_\_\_}$ Undo ______.

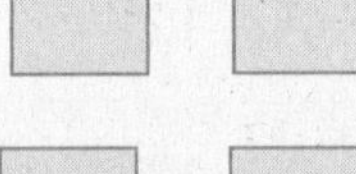

Divide each side by ______.

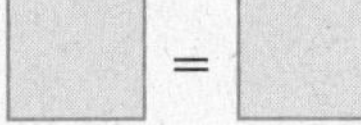

______ $=$ ______ Simplify.

FOLDABLES ORGANIZE IT

Under *Two-Step Equations* tab, write a two-step equation. Then write the order of the steps you would use to solve that equation.

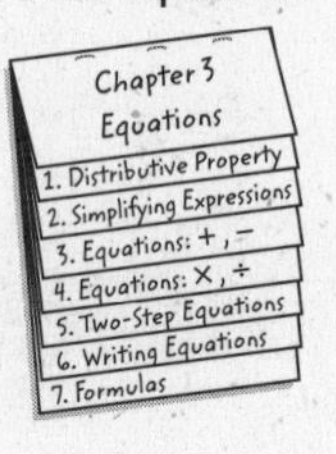

EXAMPLE Equations with Negative Coefficients

2 Solve $5 - x = 7$.

$5 -$ ______ $= 7$ Identity Property

$5 + (-1x) = 7$ Definition of subtraction

$5 + (-1x) + (-5) = 7 + (-5)$ Add -5 to each side.

$(-1x) =$ ______ Simplify.

______ $=$ ______ Divide each side by ______.

______ $=$ ______ Simplify.

EXAMPLE Combine Like Terms Before Solving

3 **Solve $b - 3b + 8 = 18$.**

$\square - 3b + 8 = 18$	Identity Property
$\square + 8 = 18$	Combine like terms.
$\square + 8 \;\square = 18 \;\square$	Subtract $\square$ from each side.
$\square = \square$	Simplify.
$\square = \square$	Divide each side by .
$\square = \square$	Simplify.

Your Turn **Solve each equation.**

a. $4x + 3 = 19$

b. 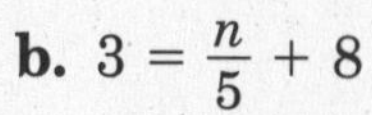 $3 = \frac{n}{5} + 8$

c. Solve $9 = -4 - m$.

d. 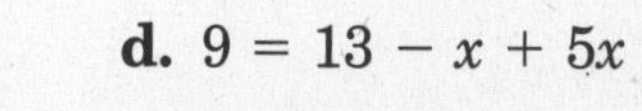 $9 = 13 - x + 5x$

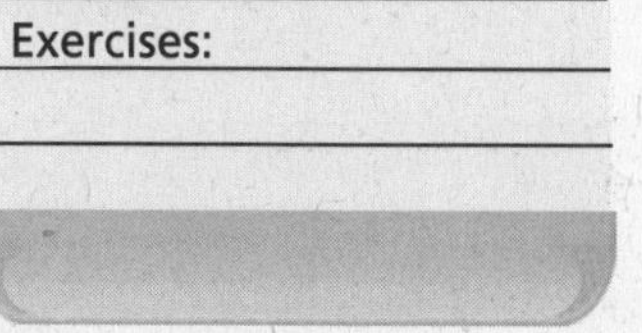

HOMEWORK ASSIGNMENT

Page(s):

Exercises:

3–6 Writing Two-Step Equations

EXAMPLE Translate Sentences into Equations

WHAT YOU'LL LEARN

- Write verbal sentences as two-step equations.
- Solve verbal problems by writing and solving two-step equations.

1 Translate *twice a number increased by 5 equals −25* into an equation.

Your Turn Translate *five times a number decreased by 9 equals –6* into an equation.

EXAMPLE Translate and Solve an Equation

2 Nine more than four times a number is 41. Find the number.

Words	Nine more than four times a number is 41.
Variable	Let n = the number.
Equation	$9 + 4n = 41$

$9 + 4n = 41$ Write the equation.

$9 + 4n - \_\_\_ = 41 - \_\_\_$ Subtract ___ from each side.

$\_\_\_ = \_\_\_$ Simplify.

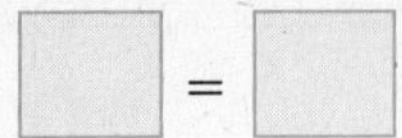

Divide each side by ___.

Your Turn Six less than three times a number is 15. Find the number.

FOLDABLES

ORGANIZE IT

Under the *Writing Equations* tab, list two words or phrases that can be translated into each of the four basic operations.

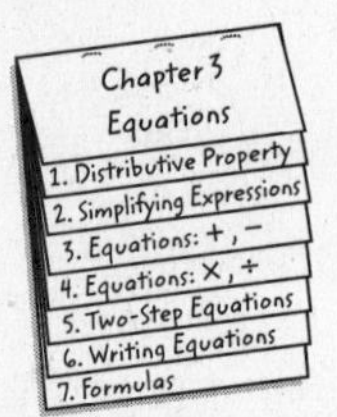

EXAMPLE Write and Solve a Two-Step Equation

3 EARNINGS **Ms. Blake earns $48,400 per year. This is $4150 more than three times as much as her daughter earns. How much does her daughter earn?**

Words: Ms. Blake earns $4150 more than three times as much as her daughter.

Variable: Let d = daughter's earnings

Equation:

Ms. Blake	earns	$4150	more than	three times as much as her daughter
______	=	$4150	+	______

______ = 4150 + ______ Write the equation.

Subtract ______ from each side.

______ − ______ = 4150 + ______ − ______

______ = ______ Simplify.

______ = ______ Divide each side by ______.

______ = ______ Simplify.

Ms. Blake's daughter earns ______.

Your Turn Tami spent $175 at the grocery store. That is $25 less than four times as much as Ted spent. How much did Ted spend?

HOMEWORK ASSIGNMENT

Page(s):

Exercises:

3–7 Using Formulas

What You'll Learn

- Solve problems by using formulas.
- Solve problems involving the perimeters and areas of rectangles.

BUILD YOUR VOCABULARY (pages 50–51)

A **formula** is an ______ that shows a relationship among certain quantities.

The ______ around a geometric figure is called the **perimeter**.

EXAMPLE Use the Distance Formula

1 TRAVEL **If you travel 135 miles in 3 hours, what is your average speed in miles per hour?**

$d = rt$ Write the formula.

____ = ____ $d =$ ____, $t =$ ____

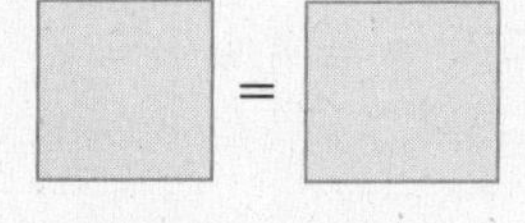

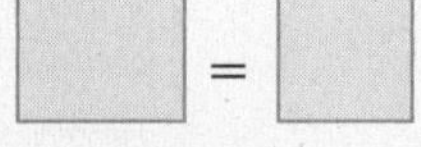

Divide each side by ____.

Simplify.

EXAMPLE Find the Perimeter of a Rectangle

2 **Find the perimeter of the rectangle.**

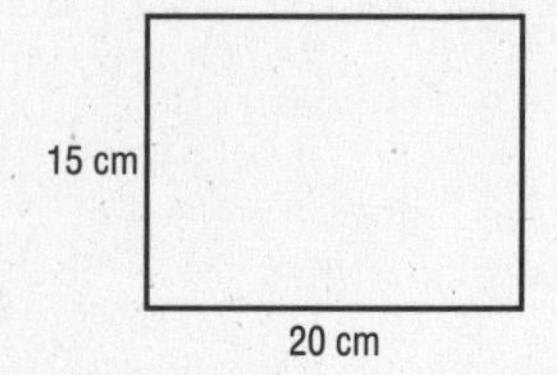

Key Concept

Perimeter of a Rectangle The perimeter of a rectangle is twice the sum of the length and width.

$P =$ ____ Write the formula.

$P =$ ____ $\ell =$ ____, $w =$ ____

$P =$ ____ Add ____ and ____.

$P =$ ____ Simplify.

Your Turn

a. If you drive 520 miles in 8 hours, what is your average speed in miles per hour?

b. Find the perimeter of the rectangle.

6 in.

14 in.

FOLDABLES™

ORGANIZE IT

Locate a rectangular object in your classroom and measure its length and width. Under the *Formulas* tab, describe how to determine its perimeter using the perimeter formula.

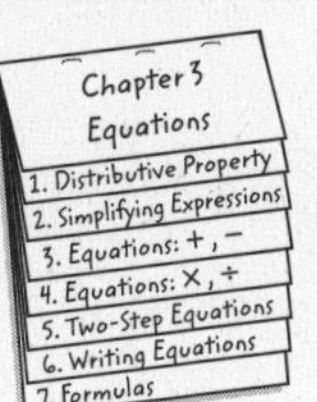

EXAMPLE Find a Missing Length

3 **The perimeter of a rectangle is 60 feet. Its width is 9 feet. Find its length.**

$P = 2(\ell + w)$ Write the formula.

$P =$ ____ Distributive Property

____ = ____ $P =$ ____, $w =$ ____

____ = ____ Simplify.

____ = ____ Subtract ____ from each side.

____ = ____ Simplify.

____ = ____ Divide each side by ____.

The length is ____ feet.

Your Turn The perimeter of a rectangle is 36 meters. Its width is 6 meters. Find its length.

Key Concept

Area of a Rectangle The area of a rectangle is the product of the length and width.

Build Your Vocabulary (page 50)

The measure of the surface enclosed by a figure is its **area**.

EXAMPLE Find the Area of a Rectangle

4 Find the area of a rectangle with length 14 feet and width 6 feet.

$A = \square$ Write the formula.

$A = \square \cdot \square$ $\ell = \square$, $w = \square$

$A = \square$ Simplify.

The area is $\square$ square feet.

EXAMPLE Find a Missing Width

5 The area of a rectangle is 40 square meters. Its length is 8 meters. Find its width.

$A = \square$ Write the formula.

$\square = \square$ $A = \square$, $\ell = \square$

$\square = \square$ Divide each side by $\square$.

The width is $\square$ meters.

Your Turn

a. Find the area of a rectangle with length 11 yards and width 6 yards.

b. The area of a rectangle is 42 square inches. Its length is 14 inches. Find its width.

Homework Assignment

Page(s):

Exercises:

CHAPTER 3

BRINGING IT ALL TOGETHER

STUDY GUIDE

	VOCABULARY PUZZLEMAKER	Build your Vocabulary
Use your **Chapter 3 Foldable** to help you study for your chapter test.	To make a crossword puzzle, word search, or jumble puzzle of the vocabulary words in Chapter 3, go to: www.glencoe.com/sec/math/t_resources/free/index.php	You can use your completed **Vocabulary Builder** (pages 50–51) to help you solve the puzzle.

3-1

The Disributive Property

Match each expression with an equivalent expression.

1. $5(4 + 7)$

2. $(5 + 4)7$

3. $-4(5 + 7)$

4. $(5 - 7)4$

5. $-4(5 - 7)$

a. $5 \cdot 7 + 4 \cdot 7$

b. $(-4) \cdot 5 + (-4) \cdot (-7)$

c. $5 \cdot 4 + 5 \cdot 7$

d. $(-4) \cdot 5 + (-4) \cdot 7$

e. $(-5) \cdot 7 + (-5) \cdot 4$

f. $5 \cdot 4 + (-7) \cdot 4$

6. In rewriting $3(x + 2)$, which term is "distributed" to the other terms in the expression?

3-2

Simplifying Algebraic Expressions

Underline the term that best completes each statement.

7. A term without a variable is a (coefficients, constant).

8. The expression $5z + 2z + 9 + 6z$ has three (like terms, terms).

Simplify each expression.

9. $6q + 2q$

10. $12y - y$

11. $5 + 7x - 3$

12. $4(b + 1) + b$

3-3

Solving Equations by Adding or Subtracting

Underline the term that best completes each statement.

13. To undo the addition of 8 in the expression $y + 8$, you would add -8. This is an example of (inverse operations, simplest form.)

14. The equations $x + 3 = 12$ and $x = 9$ are equivalent equations because they have the same (solution, variable).

Solve each equation.

15. $7 + z = 19$

16. $19 = x - 8$

17. Write and solve an equation for the sentence. The sum of –13 and a number is –16.

3-4

Solving Equations by Multiplying or Dividing

Solve each equation.

18. $3m = 39$

19. $\frac{c}{8} = -6$

20. What value of h makes $\frac{h}{-2} = 16$ a true statement?

A. –8 **B.** –32 **C.** 8 **D.** 32

3-5

Solving Two-Step Equations

Solve each equation.

21. $4y + 3 = 15$

22. $-17 = 6q + 7$

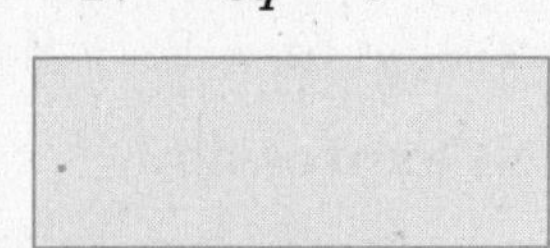

23. $9 = \frac{b}{3} - 12$

24. $31 = 2x + 6 - 7x$

3-6 Writing Two-Step Equations

Translate each sentence into an equation. Then find the number.

25. Seven more than twice a number is 15.

26. Six times a number less 14 is 16.

27. Six decreased by four times a number is 18.

28. Thirteen more than the quotient of a number and 3 is -5.

3-7 Using Formulas

29. How far does a bus driver travel if he drives 55 miles per hour for 5 hours?

30. What is the speed in miles per hour of a raft that travels 18 miles in 3 hours?

Find the perimeter and area of each rectangle.

31.

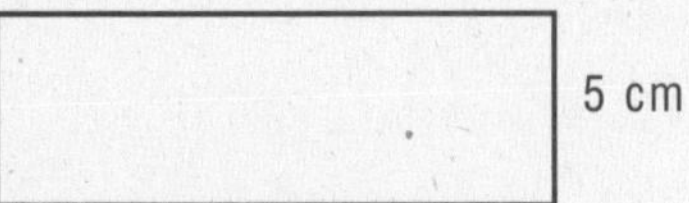

32.

ARE YOU READY FOR THE CHAPTER TEST?

Math Online

Visit **www.pre-alg.com** to access your textbook, more examples, self-check quizzes, and practice tests to help you study the concepts in Chapter 3.

Check the one that applies. Suggestions to help you study are given with each item.

☐ **I completed the review of all or most lessons without using my notes or asking for help.**

- You are probably ready for the Chapter Test.
- You may want to take the Chapter 3 Practice Test on page 141 of your textbook as a final check.

☐ **I used my Foldable or Study Notebook to complete the review of all or most lessons.**

- You should complete the Chapter 3 Study Guide and Review on pages 138–140 of your textbook.
- If you are unsure of any concepts or skills, refer back to the specific lesson(s).
- You may also want to take the Chapter 3 Practice Test on page 141.

☐ **I asked for help from someone else to complete the review of all or most lessons.**

- You should review the examples and concepts in your Study Notebook and Chapter 3 Foldable.
- Then complete the Chapter 3 Study Guide and Review on pages 138–140 of your textbook.
- If you are unsure of any concepts or skills, refer back to the specific lesson(s).
- You may want to take the Chapter 3 Practice Test on page 141.

Student Signature

Parent/Guardian Signature

Teacher Signature

Factors and Fractions

Use the instructions below to make a Foldable to help you organize your notes as you study the chapter. You will see Foldable reminders in the margin of this Interactive Study Notebook to help you in taking notes.

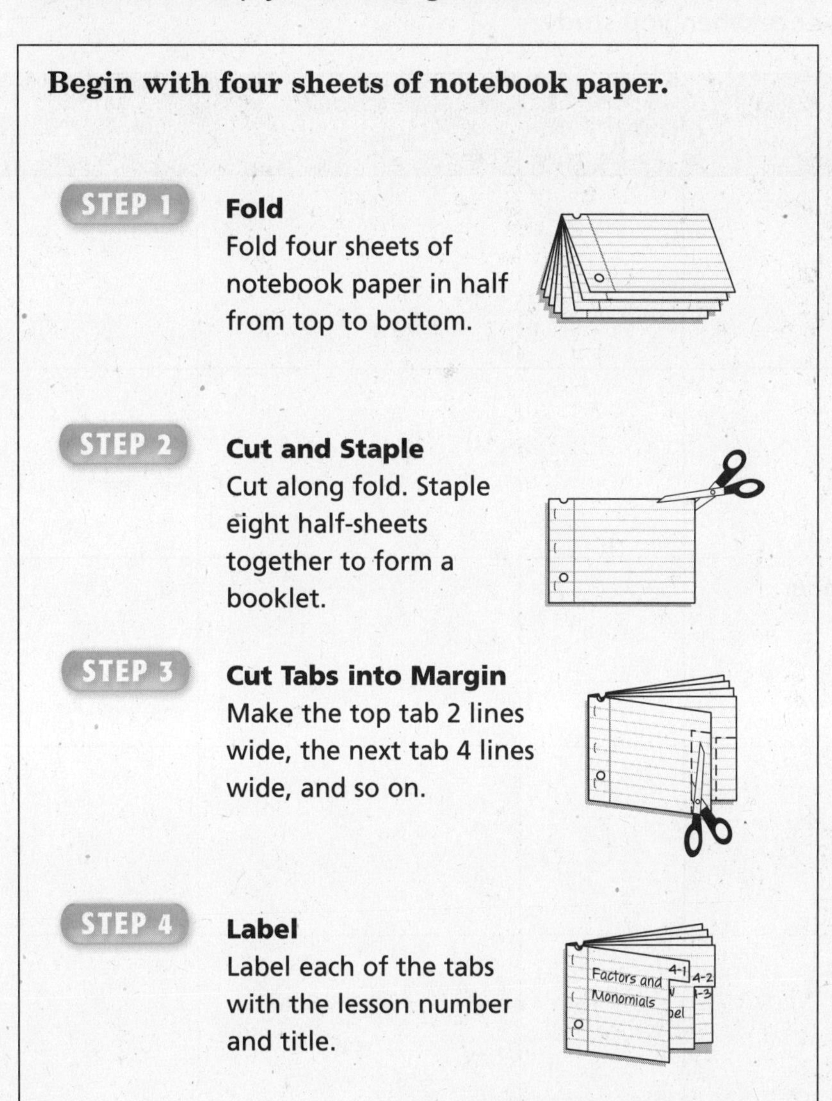

Begin with four sheets of notebook paper.

STEP 1 **Fold**
Fold four sheets of notebook paper in half from top to bottom.

STEP 2 **Cut and Staple**
Cut along fold. Staple eight half-sheets together to form a booklet.

STEP 3 **Cut Tabs into Margin**
Make the top tab 2 lines wide, the next tab 4 lines wide, and so on.

STEP 4 **Label**
Label each of the tabs with the lesson number and title.

NOTE-TAKING TIP: At the end of each lesson, write a summary of the lesson, or write in your own words what the lesson was about.

CHAPTER 4

BUILD YOUR VOCABULARY

This is an alphabetical list of new vocabulary terms you will learn in Chapter 4. As you complete the study notes for the chapter, you will see Build Your Vocabulary reminders to complete each term's definition or description on these pages. Remember to add the textbook page number in the second column for reference when you study.

Vocabulary Term	Found on Page	Definition	Description or Example
algebraic fraction			
base			
composite number			
divisible			
expanded form			
exponent			

Vocabulary Term	Found on Page	Definition	Description or Example
factor			
factor tree			
greatest common factor (GFC)			
monomial			
power			
prime factorization			
prime number			
scientific notation			
simplest form			
standard form			
Venn Diagram			

4–1 Factors and Monomials

What You'll Learn

- Determine whether one number is a factor of another.
- Determine whether an expression is a monomial.

Foldables™

Organize It

Write an example of each divisibility rule under the Lesson 4-1 tab.

Build Your Vocabulary (pages 74–75)

Two or more ______ that are multiplied to form a ______ are called **factors**.

If two numbers are ______ of a third number, we say that the third number is **divisible** by the first two numbers.

EXAMPLE Use Divisibility Rules

1 **Determine whether 435 is divisible by 2, 3, 5, 6, or 10.**

Number	Divisible	Reason
2	______	The ones digit is ______ and ______ ______ divisible by 2.
3	______	The sum of the digits is 4 + 3 + 5 or ______, and ______ ______ divisible by 3.
5	______	The ones digit is ______.
6	______	435 ______ divisible by 2, so it ______ divisible by 6.
10	______	The ones digit is not ______.

Your Turn **Determine whether 786 is divisible by 2, 3, 5, 6, or 10.**

______

EXAMPLE Find Factors of a Number

2 **List all the factors of 64.**

Using the divisibility rules to determine whether 64 is divisible by 2, 3, 5, and so on. Then use division to find other factors of 64.

Number	64 Divisible by Number?	Factor Pairs
1	yes	$1 \cdot 64$
2		
3	no	—
4		
5	no	—
6		—
7	no	—
8	yes	

So, the factors of 64 are 1, 2, 4, 8, 16, 32, and 64.

Your Turn List all the factors of 96.

BUILD YOUR VOCABULARY (page 75)

A **monomial** is a ______, a ______, or a product of ______ and/or ______.

WRITE IT

What must you do to an expression before you can determine whether or not it is a monomial?

EXAMPLE Identify Monomials

3 **a. Determine whether $4(n + 3)$ is a monomial.**

$4(n + 3) = 4$ ______ $+ 4($ ______ $)$ ______ Property

$=$ ______ Simplify.

This expression is not a monomial because in its simplest form, it involves two terms that are added.

b. Determine whether $\frac{x}{3}$ is a monomial.

This expression is a ______ because it is the product of a ______, $\frac{1}{3}$, and a ______, x.

Your Turn **Determine whether each expression is the monomial.**

a. $7ab$

b. $6(4 - x)$

HOMEWORK ASSIGNMENT

Page(s):

Exercises:

4–2 Powers and Exponents

WHAT YOU'LL LEARN

- Write expressions using exponents.
- Evaluate expressions containing exponents.

BUILD YOUR VOCABULARY (pages 74–75)

In an expression like 2^4, the **base** is the number that is ______.

The **exponent** tells how many times the base is used as a ______.

The number that can be expressed using an ______ is called a **power**.

EXAMPLE Write Expressions Using Exponents

1 **Write each expression using exponents.**

a. $6 \cdot 6 \cdot 6 \cdot 6$

The base is ______. It is a factor ______ times, so the exponent is ______.

$6 \cdot 6 \cdot 6 \cdot 6 =$ ______

b. $(-1)(-1)(-1)$

The base is ______. It is a factor ______ times, so the exponent is ______.

$(-1)(-1)(-1) =$ ______

c. $(5x + 1)(5x + 1)$

The base is ______. It is a factor ______ times, so the exponent is ______.

$(5x + 1)(5x + 1) =$ ______

d. $\frac{1}{2} \cdot x \cdot x \cdot x \cdot x \cdot y \cdot y \cdot y$

First group the factors with like bases. Then write using exponents.

$= \frac{1}{2} \cdot (x \cdot x \cdot x \cdot x) \cdot (y \cdot y \cdot y)$

$=$ ______

WRITE IT

What is the difference between $(-5)^2$ and -5^2? Explain.

______

______

______

______

______

Your Turn **Write each expression using exponents.**

a. $3 \cdot 3 \cdot 3 \cdot 3 \cdot 3 \cdot 3$

b. $(-6)(-6)(-6)(-6)$

c. $(4 - 2x)(4 - 2x)$

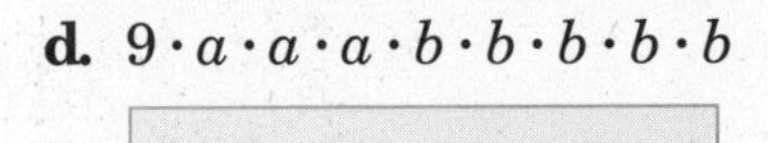

d. $9 \cdot a \cdot a \cdot a \cdot b \cdot b \cdot b \cdot b \cdot b$

EXAMPLE Evaluate Expressions

2 **Evaluate each expression. (parallels example 3 in text)**

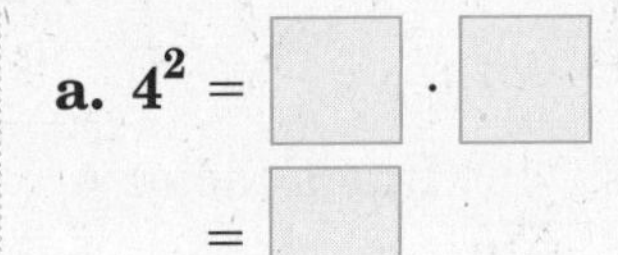

a. $4^2 =$ ____ $\cdot$ ____   ____ is a factor two times.

$=$ ____   Multiply.

b. $r^3 - 3 = (\ \ \ )^3 - 3$ Replace r with ____

$= (\ \ \ )(\ \ \ )(\ \ \ ) - 3$ Rewrite ____^{3}.

$=$ ____ $- 3$ Multiply.

$=$ ____   Subtract.

FOLDABLES ORGANIZE IT

Under the tab for Lesson 4-2, write a summary about the relationship between base, exponent, and factor. Also, give an example of a number written in standard notation.

Your Turn **Evaluate each expression.**

a. 3^4

b. $100 - x^4$ if $x = 2$

4–3 Prime Factorization

WHAT YOU'LL LEARN

- Write the prime factorization of composite numbers.
- Factor monomials.

BUILD YOUR VOCABULARY (pages 74–75)

A **prime number** is a whole number that has exactly two factors, 1 and itself.

A **composite number** is a whole number that has more than two factors.

REMEMBER IT

Zero and 1 are neither prime nor composite.

EXAMPLE Identify Numbers as Prime or Composite

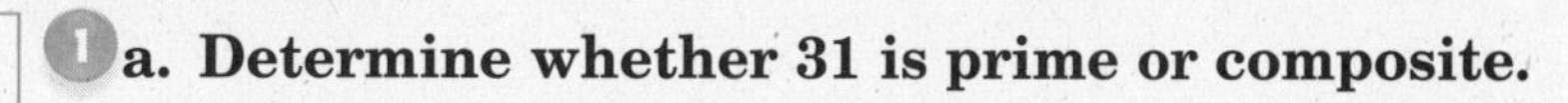

1 a. Determine whether 31 is prime or composite.

Find factors of 31 by listing the whole number pairs whose product is 31.

31 =

The number 31 has only two factors. So, it is a number.

b. Determine whether 36 is prime or composite.

Find factors of 36 by listing the whose product is 36.

36 =

36 =

36 =

36 =

36 =

The factors of 36 are .

Since the number has more than two factors, it is a number.

Your Turn **Determine whether each number is prime or composite.**

a. 49

b. 29

BUILD YOUR VOCABULARY (page 75)

When a composite number is expressed as the product of prime factors, it is called the **prime factorization** of the number.

One way to find the prime factorization of a number is to use a **factor tree**.

To **factor** a number means to write it as a product of its factors.

EXAMPLE Write Prime Factorization

2 **Write the prime factorization of 56.**

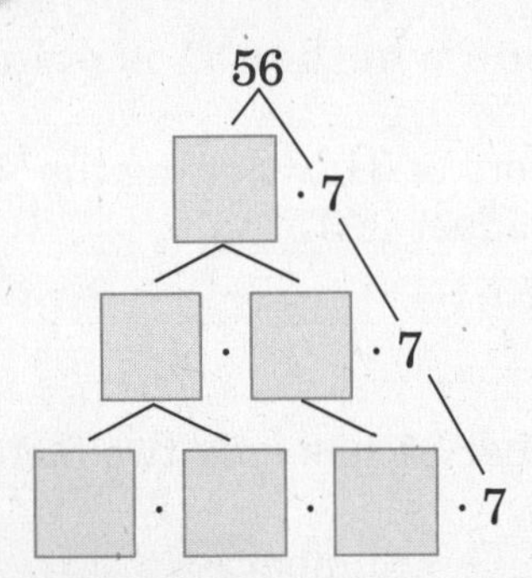

The prime factorization of 56 is ______.

EXAMPLE Factor Monomials

3 **a. Factor $16p^2q^4$.**

$16p^2q^4 = 2 \cdot 2 \cdot 2 \cdot 2 \cdot p^2 \cdot q^4$

$= \square \cdot \square \cdot \square \cdot \square \cdot \square \cdot \square \cdot \square \cdot \square \cdot \square \cdot \square$

b. Factor $-21x^2y$.

$-21x^2y = -1 \cdot \square \cdot \square \cdot x^2 \cdot y$

$= -1 \cdot \square \cdot \square \cdot \square \cdot \square \cdot \square$

ORGANIZE IT

Under the Lesson 4-3 tab, describe how to use a factor tree to find the prime factorization of a number.

HOMEWORK ASSIGNMENT

Page(s):

Exercises:

Your Turn

a. Write the prime factorization of 72.

b. Factor $12a^3b$.

c. Factor $-18mn^2$.

4-4 Greatest Common Factor (GCF)

What You'll Learn

- Find the greatest common factor of two or more numbers or monomials.
- Use the Distributive Property to factor algebraic expressions.

Build Your Vocabulary (page 75)

A **Venn diagram** shows the relationship among sets of ______ or objects by using overlapping ______ in a rectangle.

The ______ number that is a ______ of two or more number is called the **greatest common factor (GCF)**.

Foldables

Organize It

Under the Lesson 4-4 tab, describe the two methods for finding the GCF of two or more numbers.

Example Find the GCF

1 Find the GCF of 16 and 24.

Method 1 List the factors.

factors of 16: ______

factors of 24: ______

The greatest common factor of 16 and 24 is ______.

Method 2 Use prime factorization.

$16 = 2 \cdot 2 \cdot 2 \cdot 2$

$24 = 2 \cdot 2 \cdot 2 \cdot 3$

The GFC is the product of the common ______.

$2 \cdot 2 \cdot 2 =$ ______

Your Turn **Find the GCF of 18 and 30.**

EXAMPLE Find the GCF of Monomials

2 Find the GCF of $18x^3y^2$ and $42xy^2$.

Completely factor each expression.

$18x^3y^2 = \boxed{2} \cdot \boxed{3} \cdot 3 \cdot \boxed{x} \cdot x \cdot x \cdot \boxed{y} \cdot \boxed{y}$

$42xy^2 = \boxed{2} \cdot \boxed{3} \cdot 7 \cdot \boxed{x} \cdot \boxed{y} \cdot \boxed{y}$

Circle the common factors.

The GCF of $18x^3y^2$ and $42xy^2$ is ______ or ______.

Your Turn Find the GCF of $32mn^4$ and $80m^3n^2$.

______

REVIEW IT

Name the operations that are combined by the Distributive Property. (*Lesson 3-1*)

______

______

______

______

EXAMPLE Factor Expressions

3 Factor $3x + 12$.

First, find the GCF of $3x$ and 12.

$3x = \boxed{3} \cdot x$

$12 = 2 \cdot 2 \cdot \boxed{3}$

The GCF is ______.

Now, write each term as a product of the ______ and its remaining factors.

$3x + 12 = 3(\quad) + 3(\quad)$

$= 3(\quad)$ ______ Property

Your Turn Factor $4x + 20$.

______

HOMEWORK ASSIGNMENT

Page(s): ______

Exercises: ______

4–5 Simplifying Algebraic Fractions

WHAT YOU'LL LEARN

- Simplify fractions using the GCF.
- Simplify algebraic fractions.

BUILD YOUR VOCABULARY (pages 74–75)

A fraction is in **simplest form** when the GCF of the ______ and the ______ is 1.

A fraction with ______ in the ______ or ______ is called an **algebraic fraction.**

EXAMPLES Simplify Fractions

Write each fraction in simplest form.

1 $\mathbf{\frac{16}{24}}$

$16 = 2 \cdot 2 \cdot 2 \cdot 2$ Factor the ______.

$24 = 2 \cdot 2 \cdot 2 \cdot 3$ Factor the ______.

The  of 16 and 24 is 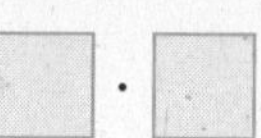 ___ · ___ · ___ or ___.

$\frac{16}{24} = \frac{16 \div \square}{24 \div \square} = \square$ Divide the numerator and denominator by the ______.

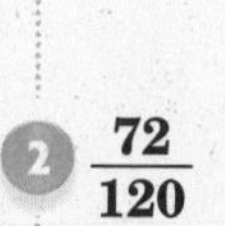

2 $\mathbf{\frac{72}{120}}$

$\frac{72}{120} = \frac{\overset{1}{\cancel{2}} \cdot \overset{1}{\cancel{2}} \cdot \overset{1}{\cancel{2}} \cdot \overset{1}{\cancel{3}} \cdot 3}{\underset{1}{\cancel{2}} \cdot \underset{1}{\cancel{2}} \cdot \underset{1}{\cancel{2}} \cdot \underset{1}{\cancel{3}} \cdot 5} = \square$ Divide the numerator and the denominator by the GCF.

Your Turn **Write each fraction in simplest form.**

a. $\frac{12}{40}$

b. $\frac{48}{80}$

WRITE IT

Describe the result if a common factor other than the GCF is used to simplify a fraction.

______

______

______

______

FOLDABLES™

ORGANIZE IT

Under the Lesson 4-5 tab, explain how to simplify both numeric and algebraic fractions.

EXAMPLE Simplify Algebraic Fractions

3 **Simplify $\frac{20m^3n^2}{65mn}$.**

$\frac{20m^3n^2}{65mn} = \frac{2 \cdot 2 \cdot \overset{1}{\cancel{5}} \cdot \overset{1}{\cancel{m}} \cdot m \cdot m \cdot \overset{1}{\cancel{n}} \cdot n}{\underset{1}{\cancel{5}} \cdot 13 \cdot \underset{1}{\cancel{m}} \cdot \underset{1}{\cancel{n}}}$ Divide the numerator and the denominator by the GCF.

= ______ Simplify.

Your Turn **Simplify $\frac{14x^4y^2}{49x^2y^2}$.**

HOMEWORK ASSIGNMENT

Page(s):

Exercises:

4–6 Multiplying and Dividing Monomials

WHAT YOU'LL LEARN

- Multiply monomials.
- Divide monomials.

EXAMPLE Multiply Powers

1 **Find $3^4 \cdot 3^6$.**

$3^4 \cdot 3^6 = 3^{\square + \square}$ The common base is $\square$.

$= 3^{\square}$ $\square$ the exponents.

Your Turn Find $4^3 \cdot 4^5$.

EXAMPLE Multiply Monomials

2 **a. Find $y^4 \cdot y$.**

$y^4 \cdot y = y^{\square + \square}$ The common base is $\square$.

$= y^{\square}$ $\square$ the exponents.

KEY CONCEPT

Product of Powers You can multiply powers with the same base by adding their exponents.

b. Find $(3p^4)(-2p^3)$.

$(3p^4)(-2p^3) = (3 \cdot \square)(p^4 \cdot \square)$ Use the Commutative and Associative Properties.

$= (\square)(p^{\square + \square})$ The common base is p.

$= \square$ $\square$ the exponents.

Your Turn **Find each product.**

a. $w^2 \cdot w^5$

b. $(-4m^3)(6m^2)$

EXAMPLE Divide Powers

KEY CONCEPT

Quotient of Powers You can divide powers with the same base by subtracting their exponents.

3 a. Find $\frac{8^{11}}{8^5}$.

$\frac{8^{11}}{8^5} = \square = \square$ The common base is $\square$.

$\square$ the exponents.

b. Find $\frac{x^{12}}{x}$.

$\frac{x^{12}}{x} = \square = \square$ The common base is $\square$.

$\square$ the exponents.

Your Turn **Find each quotient.**

a. $\frac{7^5}{7^3}$

b. $\frac{r^4}{r^1}$

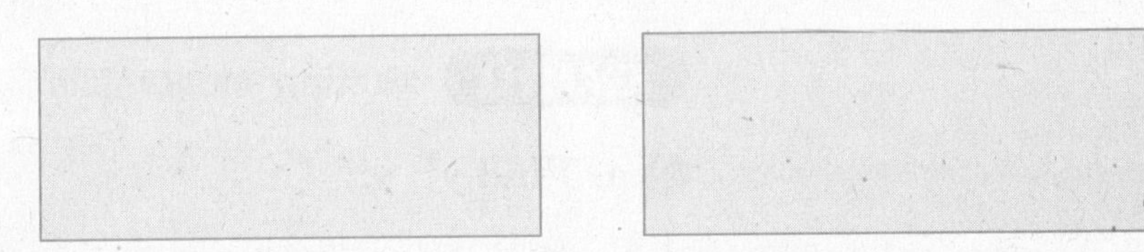

FOLDABLES

ORGANIZE IT

Under the tab for Lesson 4-6, write a summary of the way you can use exponents to multiply and divide polynomials.

HOMEWORK ASSIGNMENT

Page(s):

Exercises:

4–7 Negative Exponents

EXAMPLE Use Positive Exponents

WHAT YOU'LL LEARN

- Write expressions using negative exponents.
- Evaluate numerical expressions containing negative exponents.

1 **Write each expression using a positive exponent.**

a. 3^{-4}

$3^{-4} =$ ______ Definition of ______ exponent

b. m^{-2}

$m^{-2} =$ ______ Definition of ______ exponent

Your Turn **Write each expression using a positive exponent.**

a. 5^{-3} **b.** y^{-6}

EXAMPLE Use Negative Exponents

REMEMBER IT

A negative exponent in an expression does not change the sign of the expression.

2 **Write $\frac{1}{125}$ as an expression using a negative exponent.**

$\frac{1}{125} = \frac{1}{\text{______}}$ Find the ______ of 125.

$= \frac{1}{\text{______}}$ Definition of exponents

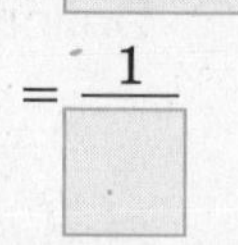

$=$ ______ Definition of negative exponent

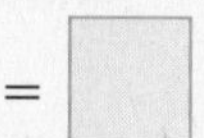

Your Turn **Write $\frac{1}{32}$ as an expression using a negative exponent.**

EXAMPLE Use Exponents to Solve a Problem

FOLDABLES™

ORGANIZE IT

Under the tab for Lesson 4-7, explain negative exponents. Give an example of a number written with a negative exponent and an equivalent expression using a positive exponent.

3 PHYSICS **An atom is an incredibly small unit of matter. The smallest atom has a diameter of approximately $\frac{1}{10}$th of a nanometer, or 0.0000000001 meter. Write the decimal as a fraction and as a power of 10.**

0.0000000001 = ______ Write the decimal as a fraction.

= ______ 10,000,000,000 = ______

= ______ Definition of negative exponent

Your Turn Write 0.000001 as a fraction and as a power of 10.

EXAMPLE Algebraic Expressions with Negative Exponents

4 **Evaluate r^{-2} if $r = -4$.**

$r^{-2} = (\ \_\_\_\ )^{-2}$ Replace r with ______.

= ______ Definition of ______ exponent

= ______ Find $(-4)^2$.

Your Turn **Evaluate d^{-3} if $d = 5$.**

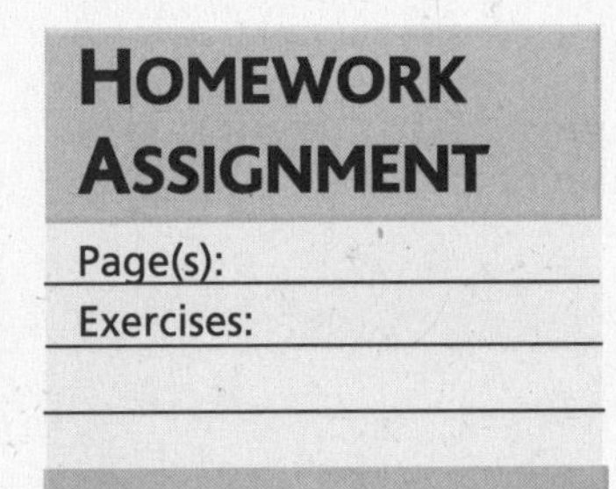

HOMEWORK ASSIGNMENT

Page(s):

Exercises:

4-8 Scientific Notation

What You'll Learn

- Express numbers in standard form and in scientific notation.
- Compare and order numbers written in scientific notation.

Key Concept

Scientific Notation A number is expressed in scientific notation when it is written as the product of a factor and a power of 10. The factor must be greater than equal to 1 and less than 10.

EXAMPLE Express Numbers in Standard Form

1 **Express each number in standard form.**

a. $\mathbf{4.395 \times 10^4}$

$4.395 \times 10^4 = 4.395 \times$ ______ $\quad 10^4 =$ ______

$= 43{,}950$ Move the decimal point ______ places to the right.

$=$ ______

b. $\mathbf{6.79 \times 10^{-6}}$

$6.79 \times 10^{-6} = 6.79 \times$ ______ $\quad 10^{-6} =$ ______

$= 0.00000679$ Move the decimal point ______ places to the ______.

$=$ ______

Your Turn **Express each number in standard form.**

a. 2.614×10^6

b. 8.03×10^{-4}

EXAMPLE Express Numbers in Scientific Notation

2 **a. Express 800,000 in scientific notation.**

______ $= 8.0 \times$ ______ The decimal point moves ______ places.

$= 8.0 \times$ ______ The exponent is

.

FOLDABLES™

ORGANIZE IT

Under the tab for Lesson 4-8, explain the significance of a positive or negative exponent in scientific notation. Give an example of a number with each, written in both standard form and in scientific notation.

Factors and Monomials

b. Express 0.0119 in scientific notation.

0.0119 = ______ × ______ The decimal point moves ______ places.

= ______ × ______ The exponent is ______.

Your Turn **Express each number in scientific notation.**

a. 3,024,000

b. 0.00042

EXAMPLE **Compare Numbers in Scientific Notation**

3 SPACE **The diameters of Mercury, Saturn, and Pluto are 4.9×10^3 kilometers, 1.2×10^5 kilometers, and 2.3×10^3 kilometers, respectively. List the planets in order of increasing diameter.**

First, order the numbers according to their exponents. Then, order the numbers with the same exponent by comparing the factors.

Step 1 Mercury and Pluto: 4.9×10^3, 2.3×10^3 ______ Saturn: 1.2×10^5

Step 2 2.3×10^3 (Pluto) ______ 4.9×10^3 (Mercury) Compare the factors: ______.

So, the order is ______.

Your Turn **Order the numbers 6.21×10^5, 2.35×10^4, 5.95×10^9, and 4.79×10^4 in decreasing order.**

HOMEWORK ASSIGNMENT

Page(s):

Exercises:

CHAPTER 4

BRINGING IT ALL TOGETHER

STUDY GUIDE

FOLDABLES™	VOCABULARY PUZZLEMAKER	BUILD YOUR VOCABULARY
Use your **Chapter 4 Foldable** to help you study for your chapter test.	To make a crossword puzzle, word search, or jumble puzzle of the vocabulary words in Chapter 4, go to: www.glencoe.com/sec/math/t_resources/free/index.php	You can use your completed **Vocabulary Builder** (pages 74–75) to help you solve the puzzle.

4-1 Factors and Monomials

Use divisibility rules to determine whether each number is divisible by 2, 3, 5, 6, or 10.

1. 45 **2.** 68 **3.** 519 **4.** 762

Determine whether each expression is a monomial. Explain why or why not.

5. $2p + q$

6. $2pq$

4-2 Powers and Exponents

Write each expression using exponents.

7. $4 \cdot 4 \cdot 4 \cdot 4$ **8.** $(-2)(-2)(-2)$ **9.** $5 \cdot r \cdot r \cdot m \cdot m \cdot m$ 

Evaluate each expression if $x = 3$, $y = 1$, and $h = -2$.

10. $3h$ **11.** hx^3 **12.** $4(2x - 4y)^3$

4-3

Prime Factorization

Write the prime factorization of each number. Use exponents for repeated factors.

13. 64 **14.** 126 **15.** 735

Factor each monomial completely.

16. $32ac$ **17.** $49s^3t$

18. $144x^3$ **19.** $25pq^4$

4-4

Greatest Common Factor (GCF)

Find the GCF of each set of numbers or monomials.

20. 25, 45 **21.** 36, 54, 66 **22.** $28a^2$, $42ab^3$

Factor each expression.

23. $7x + 14y$ **24.** $50s + 10st$

4-5

Simplifying Algebraic Fractions

25. Six ounces is what part of a pound?

26. Use a Venn diagram to explain how to simplify $\frac{18}{45}$.

4-6

Multiplying and Dividing Monomials

Find each quotient.

27. $\frac{4^6}{4^4}$

28. $\frac{(-3)^3}{(-3)}$

29. $\frac{p^2 \cdot p^3}{p^4}$

Find a match for each product.

30. $2^4 \cdot 2^3$

31. $4^3 \cdot 4^4$

32. $2^5 \cdot 2^7$

a. 4^7
b. 4^{12}
c. 2^7
d. 2^{12}

4-7

Negative Exponents

Write each expression using a positive exponent.

33. 8^{-3}

34. 5^{-10}

35. x^{-2}

Evaluate each expression if $s = 4$ and $t = 3$.

36. t^{-3}

37. $(st)^{-1}$

38. s^{-t}

4-8

Scientific Notation

Tell direction and the number of places you need to move the decimal point to write each number in standard notation.

39. 2.3×10^4

40. 1.5×10^{-7}

41. 7.1×10^{11}

42. The table at the right shows the average wave lengths of 3 types of radiation. Write the radiation types in order from longest to shortest wave length.

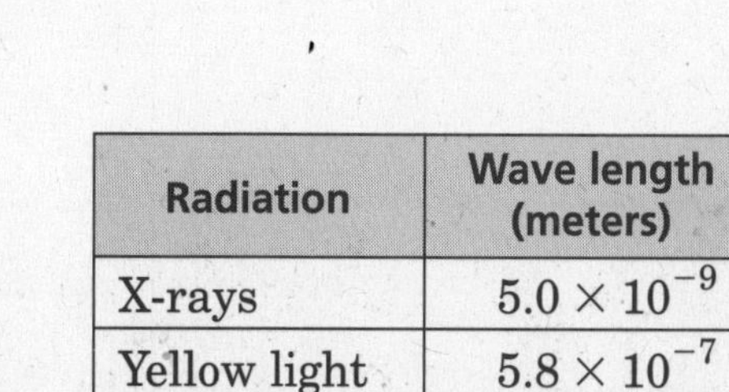

Radiation	Wave length (meters)
X-rays	5.0×10^{-9}
Yellow light	5.8×10^{-7}
Blue light	4.7×10^{-7}

Checklist

ARE YOU READY FOR THE CHAPTER TEST?

Visit **pre-alg.com** to access your textbook, more examples, self-check quizzes, and practice tests to help you study the concepts in Chapter 4.

Check the one that applies. Suggestions to help you study are given with each item.

☐ **I completed the review of all or most lessons without using my notes or asking for help.**

- You are probably ready for the Chapter Test.
- You may want take the Chapter 4 Practice Test on page 195 of your textbook as a final check.

☐ **I used my Foldable or Study Notebook to complete the review of all or most lessons.**

- You should complete the Chapter 4 Study Guide and Review on pages 191–194 of your textbook.
- If you are unsure of any concepts or skills, refer back to the specific lesson(s).
- You may also want to take the Chapter 4 Practice Test on page 195.

☐ **I asked for help from someone else to complete the review of all or most lessons.**

- You should review the examples and concepts in your Study Notebook and Chapter 4 Foldable.
- Then complete the Chapter 4 Study Guide and Review on pages 191–194 of your textbook.
- If you are unsure of any concepts or skills, refer back to the specific lesson(s).
- You may also want to take the Chapter 4 Practice Test on page 195.

Student Signature

Parent/Guardian Signature

Teacher Signature

Rational Numbers

Use the instructions below to make a Foldable to help you organize your notes as you study the chapter. You will see Foldable reminders in the margin of this Interactive Study Notebook to help you in taking notes.

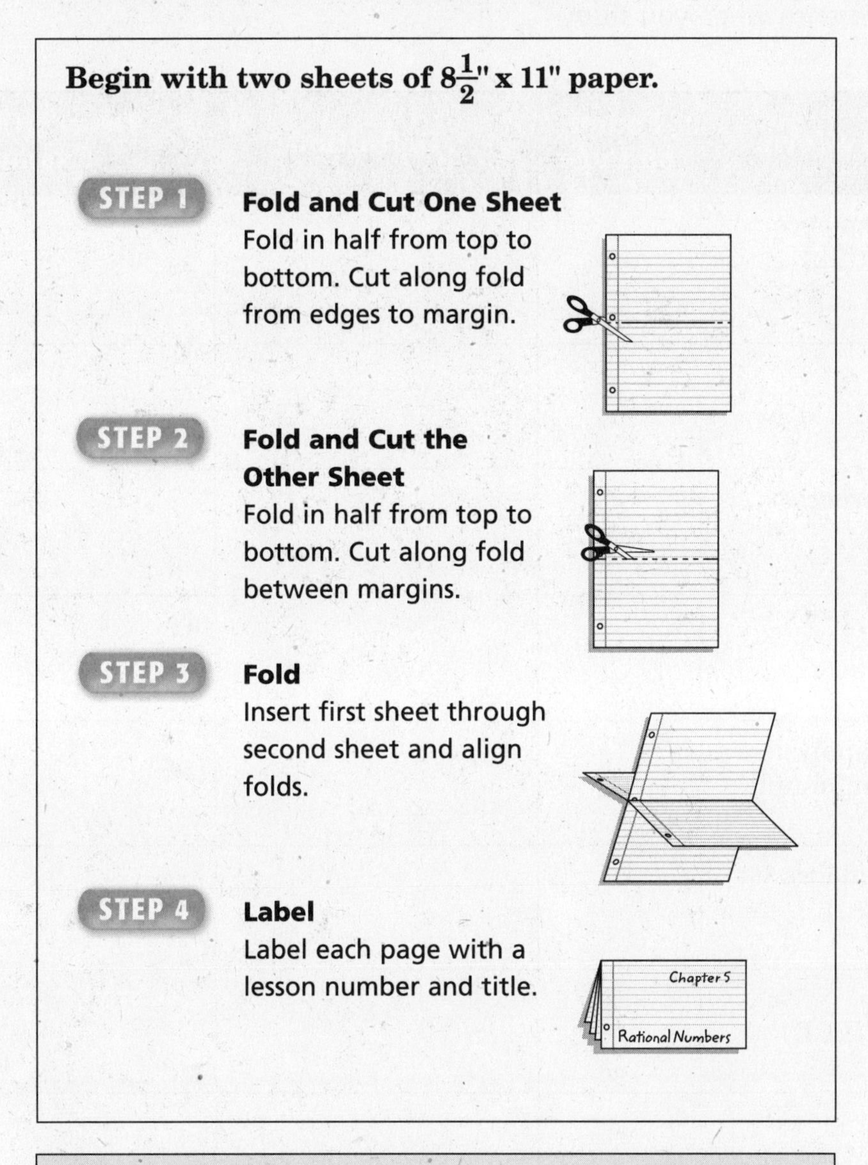

Begin with two sheets of $8\frac{1}{2}$" x 11" paper.

STEP 1 **Fold and Cut One Sheet**
Fold in half from top to bottom. Cut along fold from edges to margin.

STEP 2 **Fold and Cut the Other Sheet**
Fold in half from top to bottom. Cut along fold between margins.

STEP 3 **Fold**
Insert first sheet through second sheet and align folds.

STEP 4 **Label**
Label each page with a lesson number and title.

NOTE-TAKING TIP: As you read each lesson, list examples of ways the new knowledge has been or will be used in your daily life.

BUILD YOUR VOCABULARY

This is an alphabetical list of new vocabulary terms you will learn in Chapter 5. As you complete the study notes for the chapter, you will see Build Your Vocabulary reminders to complete each term's definition or description on these pages. Remember to add the textbook page number in the second column for reference when you study.

Vocabulary Term	Found on Page	Definition	Description or Example
arithmetic sequence [ar-ihth-MEH-tihk]			
bar notation			
common difference			
common ratio			
dimensional analysis [duh-MEHN-chuhn-uhl]			
geometric sequence [JEE-uh-MEH-trihk SEE-kwuhn(t)s]			
least common denominator (LCD)			
least common multiple (LCM)			
mean			

Vocabulary Term	Found on Page	Definition	Description or Example
measure of central tendency			
median			
mode			
multiple			
multiplicative inverse [muhl-tuh-PLIH-kuh-tihv IHN-vuhrs]			
period			
rational number [RASH-nuhl]			
reciprocal [rih-SIHP-ruh-kuhl]			
repeating decimal			
sequence			
term			

5–1 Writing Fractions as Decimals

WHAT YOU'LL LEARN

- Write fractions as terminating or repeating decimals.
- Compare fractions and decimals.

EXAMPLE Write a Fraction as a Terminating Decimal

1 **Write $\frac{1}{16}$ as a decimal.**

Method 1 Use paper and pencil. Divide 1 by 16.

Method 2 Use a calculator.

 ÷  ENTER 0.0625

So, $\frac{1}{16}$ = ______

FOLDABLES

ORGANIZE IT

Under the tab for Lesson 5-1, write an example of when you might want to change two fractions to decimals in order to determine which is larger.

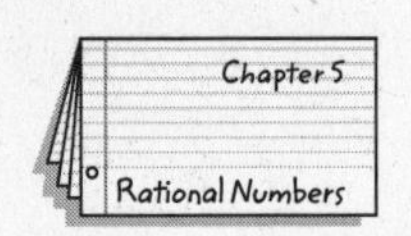

EXAMPLE Write a Mixed Number as a Decimal

2 **Write $1\frac{1}{4}$ as a decimal.**

$1\frac{1}{4}$ = ______ + ______ Write as the sum of an integer and a ______.

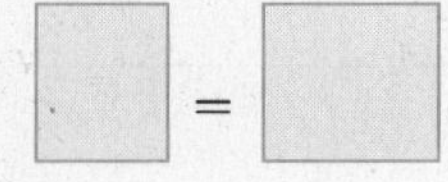

= ______ + ______ ______ = ______

= ______ Add.

Your Turn **Write each fraction or mixed number as a decimal.**

a. $\frac{3}{8}$

b. $2\frac{3}{5}$

BUILD YOUR VOCABULARY (pages 98–99)

A decimal with one or more ______ that repeat forever is called a **repeating decimal**.

You can use **bar notation** to indicate that a digit **repeats** forever.

The **period** of a repeating decimal is the digit or digits that ______.

EXAMPLE Write Fractions as Repeating Decimals

3 Write each fraction as a decimal.

a. $-\frac{4}{33} \rightarrow 33\overline{)-4.0000...}$ with quotient $0.1212...$

The digits ______ repeat.

$-\frac{4}{33} =$ ______

b. $\frac{2}{11} \rightarrow 11\overline{)2.0000...}$

The digits ______ repeat.

$\frac{2}{11} =$ ______

Your Turn **Write each fraction as a decimal.**

a. $-\frac{2}{3}$ ______ **b.** $\frac{4}{15}$ ______

EXAMPLE Compare Fractions and Decimals

4 Replace ● with <, >, or = to make $0.7 \bullet \frac{13}{20}$ a true sentence.

$0.7 \bullet \frac{13}{20}$ — Write the sentence.

$0.7 \bullet$ ______ — Write $\frac{13}{20}$ as a decimal.

0.7 ______ ______ — In the tenths place, ______.

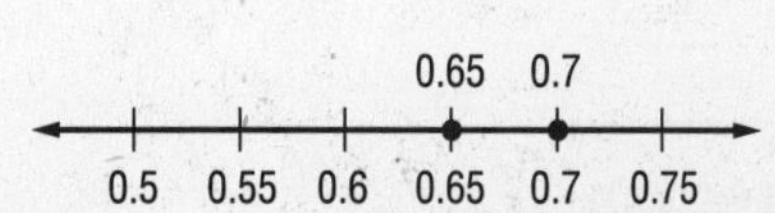

On a number line, 0.7 is to the right of 0.65, so 0.7 ______ $\frac{13}{20}$.

Your Turn Replace • with <, >, or = to make $\frac{3}{8}$ • 0.4 a true sentence.

EXAMPLE Compare Fractions to Solve a Problem

5 GRADES **Jeremy got a score of $\frac{16}{20}$ on his first quiz and $\frac{20}{25}$ on his second quiz. Which grade was the higher score?**

Write the fractions as ______ and then compare the ______.

Quiz #1: $\frac{16}{20}$ = ______ Quiz #2: $\frac{20}{25}$ = ______

The scores were the same, ______.

Your Turn One recipe for cookies requires $\frac{5}{8}$ of a cup of butter and a second recipe for cookies requires $\frac{3}{5}$ of a cup of butter. Which recipe uses less butter?

HOMEWORK ASSIGNMENT

Page(s):

Exercises:

5–2 Rational Numbers

WHAT YOU'LL LEARN

- Write rational numbers as fractions.
- Identify and classify rational numbers.

BUILD YOUR VOCABULARY (page 99)

A number that can be written as a ______ is called a **rational number**.

FOLDABLES™

ORGANIZE IT

In your notes, describe the fractions you use during a normal day at school and at home.

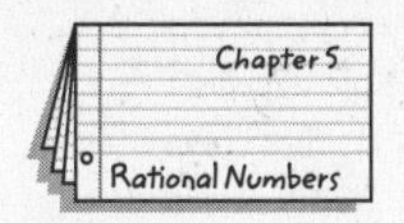

EXAMPLE Write Mixed Numbers and Integers as Fractions

1 **Write each number as a fraction.**

a. $-4\frac{3}{8} =$ ______ Write $-4\frac{3}{8}$ as an ______ fraction.

b. $10 =$ ______

EXAMPLE Write Terminating Decimals as Fractions

2 **Write 0.26 as a fraction or mixed number in simplest form.**

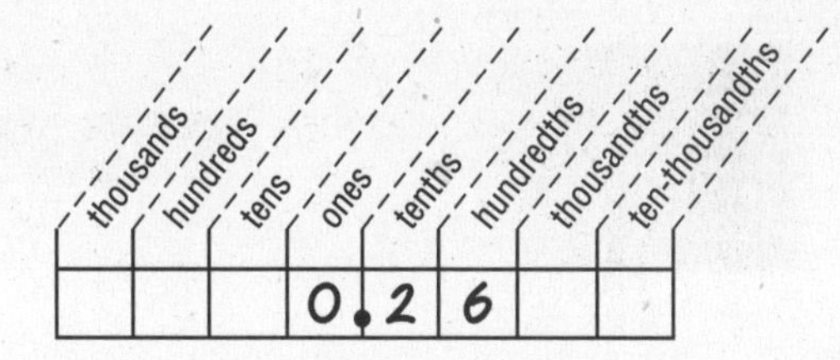

$0.26 =$ ______ 0.26 is 26 ______.

$=$ ______ Simplify. The GCF of ______ and ______ is ______.

Your Turn **Write each number as a fraction or mixed number in simplest form.**

a. $2\frac{3}{5}$ **b.** -6 **c.** 0.84

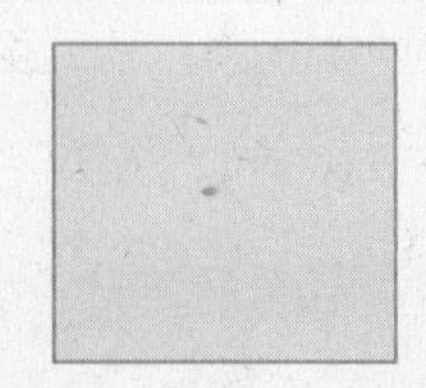

EXAMPLE Write Repeating Decimals as Fractions

WRITE IT

What would you multiply each side by if *three* digits repeat? Explain.

____

____

____

____

____

____

____

3 Write $0.\overline{39}$ as a fraction in simplest form.

$N = 0.3939...$ Let *N* represent the number.

____ $N =$ ____ $(0.3939...)$ Multiply each side by ____.

____ $=$ ____

Subtract *N* from ____ to eliminate the repeating part, 0.3939... .

____ $= 39.3939...$

$-(N = \quad 0.3939...)$

____ $=$ ____ ____

____ $=$ ____ Divide each side by ____.

$N =$ ____ or ____ Simplify.

Your Turn Write $0.\overline{4}$ as a fraction in simplest form.

EXAMPLE Classify Numbers

4 Identify all sets to which the number 15 belongs.

15 is a ____ number, an ____, a natural number, and a rational number.

Your Turn Identify all sets to which each number −7 belongs.

HOMEWORK ASSIGNMENT

Page(s):

Exercises:

5-3 Multiplying Rational Numbers

WHAT YOU'LL LEARN

- Multiply fractions.
- Use dimensional analysis to solve problems.

EXAMPLE Multiply Fractions

1 **Find $\frac{2}{5} \cdot \frac{5}{8}$. Write the product in simplest form.**

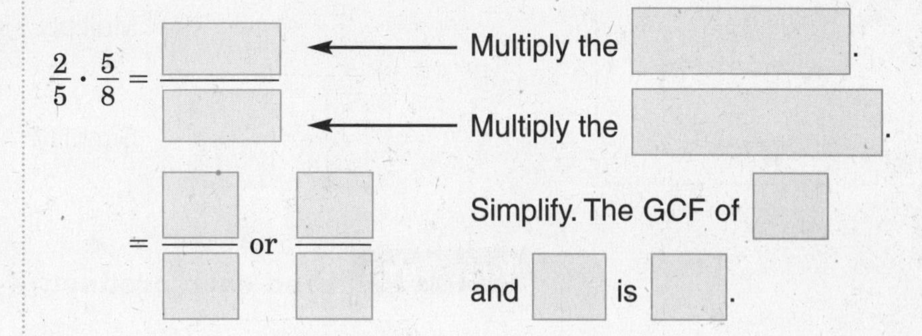

KEY CONCEPT

Multiplying Fractions
To multiply fractions, multiply the numerators and multiply the denominators.

EXAMPLE Simplify Before Multiplying

2 **Find $\frac{8}{9} \cdot \frac{5}{6}$.**

$\frac{8}{9} \cdot \frac{5}{6} = \frac{\overset{4}{\cancel{8}}}{9} \cdot \frac{5}{\underset{3}{\cancel{6}}}$ Divide 8 and 6 by their GCF, ☐.

$= \frac{4 \cdot 5}{9 \cdot 3}$ Multiply the numerators and multiply the denominators.

$=$ ☐ Simplify.

EXAMPLE Multiply Negative Fractions

3 **Find $-\frac{1}{4} \cdot \frac{2}{7}$.**

$-\frac{1}{4} \cdot \frac{2}{7} = \frac{1}{\underset{2}{\cancel{4}}} \cdot \frac{\overset{1}{\cancel{2}}}{7}$ Divide 2 and 4 by their GCF, ☐.

$= -\frac{1 \cdot 1}{2 \cdot 7}$ Multiply the numerators and multiply the denominators.

Simplify.

EXAMPLE Multiply Mixed Numbers

4 **Find $1\frac{1}{2} \cdot 3\frac{2}{3}$.**

$1\frac{1}{2} \cdot 3\frac{2}{3} = \frac{3}{2} \cdot \frac{11}{3}$ Rename the mixed numbers.

$= \frac{\overset{1}{\cancel{3}}}{2} \cdot \frac{11}{\underset{1}{\cancel{3}}}$ Divide by the GCF, ____.

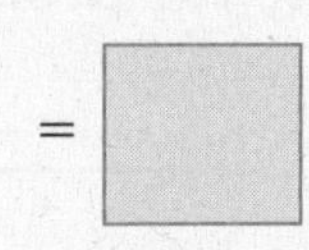

$=$ ____ Multiply.

$=$ ____ or ____ Simplify.

FOLDABLES

ORGANIZE IT

Under the tab for Lesson 5-3, write an expression in which you would multiply rational numbers and explain what it means.

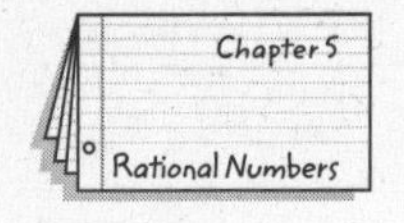

Your Turn **Find each product. Write in simplest form.**

a. $\frac{3}{8} \cdot \frac{2}{9}$ ____

b. $\frac{3}{8} \cdot \frac{2}{15}$ ____

c. $\frac{6}{14} \cdot -\frac{21}{40}$ ____

d. $2\frac{2}{7} \cdot 3\frac{1}{4}$ ____

EXAMPLE Multiply Algebraic Fractions

5 $\frac{3p^2}{q} \cdot \frac{q^2}{r} = \frac{3p \cdot p}{\underset{1}{\cancel{q}}} \cdot \frac{\overset{1}{\cancel{q}} \cdot q}{r}$ The GCF of q^2 and q is ____.

$=$ ____ Simplify.

Your Turn Find $\frac{5mn^3}{p^2} \cdot \frac{mp}{n^2}$.

____

BUILD YOUR VOCABULARY (page 98)

Dimensional analysis is the process of including ____ when you compute. You can use dimensional analysis to check whether your answers are reasonable.

EXAMPLE Use Dimensional Analysis

6 TRACK **The track at Cole's school is $\frac{1}{4}$ mile around. If Cole runs one lap in two minutes, how far (in miles) does he run in 30 minutes?**

Words: Distance equals the rate multiplied by the time.

Variables: Let d = distance, r = rate, and t = time.

Formula: $d = rt$

d = ☐ mile per 2 minutes · ☐ minutes

= $\frac{☐ \text{ mile}}{\cancel{2}_1 \cancel{\text{min}}} \times \overset{15}{\cancel{30}} \cancel{\text{min}}$ Divide by the common factors and units.

= ☐ · ☐ miles Multiply.

= ☐ or ☐ miles Simplify.

Cole runs ☐ miles in ☐ minutes.

Your Turn Walking Bob walks $\frac{2}{3}$ mile in 12 minutes. How far does he walk in 30 minutes?

HOMEWORK ASSIGNMENT

Page(s):

Exercises:

5–4 Dividing Rational Numbers

WHAT YOU'LL LEARN

- Divide fractions using multiplicative inverses.
- Use dimensional analysis to solve problems.

BUILD YOUR VOCABULARY (page 99)

Two numbers whose ______ is ______ are called **multiplicative inverses** or **reciprocals**.

KEY CONCEPTS

Inverse Property of Multiplication The product of a number and its multiplicative inverse is 1.

Dividing Fractions To divide by a fraction, multiply by its multiplicative inverse.

EXAMPLE Find Multiplicatives Inverses

1 Find the multiplicative inverse of $\frac{6}{7}$.

$\frac{6}{7} \cdot$ ______ $= 1$ The product is 1.

The multiplicative inverse or reciprocal of $\frac{6}{7}$ is ______.

EXAMPLE Divide by a Fraction

2 Find $\frac{4}{5} \div \frac{3}{10}$.

$\frac{4}{5} \div \frac{3}{10} = \frac{4}{5} \cdot$ ______ Multiply by the multiplicative  of $\frac{3}{10}$.

$= \frac{4}{\cancel{5}_1} \cdot \frac{\cancel{10}^2}{3}$ Divide ______ and ______ by their GCF, ______.

$=$ ______ or ______ Simplify.

EXAMPLE Divide by a Whole Number

3 Find $\frac{5}{6} \div 3$.

$\frac{5}{6} \div 3 = \frac{5}{6} \div$ ______ Write 3 as 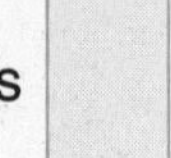.

$= \frac{5}{6} \cdot$ ______ Multiply by the multiplicative inverse of $\frac{3}{1}$.

$=$ ______ Simplify.

EXAMPLE Divide by a Mixed Number

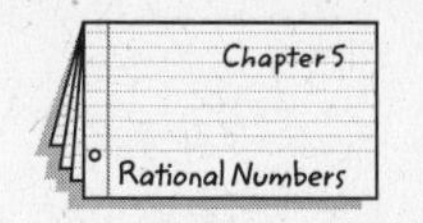

FOLDABLES

ORGANIZE IT

Under the tab for Lesson 5-4, write a word problem in which you would divide rational numbers to solve the problem.

4 **Find $4\frac{2}{3} \div -3\frac{1}{9}$.**

$4\frac{2}{3} \div -3\frac{1}{9} = \square \div \square$ Rename the mixed numbers as $\square$.

$= \square \cdot \square$ Multiply by the multiplicative inverse of $\square$.

$= \frac{\overset{1}{\cancel{14}}}{\underset{1}{\cancel{3}}} \cdot -\frac{\overset{3}{\cancel{9}}}{\underset{2}{\cancel{28}}}$ Divide out common factors.

$= \square$ or $\square$ Simplify.

EXAMPLE Divide by an Algebraic Fraction

5 **Find $\frac{5x}{8y} \div \frac{10}{16y}$.**

$\frac{5x}{8y} \div \frac{10}{16y} = \frac{5x}{8y} \times \square$ Multiply by the multiplicative

inverse of $\square$.

$= \frac{\overset{1}{\cancel{5}}x}{\underset{1}{\cancel{8}}\underset{1}{\cancel{y}}} \cdot \frac{\overset{2}{\cancel{16}}\overset{1}{\cancel{y}}}{\underset{2}{\cancel{10}}}$ Divide out common factors.

$= \square$ or $\square$ Simplify.

Your Turn

a. Find the multiplication inverse of $\frac{4}{9}$ ·

Find each quotient. Write in simplest form.

b. $\frac{3}{8} \div \frac{5}{6}$

c. $\frac{5}{12} \div 10$

d. $3\frac{3}{4} \div 2\frac{5}{8}$

e. $\frac{6m}{10p} \div \frac{9m}{4}$

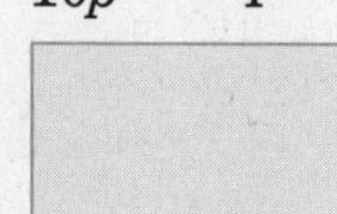

EXAMPLE Use Dimensional Analysis

6 TRAVEL **How many gallons of gas are needed to travel $78\frac{3}{4}$ miles if a car gets $25\frac{1}{2}$ miles per gallon?**

To find how many gallons, divide ______ by ______.

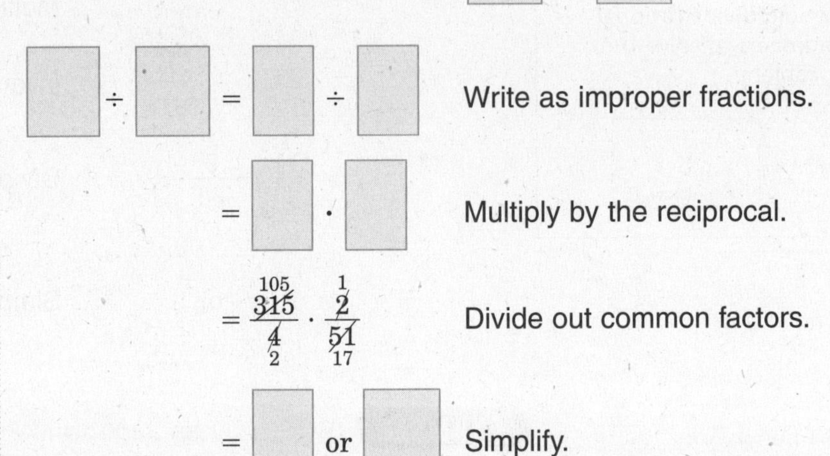

______ ÷ ______ = ______ ÷ ______ Write as improper fractions.

= ______ · ______ Multiply by the reciprocal.

$= \frac{\overset{105}{\cancel{315}}}{\underset{2}{\cancel{4}}} \cdot \frac{\overset{1}{\cancel{2}}}{\underset{17}{\cancel{51}}}$ Divide out common factors.

= ______ or ______ Simplify.

So, ______ gallons of gas are needed.

Your Turn Emily has $32\frac{2}{3}$ yards of fabric. She wants to make pillows which each require $3\frac{5}{6}$ yards of fabric to complete. How many pillows can Emily make?

HOMEWORK ASSIGNMENT

Page(s):

Exercises:

5–5 Adding and Subtracting Like Fractions

WHAT YOU'LL LEARN

Add like fractions.
Subtract like fractions.

KEY CONCEPTS

Adding Like Fractions To add fractions with like denominators, add the numerators and write the sum over the denominator.

Subtracting Like Fractions To subtract fractions with like denominators, subtract the numerators and write the difference over the denominator.

EXAMPLE Add Fractions

1 **Find $\frac{3}{4} + \frac{3}{4}$.**

$\frac{3}{4} + \frac{3}{4} = \frac{\square}{\square}$ The denominators are the same. ______ the numerators.

$= \square$ or $\square$ or $\square$ Simplify and rename as a mixed number.

EXAMPLE Add Mixed Numbers

2 **Find $3\frac{4}{9} + 8\frac{2}{9}$.**

$3\frac{4}{9} + 8\frac{2}{9} = (\square) + (\square)$ Add the whole numbers and fractions separately.

$= \square + \frac{\square}{\square}$ Add the numerators.

$= \square$ or $\square$ Simplify.

EXAMPLE Subtract Fractions

3 **Find the difference and write it in simplest form.**

$\frac{11}{12} - \frac{5}{12} = \frac{\square}{\square}$ The denominators are the same. Subtract the numerators.

$= \square$ or $\square$ Simplify.

FOLDABLES

ORGANIZE IT

Under the tab for Lesson 5-5, describe real-life situations in which you would add or subtract rational numbers.

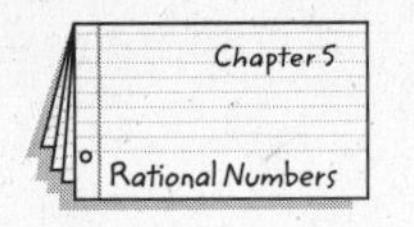

Your Turn **Add or subtract.**

a. $\frac{2}{9} + \frac{8}{9}$ **b.** $5\frac{3}{14} + 2\frac{5}{14}$ **c.** $\frac{17}{20} - \frac{11}{20}$

EXAMPLE Subtract Mixed Numbers

4 **Evaluate $r - q$ if $r = 7\frac{3}{5}$ and $q = 9\frac{1}{5}$.**

$r - q = \square - \square$ $r = \square$, $q = \square$

$= \square - \square$ Write the mixed numbers as improper fractions.

$= \square$ Subtract the numerators.

$= \square$ Simplify.

EXAMPLE Add Algebraic Fractions

5 **Find $\frac{5}{2b} + \frac{3}{2b}$. Write the sum in simplest form.**

$\frac{5}{2b} + \frac{3}{2b} = \frac{\square}{\square}$ The denominators are the same. Add the numerators.

$=$ 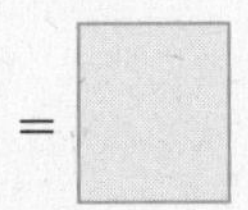 Add the numerators.

$=$ 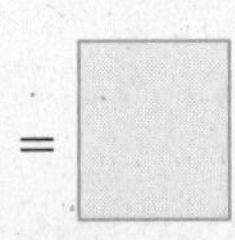 Simplify.

Your Turn

a. Evaluate $m - n$ if $m = 4\frac{7}{9}$ and $n = 8\frac{2}{9}$.

b. Find $\frac{3x}{16} + \frac{5x}{16}$. Write the sum in simplest form.

HOMEWORK ASSIGNMENT

Page(s):

Exercises:

5–6 Least Common Multiple

WHAT YOU'LL LEARN

- Find the least common multiple of two or more numbers.
- Find the least common denominator of two or more fractions.

BUILD YOUR VOCABULARY (pages 98–99)

A **multiple** of a number is a ______ of that number and a whole number.

Sometimes numbers have some of the ______ multiples. These are called **common multiples**.

The **least** of the *nonzero* common multiples of two or more numbers is called the **least common multiple (LCM)**.

EXAMPLE Find the LCM

1 **Find the LCM of 168 and 180.**

Number	Prime Factorization	Exponential Form
168		
180		

The prime factors of both numbers are ______.

Multiply the greatest powers of ______ appearing in either factorization.

LCM = ______ = ______

FOLDABLES

ORGANIZE IT

Under the tab for Lesson 5-6, describe what a least common multiple is. Give two numbers and their least common multiple.

Chapter 5

Rational Numbers

EXAMPLE The LCM of Monomials

2 **Find the LCM of $12x^2y^2$ and $6y^3$.**

$12x^2y^2 =$ ______

$6y^3 =$ ______

LCM = ______ Multiply the greatest power of each prime factor or variable.

= ______

Your Turn **Find the least common denominator (LCM) of each set of numbers or monomials.**

a. 144, 96

b. $18ab^3$, $24a^2b$

BUILD YOUR VOCABULARY (page 98)

The **least common denominator (LCD)** of two or more fractions is the ______ of the ______.

EXAMPLE Find the LCD

3 **Find the LCD of $\frac{7}{8}$ and $\frac{13}{20}$.**

8 = ______

20 = ______

Write the prime factorization of 8 and 20. Highlight the greatest power of each prime factor.

LCM = ______ or ______

Multiply.

The LCD of $\frac{7}{8}$ and $\frac{13}{20}$ is ______.

EXAMPLE Find the LCD of Algebraic Fractions

4 **Find the LCD of $\frac{9}{36a^2b}$ and $\frac{16}{27ab^2}$.**

$36a^2b$ = ______

$27ab^2$ = ______

LCM = ______ or ______.

The LCD of $\frac{9}{36a^2b}$ and $\frac{16}{27ab^2}$ is ______.

Your Turn **Find the least common denominator (LCD) of each pair of fractions.**

a. $\frac{5}{9}, \frac{11}{12}$

b. $\frac{7}{24m^3n}, \frac{9}{40mn^5}$

EXAMPLE Compare Fractions

5 **Replace • with <, >, or = to make $\frac{7}{15} \bullet \frac{3}{7}$ a true statement.**

The LCD of the fractions is ______ or ______.

Rewrite the fractions using the LCD and then compare the

$\frac{7}{15} = \frac{7 \cdot \square}{3 \cdot 5 \cdot \square} = \square$ Multiply the fraction by ______ to make the denominator 105.

$\frac{3}{7} = \frac{3 \cdot}{7 \cdot}\square = \square$ Multiply the fraction by ______ to make the denominator 105.

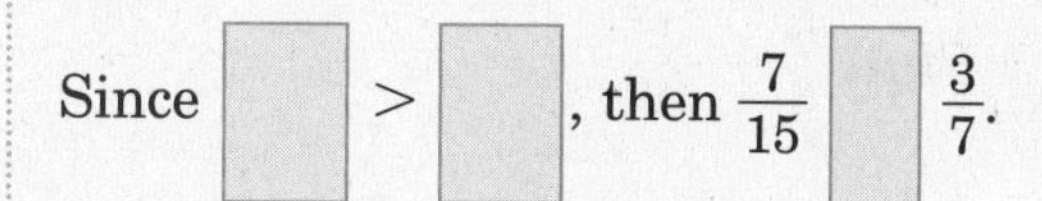

Since ______ > ______, then $\frac{7}{15}$ ______ $\frac{3}{7}$.

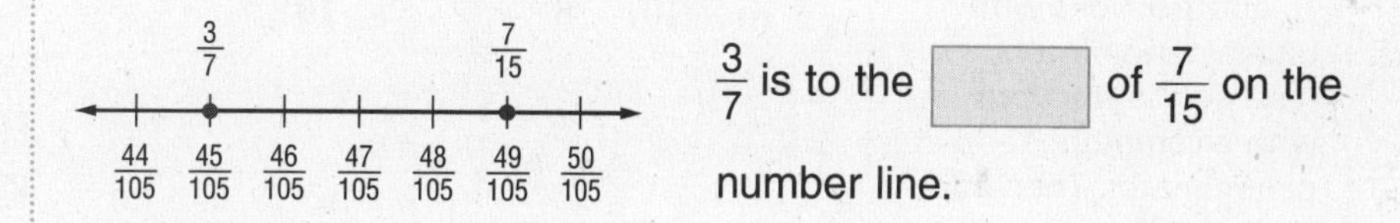

$\frac{3}{7}$ is to the ______ of $\frac{7}{15}$ on the number line.

Your Turn Replace • with <, >, or = to make $\frac{5}{21} \bullet \frac{9}{14}$ a true statement.

HOMEWORK ASSIGNMENT

Page(s):

Exercises:

5–7 Adding and Subtracting Unlike Fractions

WHAT YOU'LL LEARN

- Add unlike fractions.
- Subtract unlike fractions.

EXAMPLE Add Unlike Fractions

1 **Find $\frac{3}{4} + \frac{1}{7}$.**

$\frac{3}{4} + \frac{1}{7} = \frac{3}{4} \cdot \square + \frac{1}{7} \cdot \square$ — Use 4 · 7 or $\square$ as the common denominator.

$= \square + \square$ — Rename each fraction with the common denominator. Add the numerators.

$= \square$

KEY CONCEPT

Adding Unlike Fractions To add fractions with unlike denomintors, rename the fractions with a common denominator. Then add and simplify.

EXAMPLE Add Fractions

2 **Find $\frac{5}{6} + \frac{3}{10}$.**

$\frac{5}{6} + \frac{3}{10} = \frac{5}{6} \cdot \square + \frac{3}{10} \cdot \square$ — The LCD is $\square$.

$= \square + \square$ — Rename each fraction with the LCD.

$= \square$ or $\square$ — Add the numerators. Simplify.

EXAMPLE Add Mixed Numbers

3 **Find $2\frac{1}{8} + (-3\frac{2}{3})$.**

$2\frac{1}{8} + \left(-3\frac{2}{3}\right) = \frac{\square}{8} + \left(-\frac{\square}{3}\right)$ — Write the mixed numbers as improper fractions.

$= \square + \left(\square\right)$ — Rename fractions using the LCD, $\square$.

$= \square$ — Add the numerators.

$= \square$ — Simplify.

Your Turn **Find each sum. Write in simplest form.**

a. $\frac{2}{3} + \frac{1}{8}$

b. $\frac{5}{12} + \frac{5}{9}$

c. $4\frac{2}{5} + \left(-6\frac{2}{3}\right)$

KEY CONCEPT

Subtracting Unlike Fractions To subtract fractions with unlike denominators, rename the fractions with a common denominator. Then subtract and simplify.

FOLDABLES Under the tab for Lesson 5-7, describe a situation in which you would add or subtract unlike fractions.

EXAMPLE Subtract Fractions

4 **Find $\frac{9}{16} - \frac{5}{8}$.**

$\frac{9}{16} - \frac{5}{8} = \frac{9}{16} - \frac{5}{8} \cdot$ ☐ The LCD is ☐.

$= \frac{9}{16} -$ ☐ Rename $\frac{5}{8}$ using the LCD.

$=$ ☐ Subtract the numerators.

EXAMPLE Subtract Mixed Numbers

5 **Find $4\frac{2}{3} - 3\frac{6}{7}$.**

$4\frac{2}{3} - 3\frac{6}{7} = \frac{☐}{3} - \frac{☐}{7}$ Write the mixed numbers as improper fractions.

$=$ ☐ $-$ ☐ Rename the fractions using the LCD. Subtract the numerators.

$=$ ☐ Simplify.

Your Turn **Find each difference. Write in simplest form.**

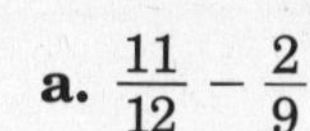

a. $\frac{11}{12} - \frac{2}{9}$

b. $3\frac{5}{6} - 2\frac{1}{8}$

HOMEWORK ASSIGNMENT

Page(s):

Exercises:

5–8 Measures of Central Tendency

WHAT YOU'LL LEARN

- Use the mean, median, and mode as measures of central tendency.
- Analyze data using mean, median, and mode.

KEY CONCEPTS

Measures of Central Tendency

mean the sum of the data divided by the number of items in the data set

median the middle number of the ordered data, or the mean of the middle two numbers

mode the number or numbers that occur most often

EXAMPLE Find the Mean, Median, and Mode

1 MOVIES **The revenue of the 10 highest grossing movies as of June 2000 are given in the table. Find the mean, median, and mode of the revenues.**

Top 10 Movie Revenues (millions of $)	
601	330
461	313
431	309
400	306
357	290

$$\text{mean} = \frac{\text{sum of revenues}}{\text{number of movies}}$$

$$= \frac{601 + 461 + 431 + 400 + 357 + 330 + 313 + 309 + 306 + 290}{\square}$$

$$= \frac{\square}{\square} \text{ or } \square$$

The mean revenue is ______

To find the median, order the numbers from least to greatest.

290, 306, 309, 313, 330, 357, 400, 431, 461, 601

$$\text{median} = \frac{\square}{\square} = \square$$

The mean revenue is ______.

There is ______ because each number in the set occurs ______.

FOLDABLES

ORGANIZE IT

Under the tab for Lesson 5–8, explain the differences between mean, median, and mode.

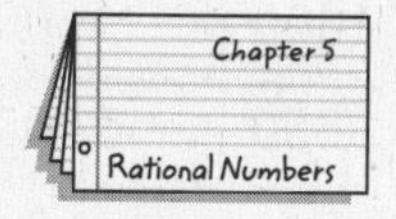

Your Turn The test scores for a class of nine students are 85, 93, 78, 99, 62, 83, 90, 75, and 85. Find the mean, median, and mode of the test scores.

EXAMPLE Use a Line Plot

2 **OLYMPICS** **The line plot shows the number of gold medals earned by each country that participated in the 1998 Winter Olympic games in Nagano, Japan. Find the mean, median, and mode for the gold medals won.**

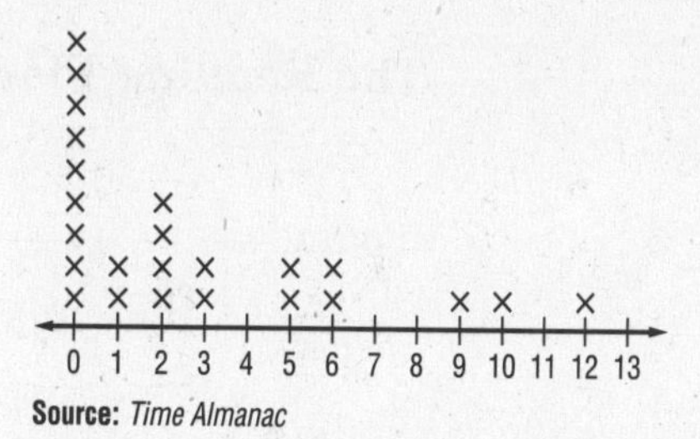

Source: *Time Almanac*

$$\text{mean} = \frac{9(0) + 2(1) + 4(2) + 2(3) + 2(5) + 2(6) + 1(9) + 1(10) + 1(12)}{24} = 2.875$$

There are ______ numbers. The median number is the average of the ______ and ______ numbers. The median is ______.

The number ______ occurs most frequently in the set of data.

The mode is ______.

Your Turn A survey of school-age children shows the family sizes displayed in the line plot. Find the mean, median, and mode.

1 2 3 4 5 6 7 8 9 10

EXAMPLE Use Mean, Median, and Mode to Analyze Data

3 **The table shows the monthly salaries of the employees at two bookstores. Find the mean, median, and mode for each set of data. Based on the averages, which bookstore pays it employees better?**

Bob's Books	The Reading Place
1290	1400
1400	1450
1400	1550
1600	1600
3650	2000

Bob's Books

mean: ______ = ______

median: ______ mode: ______

The Reading Place

mean: ______ = ______

median: ______ mode: ______

The $3650 salary at Bob's Books is an extreme value that increases the mean salary. The employees at The Reading Place are generally better paid as shown by the median.

Your Turn The number of hours spent exercising each week by men and women are given in the table. Find the mean, median, and mode for each set of data. Based on the averages, which gender exercises more?

Men	Women
5	1
2	6
3	4
8	2
12	1
4	8

HOMEWORK ASSIGNMENT

Page(s):

Exercises:

5–9 Solving Equations with Rational Numbers

What You'll Learn

- Solve equations containing rational numbers.

Review It

Which properties allow you to add or subtract the same number from each side of an equation? *(Lesson 3–3)*

____

____

____

____

____

EXAMPLE Solve by Using Addition

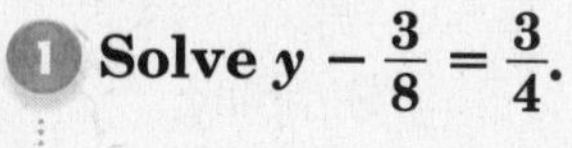

1 **Solve $y - \frac{3}{8} = \frac{3}{4}$.**

$y - \frac{3}{8} = \frac{3}{4}$ Write the equation.

$y - \frac{3}{8} + \square = \frac{3}{4} + \square$ Add $\square$ to each side.

$y = \frac{3}{4} + \square$ Simplify.

$y = \square + \square$ Rename the fractions using the $\square$ and add.

$y = \square$ Simplify.

$y = \square$ Write as a mixed number.

EXAMPLE Solve by Using Subtraction

2 **Solve $m + 8.6 = 11.2$.**

$m + 8.6 = 11.2$ Write the equation.

$m + 8.6 - \square = 11.2 - \square$ Subtract $\square$ from each side.

$m = \square$ Simplify.

Your Turn **Solve.**

a. $\frac{2}{3} = x - \frac{1}{2}$

b. $15.4 = b + 9.3$

FOLDABLES

ORGANIZE IT

Under the tab for Lesson 5–9, write an equation involving fractions that can be solved using division. Solve your problem.

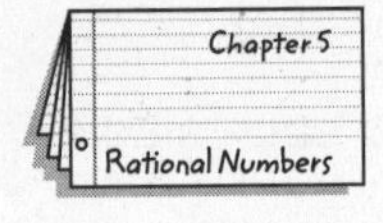

EXAMPLE Solve by Using Division

3 **Solve $9a = 3.6$.**

$9a = 3.6$ Write the equation.

$\frac{9a}{\square} = \frac{3.6}{\square}$ Divide each side by $\square$.

$a = \square$ Simplify.

EXAMPLE Solve by Using Multiplication

4 **Solve $8 = \frac{1}{6}x$.**

$8 = \frac{1}{6}x$ Write the equation.

$\square(8) = \square\left(\frac{1}{6}x\right)$ Multiply each side by $\square$.

$\square = \square$ Simplify.

Your Turn **Solve.**

a. $-6m = -4.8$

b. $9 = \frac{1}{5}m$

HOMEWORK ASSIGNMENT

Page(s):

Exercises:

5–10 Arithmetic and Geometric Sequences

What You'll Learn

- Find the terms of arithmetic sequences.
- Find the terms of geometric sequences.

BUILD YOUR VOCABULARY (pages 98–99)

A **sequence** is an ordered list of numbers.

An **arithmetic sequence** is a sequence in which the difference between any two consecutive terms is the same. Each number in the sequence is called a **term**. The difference between consecutive terms is called the **common difference**.

EXAMPLE Identify an Arithmetic Sequence

1 **State whether −5, −1, 3, 7, 11 is arithmetic. If it is, state the common difference and write the next three terms.**

−5, −1, 3, 7, 11

Notice that −1 − (−5) = ______, 3 − (−1) = ______, and so on.

The terms have a common difference of ______, so the sequence is ______.

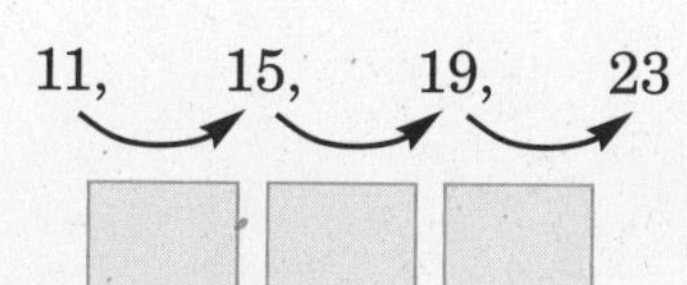

11, 15, 19, 23

Continue the ______ to find the next three terms.

EXAMPLE Identify an Arithmetic Sequence

2 **State whether 0, 2, 6, 12, 20 is arithmetic. If it is, state the common difference and write the next three terms.**

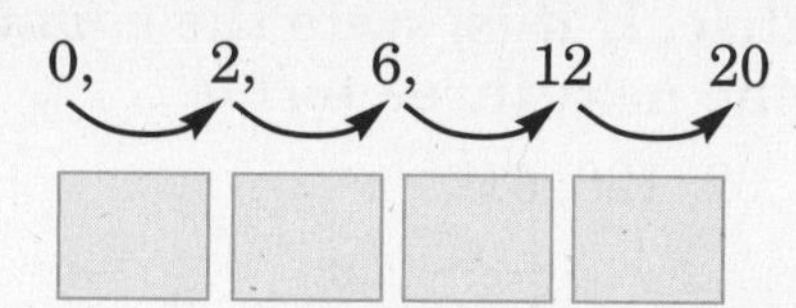

0, 2, 6, 12 20

The terms do not have a common difference.

The sequence is ______.

FOLDABLES

ORGANIZE IT

Under the tab for Lesson 5-10, write an example of an arithmetic sequence and a geometric sequence.

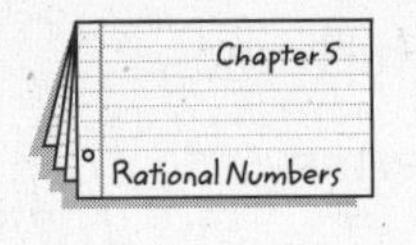

Your Turn **State whether each sequence is arithmetic. If it is, state the common difference and write the next three terms.**

a. 12, 7, 2, -3, -8, ...

b. 20, 17, 11, 2, −10, ...

BUILD YOUR VOCABULARY (page 98)

A **geometric sequence** is a sequence in which the ______ of any two ______ terms is the same. The ______ of consecutive terms is called the **common ratio.**

EXAMPLE Identify Geometric Sequences

3 **State whether 27, −9, 3, −1, $\frac{1}{3}$ is geometric. If it is, state the common ratio and write the next three terms.**

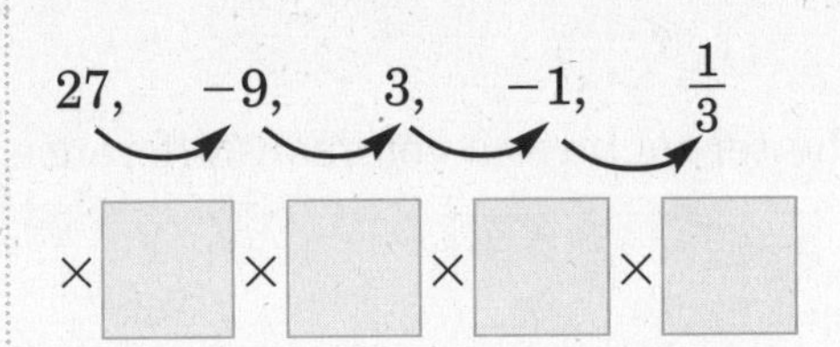

The common ratio is ______ so the sequence is ______.

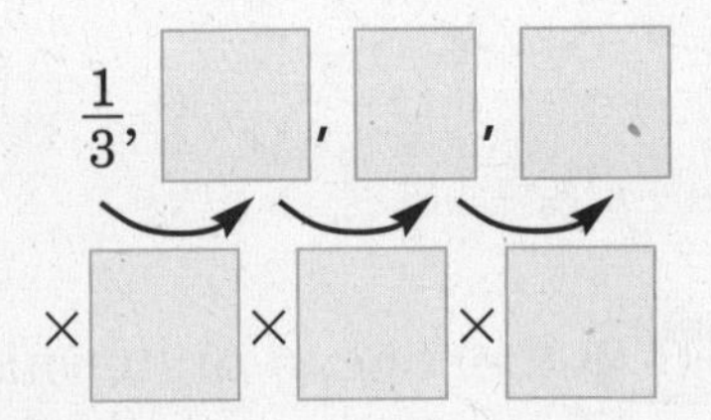

Continue the ______ to find the next three terms.

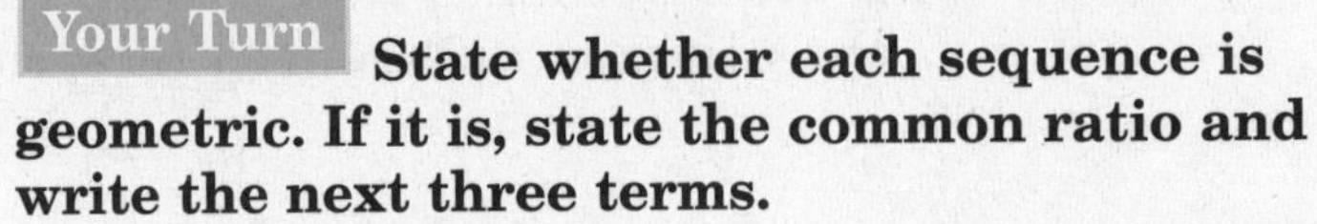

Your Turn **State whether each sequence is geometric. If it is, state the common ratio and write the next three terms.**

a. 2, 8, 32, 128, 512, ...

b. 100, 20, 4, $\frac{4}{5}$, $\frac{4}{25}$, ...

HOMEWORK ASSIGNMENT

Page(s):
Exercises:

CHAPTER 5

BRINGING IT ALL TOGETHER

STUDY GUIDE

FOLDABLES™	VOCABULARY PUZZLEMAKER	BUILD YOUR VOCABULARY
Use your **Chapter 5 Foldable** to help you study for your chapter test.	To make a crossword puzzle, word search, or jumble puzzle of the vocabulary words in Chapter 5, go to: www.glencoe.com/sec/math/t_resources/free/index.php	You can use your completed **Vocabulary Builder** (pages 98–99) to help you solve the puzzle.

5-1 Fractions as Decimals

Write each fraction or mixed number as a decimal. Use a bar to show a repeating decimal.

1. $\frac{5}{6}$

2. $\frac{7}{8}$

3. $-4\frac{1}{11}$

Replace each ● with <, >, or = to make a true sentence.

4. -5.43 ● -5.62

5. $\frac{4}{5}$ ● $\frac{9}{11}$

6. 0.76 ● $\frac{23}{29}$

5-2 Rational Numbers

Write each decimal as a fraction or mixed number in simplest form.

7. 0.62

8. 3.48

9. $1.\overline{7}$

5-3 Multiplying Rational Numbers

Find each product. Write in simplest form.

10. $\frac{3}{5}\left(-\frac{2}{3}\right)$

11. $-\frac{4}{15}\left(-\frac{55}{6}\right)$

12. $\frac{p}{15} \cdot \frac{3}{p^2}$

5-4

Dividing Rational Numbers

Find each quotient. Write in simplest form.

13. $\frac{2}{9} \div \frac{1}{8}$

14. $-\frac{3}{11} \div \frac{7}{22}$

15. $\frac{5pq}{t} \div \frac{6q}{t}$

16. Holly is wallpapering her kitchen. How many $8\frac{1}{2}$ feet lengths of wallpaper can she cut from a roll of wallpaper that is $59\frac{1}{2}$ feet long?

5-5

Adding and Subtracting Like Fractions

Find each sum or difference. Write in simplest form.

17. $\frac{3}{10} + \frac{6}{10}$

18. $-\frac{3}{11} - \frac{9}{11}$

19. $-3\frac{7}{18}m + 5\frac{5}{18}m$

5-6

Least Common Multiple

Find the least common multiple (LCM) of each set of numbers of monomials.

20. 12, 42

21. 8, 12, 18

22. 14, 63

Find the least common denominator (LCD) of each pair of fractions.

23. $\frac{5}{6}, \frac{7}{15}$

24. $\frac{11}{18}, \frac{23}{32}$

25. $\frac{1}{6xy}, \frac{7}{9y}$

5-7

Adding and Subtracting Unlike Fractions

Find each sum or difference. Write in simplest form.

26. $\frac{4}{7} + \frac{2}{5}$

27. $\frac{5}{8} - \frac{9}{20}$

28. $6\frac{1}{9} + 4\frac{5}{12}$

5-8 Measures of Central Tendency

Find the mean, median, and mode for each set of data. If necessary, round to the nearest tenth.

29. 6, 8, 12, 7, 6, 11, 20

30. 9.2, 9.7, 8.6, 9.8, 9.9, 8.9, 9.0, 8.5

31. Which measure of central tendency is most affected by an extreme value?

5-9 Adding and Subtracting Unlike Fractions

Match each equation with the appropriate first step of its solution.

32. $y - 6 = 11.8$

33. $6 + x = -9$

34. $\frac{c}{6} = \frac{1}{2}$

35. $-\frac{1}{6}p = -\frac{1}{6}$

a. Multiply each side by 6.
b. Add 6 to each side.
c. Subtract 6 from each side.
d. Divide each side by -6.
e. Multiply each side by -6.

36. Dividing by a fraction is the same as multiplying by the ______.

5-10 Arithmetic and Geometric Sequences

State whether each sentence is true or false. If false, replace the underlined word to make a true sentence.

37. A <u>sequence</u> is an ordered list of numbers.

38. An <u>arithmetic</u> sequence is a sequence in which the quotient of any two consecutive terms is the same.

39. Find the seventh term in the sequence $-18, -14, -10, -6, \ldots$.

40. What is the common ratio in the sequence $6, 2, \frac{2}{3}, \frac{2}{9}, \ldots$? $\frac{1}{3}$

ARE YOU READY FOR THE CHAPTER TEST?

Visit **pre-alg.com** to access your textbook, more examples, self-check quizzes, and practice tests to help you study the concepts in Chapter 5.

Check the one that applies. Suggestions to help you study are given with each item.

☐ **I completed the review of all or most lessons without using my notes or asking for help.**

- You are probably ready for the Chapter Test.
- You may want take the Chapter 5 Practice Test on page 259 of your textbook as a final check.

☐ **I used my Foldable or Study Notebook to complete the review of all or most lessons.**

- You should complete the Chapter 5 Study Guide and Review on pages 254–258 of your textbook.
- If you are unsure of any concepts or skills, refer back to the specific lesson(s).
- You may also want to take the Chapter 5 Practice Test on page 259.

☐ **I asked for help from someone else to complete the review of all or most lessons.**

- You should review the examples and concepts in your Study Notebook and Chapter 5 Foldable.
- Then complete the Chapter 5 Study Guide and Review on pages 254–258 of your textbook.
- If you are unsure of any concepts or skills, refer back to the specific lesson(s).
- You may also want to take the Chapter 5 Practice Test on page 259.

Student Signature

Parent/Guardian Signature

Teacher Signature

Ratio, Proportion, and Percent

Use the instructions below to make a Foldable to help you organize your notes as you study the chapter. You will see Foldable reminders in the margin of this Interactive Study Notebook to help you in taking notes.

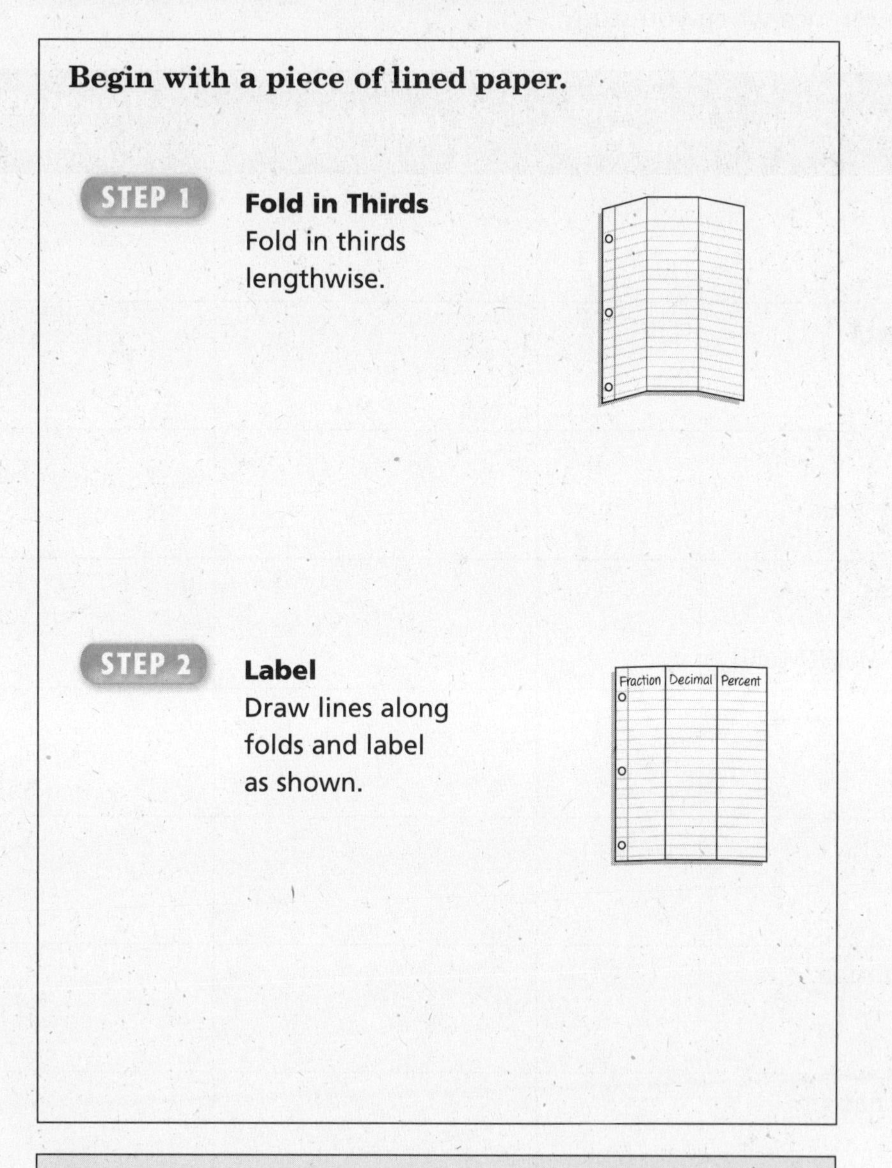

Begin with a piece of lined paper.

STEP 1 **Fold in Thirds**
Fold in thirds lengthwise.

STEP 2 **Label**
Draw lines along folds and label as shown.

NOTE-TAKING TIP: When you take notes, record real-life examples of how you can use fractions, decimals, and percent, such as telling time and making change.

Chapter 6

Build Your Vocabulary

This is an alphabetical list of new vocabulary terms you will learn in Chapter 6. As you complete the study notes for the chapter, you will see Build Your Vocabulary reminders to complete each term's definition or description on these pages. Remember to add the textbook page number in the second column for reference when you study.

Vocabulary Term	Found on Page	Definition	Description or Example
base			
cross products			
discount			
experimental probability [ik-spehr-uh-MEHN-tuhl]			
outcomes			
percent			
percent equation			
percent of change			
percent proportion			

Vocabulary Term	Found on Page	Definition	Description or Example
probability			
proportion			
ratio			
sample space			
scale			
scale drawing or scale model			
scale factor			
simple event			
simple interest			
theoretical probability [thee-uh-REHT-ih-kuhl]			
unit rate			

6–1 Ratios and Rates

BUILD YOUR VOCABULARY (page 131)

A **ratio** is a ______ of two numbers by ______

A **rate** is a ______ of two ______ having different kinds of units.

When a rate is simplified so that it has a denominator of ______, it is called a **unit rate**.

WHAT YOU'LL LEARN

- Write ratios as fractions in simplest form.
- Determine unit rates.

REVIEW IT

What does it mean for a fraction to be in simplest form? *(Lesson 4-5)*

______

______

______

______

______

______

EXAMPLE Write Ratios as Fractions

1 **Express the ratio *10 roses out of 12 flowers* as a fraction in simplest form.**

$\frac{10}{12} =$ ______ Divide the numerator and denominator by the ______, ______.

The ratio of roses to flowers is ______ to ______. This means that for every ______ flowers, ______ of them are roses.

EXAMPLE Write Ratios as Fractions

2 **Express the ratio *21 inches to 2 yards* as a fraction in simplest form.**

$\frac{21 \text{ inches}}{2 \text{ yards}} = \frac{21 \text{ inches}}{\square \text{ inches}}$ Convert ______ yards to inches.

$= \frac{\square \text{ inches}}{\square \text{ inches}}$ Divide the numerator and denominator by the ______, ______.

Written in simplest form, the ratio is ______.

Your Turn **Express each ratio as a fraction in simplest form.**

a. 8 golden retrievers out of 12 dogs

b. 4 feet to 18 inches

EXAMPLE Find Unit Rate

3 SHOPPING **A 12-oz bottle of cleaner costs $4.50. A 16-oz bottle of cleaner costs $6.56. Which costs less per ounce?**

Find and compare the unit rates of the bottles.

$\frac{\$4.50}{12 \text{ ounces}} = \frac{\square}{1 \text{ ounce}}$ $\qquad$ $\frac{\$6.56}{16 \text{ ounces}} = \frac{\square}{1 \text{ ounce}}$

The ______ bottle has the lower ______.

EXAMPLE Convert Rates

4 ANIMALS **A snail moved 30 feet in 2 hours. How many inches per minute did the snail move?**

You need to convert feet to inches and hours to minutes.

$\frac{30 \text{ ft}}{2 \text{ hr}} = \frac{30 \text{ ft}}{2 \text{ hr}} \cdot \frac{12 \text{ in}}{1 \text{ ft}} \div \frac{60 \text{ min}}{1 \text{ hr}}$

$= \frac{30 \text{ ft}}{2 \text{ hr}} \cdot \frac{12 \text{ in}}{1 \text{ ft}} \cdot \square$ $\qquad$ Write the reciprocal of $\frac{60 \text{ min}}{1 \text{ hr}}$.

$=$  $= \square$ $\qquad$ Divide the common factors and units. Simplify.

Your Turn

a. A 6-pack of a soft drink costs $1.50. A 12 pack of a soft drink costs $2.76. Which pack costs less per can?

b. Dave jogs 2 miles in 22 minutes. How many feet per second is this?

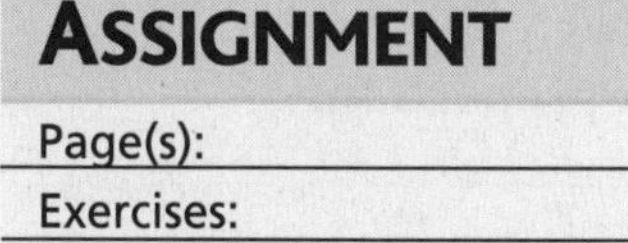

HOMEWORK ASSIGNMENT

Page(s):

Exercises:

6–2 Using Proportions

WHAT YOU'LL LEARN

- Solve proportions.
- Use proportions to solve real-world problems.

BUILD YOUR VOCABULARY (pages 130–131)

A **proportion** is a statement of ______ of two ______.

In the proportion $\frac{a}{b} = \frac{c}{d}$, the ______ ad and cb are called the **cross products** of the proportion.

KEY CONCEPTS

Proportion A proportion is an equation stating that two ratios are equal.

Property of Proportions The cross products of a proportion are equal.

EXAMPLE Identify Proportion

1 **Determine whether $\frac{2}{3}$ and $\frac{12}{20}$ form a proportion.**

$\frac{2}{3} \stackrel{?}{=} \frac{12}{20}$ Write a proportion.

______ $\stackrel{?}{=}$ ______ ______ products

______ Simplify.

So, $\frac{2}{3}$ ______ $\frac{12}{20}$.

EXAMPLE Solve Proportions

2 **Solve $\frac{c}{36} = \frac{9}{15}$.**

$\frac{c}{36} = \frac{9}{15}$

$c \cdot 15 = 36 \cdot 9$ Cross products

______ = ______ Multiply.

______ = ______ Divide.

$c =$ ______

Your Turn

a. Determine whether $\frac{3}{4}$ and $\frac{9}{10}$ form a proportion.

b. Solve $\frac{x}{12} = \frac{3}{8}$. ______

EXAMPLE Use a Proportion to Solve a Problem

3 ARCHITECTURE **An architect builds a model of a building before the actual building is built. The model is 8 inches tall and the actual building will be 22 feet tall. The model is 20 inches wide. Find the width of the actual building.**

Write and solve a proportion using ratios that compare actual height to model height.

$\frac{\text{actual height}}{\text{model height}} = \frac{\text{actual width}}{\text{model width}}$

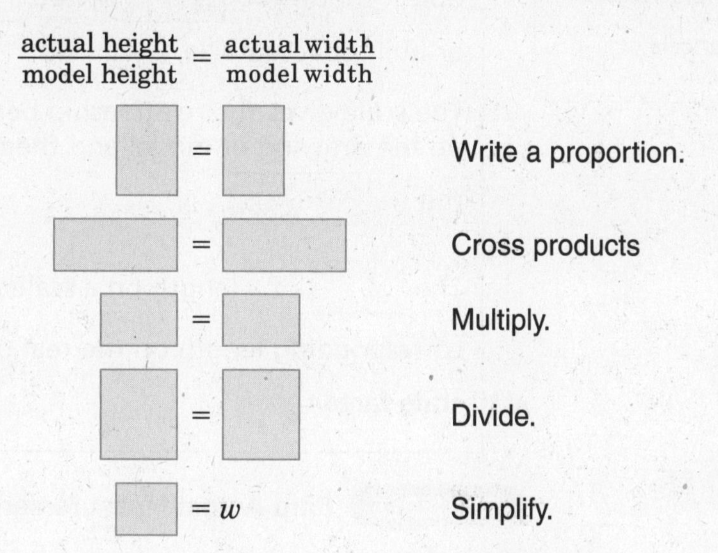

The actual width of the building is ______.

Your Turn A model of a jet airplane has a length of 9 inches and a wingspan of 6 inches. Find the wingspan of the actual plane if the length is 120 feet.

HOMEWORK ASSIGNMENT

Page(s):

Exercises:

6–3 Scale Drawings and Models

BUILD YOUR VOCABULARY (page 131)

A **scale drawing** or **scale model** is used to represent an object that is too ______ or too ______ to be drawn or built at actual size.

The **scale** gives the relationship between the measurements on the drawing or model and the measurements of the ______.

The ______ of a length on a scale drawing or model to the corresponding length on the real object is called the **scale factor**.

WHAT YOU'LL LEARN

- Use scale drawings.
- Construct scale drawings.

EXAMPLE Find Actual Measurements

1 MAP **A map has a scale of 1 inches = 8 miles. Two towns are 3.25 inches apart on the map.**

a. What is the actual distance between the two towns?

Let x represent the actual distance between the two towns. Write and solve a proportion.

map distance → 1 inch / actual distance → 8 miles = ______ inches ← map distance / ______ miles ← actual distance

$$\frac{1 \text{ inch}}{8 \text{ miles}} = \frac{\text{\_\_\_\_\_\_ inches}}{\text{\_\_\_\_\_\_ miles}}$$

______ = ______ Find the cross products.

$x =$ ______ Simplify.

The actual distance between the two towns is ______.

b. What is the scale factor?

To find the scale factor, write the ratio of 1 inch to 8 miles in simplest form.

$$\frac{1 \text{ inch}}{8 \text{ miles}} = \frac{1 \text{ inch}}{\text{\_\_\_\_\_\_ inches}}$$ Convert 8 miles to inches.

The scale factor is ______.

REMEMBER IT

When finding the scale factor, be sure to use the same units of measure.

Your Turn **A scale drawing of a new house has a scale of 1 inch = 4 feet. The height of the living room ceiling is 2.75 inches on the scale drawing.**

a. What is the actual height of the ceiling?

b. What is the scale factor?

EXAMPLE Determine the Scale

2 MODEL CAR **A model car is 4 inches long. The actual car is 12 feet long. What is the scale of the model?**

model length → $\frac{4 \text{ inches}}{12 \text{ feet}} = \frac{\square}{\square}$ ← model length
actual length → ← actual length

$\square = \square$ Find the cross products.

$\square = \square$ Simplify.

$\square = \square$ Divide each side by 4.

$x = \square$

The scale is ______.

Your Turn A model log cabin is 12 inches high. The actual log cabin is 42 feet high. What is the scale of the model?

EXAMPLE Construct a Scale Drawing

③ PATIO DESIGN **Sheila is designing a patio that is 16 feet long and 14 feet wide. Make a scale drawing of the patio. Use a scale of 0.5 inch = 5 feet.**

Step 1 Find the measure of the patio's length on the drawing.

drawing length → $\frac{0.5 \text{ inch}}{4 \text{ feet}} = \frac{x \text{ inches}}{16 \text{ feet}}$ ← drawing length
actual length → ← actual length

$0.5 \cdot 16 = 4 \cdot x$ Cross products

$8 = 4x$ Simplify.

$\square = x$ Divide.

On the drawing, the length is ☐ inches.

Step 2 Find the measure of the patio's width on the drawing.

drawing length → $\frac{0.5 \text{ inch}}{4 \text{ feet}} = \frac{w \text{ inches}}{14 \text{ feet}}$ ← drawing width
actual length → ← actual width

$0.5 \cdot 14 = 4 \cdot w$ Cross products

$7 = 4w$ Simplify.

$\square = w$ Divide.

On the drawing, the width is ☐ or $1\frac{3}{4}$ inches.

Step 3 Make the scale drawing.
Use $\frac{1}{4}$-inch grid paper.

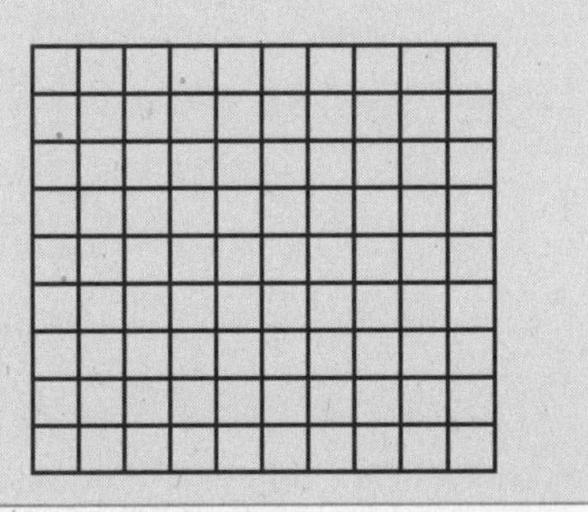

WRITE IT

What two numbers do you need to construct a scale drawing of an object?

______

______

______

______

Your Turn A garden is 18 feet long and 14 feet wide. Make a scale drawing of the garden. Use a scale of 0.5 inch = 4 feet.

18 ft

14 ft

HOMEWORK ASSIGNMENT

Page(s):
Exercises:

6–4 Fractions, Decimals, and Percents

What You'll Learn

- Express percents as fractions and vice versa.
- Express percents as decimals and vice versa.

BUILD YOUR VOCABULARY (page 130)

A **percent** is a ratio that compares a number to 100.

EXAMPLE Percents as Fractions

1 **Express each percent as a fraction in simplest form.**

a. $60\% = \frac{60}{100}$

$= \square$

b. $104\% = \frac{104}{100}$

$= \square$ or $\square$

c. $0.3\% = \frac{0.3}{100}$

$= \frac{0.3}{100} \cdot \square$

$= \square$

d. $56\frac{1}{4}\% = \frac{56\frac{1}{4}}{100}$

$= 56\frac{1}{4} \div \square$

$= \frac{\overset{9}{\cancel{225}}}{4} \cdot \frac{1}{\underset{4}{\cancel{100}}}$ or $\square$

EXAMPLE Fractions as Percents

2 **Express each fraction as a percent.**

a. $\frac{19}{20} = \square$ or $\square$

b. $\frac{8}{5} = \square$ or $\square$

Your Turn

Express each percent as a fraction in simplest form.

a. 35% $\square$

b. 160% $\square$

c. 0.8% $\square$

d. $32\frac{1}{2}\%$ $\square$

Express each fraction as a percent.

e. $\frac{17}{25}$ $\square$

f. $\frac{14}{10}$ $\square$

Key Concepts

Percents and Decimals

To write a percent as a decimal, divide by 100 and remove the percent symbol.

To write a decimal as a percent, multiply by 100 and add the percent symbol.

FOLDABLES

ORGANIZE IT

Under each tab of your Foldable, describe a real-life situation where it would be helpful to covert to a fraction, decimal, or percent.

Fraction	Decimal	Percent

EXAMPLE Percents as Decimals

3 **Express each percent as a decimal.**

Divide by 100 and remove the %.

a. $\mathbf{60\%} = 60\% =$ ______ **b.** $\mathbf{2\%} = 02\% =$ ______

c. $\mathbf{658\%} = 6.58 =$ ______ **d.** $\mathbf{0.4\%} = 00.4 =$ ______

EXAMPLE Decimals as Percents

4 **Express each decimal as a percent.**

Multiply by 100 and add the %.

a. $\mathbf{0.4} = 0.40 =$ ______ **b.** $\mathbf{0.05} = 0.05 =$ ______

EXAMPLE Fractions as Percents

5 **Express each fraction as a percent. Round to the nearest tenth percent, if necessary.**

a. $\frac{5}{8} = 0.625 =$ ______ **b.** $\frac{1}{3} = 0.333... \approx$ ______

c. $\frac{9}{1000} = 0.009 =$ ______ **d.** $\frac{23}{14} \approx 1.642 =$ ______

Your Turn

Write each percent as a decimal.

a. 84% ______ **b.** 7% ______

c. 302% ______ **d.** 0.9% 

Write each decimal as a percent.

e. 0.84 ______ **f.** 0.01 

Write each fraction as a percent. Round to the nearest tenth percent, if necessary.

g. $\frac{3}{8}$ ______ **h.** $\frac{5}{12}$ ______

i. $\frac{13}{1000}$ ______ **j.** $\frac{21}{17}$ ______

HOMEWORK ASSIGNMENT

Page(s):

Exercises:

6–5

Using the Percent Proportion

What You'll Learn

- Use the percent proportion to solve problems.

Build Your Vocabulary (page 130)

In a **percent proportion**, one of the numbers, called the part, is being ______ to the ______, called the **base**.

Key Concept

Percent Proportion

$\frac{\text{part}}{\text{base}} = \frac{\text{percent}}{100}$

EXAMPLE Find the Percent

1 Twenty is what percent of 25?

Twenty is being compared to 25. So, ______ is the part and ______ is the base. Let p represent the ______.

$\frac{a}{b} = \frac{p}{100} \rightarrow \frac{\quad}{\quad} = \frac{p}{100}$ Replace a with ______ and b with ______.

______ = ______ Find the cross products.

______ $= p$ Simplify.

So, 20 is ______ of 25.

EXAMPLE Find the Percent

2 What percent of 8 is 12?

Twelve is being compared to 8. So, ______ is the part and ______ is the base. Let p represent the ______.

$\frac{a}{b} = \frac{p}{100} \rightarrow \frac{\quad}{\quad} = \frac{p}{100}$ Replace a with ______ and b with ______.

______ = ______ Find the cross products.

______ $= p$ Simplify.

So, 150% of 8 is ______.

EXAMPLE Find the Part

3 **What number is 8.8% of 20?**

The percent is ____, and the base is ____.

Let a represent the ____.

$\frac{a}{b} = \frac{p}{100} \rightarrow \frac{a}{\square} = \frac{\square}{100}$ Replace b with ____ and p with ____.

$\square = \square$ Find the cross products.

$a = \square$

So, 8.8% of 20 is ____.

EXAMPLE Find the Base

4 **Seventy is 28% of what number?**

The percent is ____ and the part is ____.

Let b represent the ____.

$\frac{a}{b} = \frac{p}{100} \rightarrow \frac{\square}{b} = \frac{\square}{100}$ Replace a with ____ and p with ____.

$\square = \square$ Find the cross products.

$\square = b$

So, 70 is 28% of ____.

Your Turn **Use the percent proportion to solve each problem. Round to the nearest tenth.**

a. Twelve is what percent of 40?

b. What percent of 20 is 35?

c. What number is 42.5% of 90?

d. Ninety is 24% of what number?

HOMEWORK ASSIGNMENT

Page(s):

Exercises:

6–6 Finding Percents Mentally

WHAT YOU'LL LEARN

- Compute mentally with percents.
- Estimate with percents.

FOLDABLES

ORGANIZE IT

Under the percents tab of your Foldable, write the percent-fraction equivalents found on page 293 of your textbook.

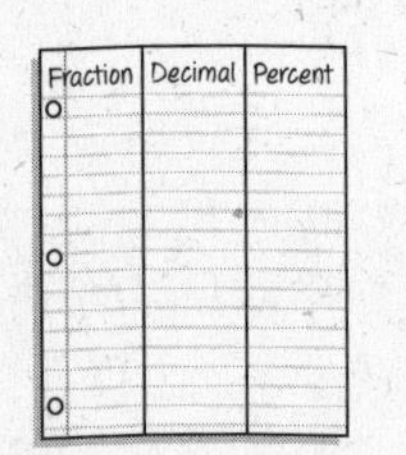

EXAMPLE Find Percent of a Number Mentally

1 Find the percent of each number mentally.

a. 50% of 46

50% of 46 = ____ of 46 Think: 50% = ____.

= ____ Think: ____ of 46 is ____.

So, 50% of 46 is ____.

b. 70% of 110

70% of 110 = ____ of 110 Think: 70% = ____.

= ____ Think: ____ of 110 is ____.

So, 70% of 110 is ____.

Your Turn **Find the percent of each number mentally.**

a. 50% of 82 **b.** 25% of 36 **c.** 80% of 60

EXAMPLE Estimate Percents

2 a. Estimate 22% of 494.

22% is about ____ or ____.

494 is about ____.

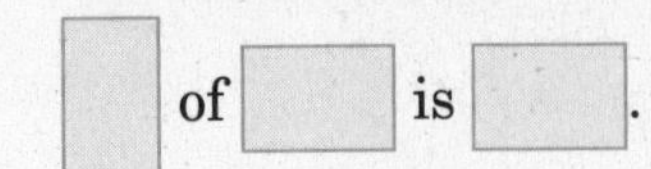

____ of ____ is ____.

So, 22% of 494 is about ____.

b. Estimate 63% of 788.

63% is about ____ or ____.

788 is about ____.

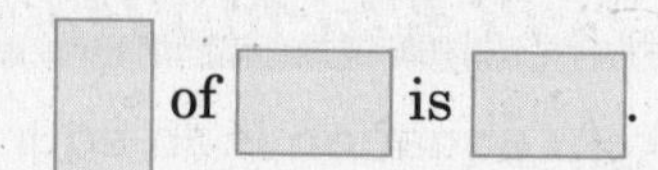

So, 63% of 788 is about ____.

c. Estimate $\frac{1}{4}$% of 1219.

$\frac{1}{4}\% = \frac{1}{4} \times$ ____.

1219 is almost ____.

So, $\frac{1}{4}$% of 1219 is about ____ × ____ or ____.

d. Estimate 155% of 38.

155% means about ____ for every 100

or about ____ for every 10.

38 has about ____ tens.

____ × ____ = ____.

So, 155% of 38 is about ____.

Your Turn **Estimate.**

a. 38% of 400

b. 72% of 318

c. $\frac{1}{5}$% of 2482

d. 183% of 93

HOMEWORK ASSIGNMENT

Page(s):

Exercises:

6–7 Using Percent Equations

BUILD YOUR VOCABULARY (page 130)

The **percent equation** is an equivalent form of the percent ______ in which the percent is written as a decimal.

WHAT YOU'LL LEARN

- Solve percent problems using percent equations.
- Solve real-life problems involving discount and interest.

EXAMPLE Find the Part

1 Find 38% of 22.

You know that the base is ______ and the percent is ______.
Let n represent the part.

$n =$ ______ Write 38% as the decimal ______.

$n =$ ______ Simplify.

So, 38% of 22 is ______.

REMEMBER IT

To determine whether your answer is reasonable, estimate *before* finding the exact answer.

EXAMPLE Find the Percent

2 19 is what percent of 25?

You know that the base is ______ and the part is ______.

Let n represent the percent.

______ $= n($ ______ $)$

______ $= n$ Divide each side by ______.

______ $= n$ Simplify.

So, 19 is ______ of 25.

Your Turn

a. Find 64% of 48.

______

b. 8 is what percent of 25?

______

EXAMPLE Find the Base

3 **84 is 16% of what number?**

You know that the part is ____ and the percent is ____. Let n represent the base.

$\square = \square n$	Write 16% as the decimal ____.
$\frac{\square}{0.16} = \frac{\square}{0.16}$	Divide each side by ____.
$\square = n$	Simplify.

So, 84 is 16% of ____.

Your Turn 315 is 42% of what number?

____

BUILD YOUR VOCABULARY (page 130)

Discount is the amount by which the regular price of an item is reduced.

EXAMPLE Find Discount

4 JEWELRY **The regular price of a ring is $495. It is on sale at a 20% discount. What is the sale price of the ring?**

Method 1

First, use the percent equation to find 20% of 495. Let d represent the discount.

$d = \square$	The base is ____ and the percent is ____.
$d = \square$	Simplify.

Then, find the sale price.

$495 - \square = \square$	Subtract the discount from the original price.

The sale price is ____.

Method 2

A discount of 20% means the ring will cost ____ – ____ 

or ____ of the original price. Use the percent equation to find 80% of 495. Let s represent the sale price.

$s = 0.80(495)$ The base is ____ and the percent is ____.

$s =$ ____ The sale price of the ring will be ____.

Your Turn The regular price of a stereo system is $1295. The system is on sale at a 15% discount. Find the sale price of the stereo system.

BUILD YOUR VOCABULARY (page 131)

Simple interest is the amount of money paid or earned for the use of money.

EXAMPLE Apply Simple Interest Formula

5 BANKING **Suppose you invest $2000 at an annual interest rate of 4.5%. How long will it take for it to earn $495?**

$I = prt$ Simple interest formula

$495 = 2000(0.045)t$ $I = 495$, $p = 2000$, $r = 0.045$

____ = ____ Simplify.

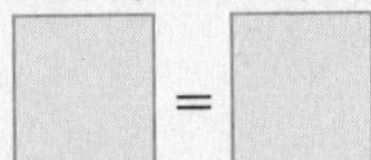

____ = ____ Divide each side by ____.

____ $= t$ Simplify.

It will take ____ years to earn $495.

Your Turn Suppose you invest $3500 at an annual interest rate of 6.25%. How long will it take for it to earn $875?

HOMEWORK ASSIGNMENT

Page(s):

Exercises:

6–8 Percent of Change

WHAT YOU'LL LEARN

- Find percent of increase.
- Find percent of decrease.

BUILD YOUR VOCABULARY (page 130)

A **percent of change** tells the ______ an amount has increased or decreased in relation to the ______ amount.

EXAMPLE Find Percent of Change

1 **Find the percent of change from 325 to 390.**

Step 1 Subtract to find the amount of change.

______ – ______ = ______ new amount – original amount

Step 2 percent of change = $\frac{\text{amount of change}}{\text{original amount}}$

= ______ = ______ or ______

The percent of change from 325 to 390 is ______.

EXAMPLE Find the Percent of Increase

2 TUITION **In 1965, when John entered college, the tuition per year was $7500. In 2000, when his daughter went to the same school, the tuition was $25,500. Find the percent of change.**

Step 1 Subtract to find the amount of change.

______ – ______ = ______

Step 2 percent of change = $\frac{\text{amount of change}}{\text{original tuition}}$

= ______ = ______ or ______

The percent of change is ______.

Your Turn

a. Find the percent of change from 84 to 105.

a. In 1990, the price of a textbook was $38. In 2000, the price of the same textbook was $81. Find the percent of change.

FOLDABLES™

ORGANIZE IT

On the back of your Foldable, describe how to find a percent of increase and a percent of decrease.

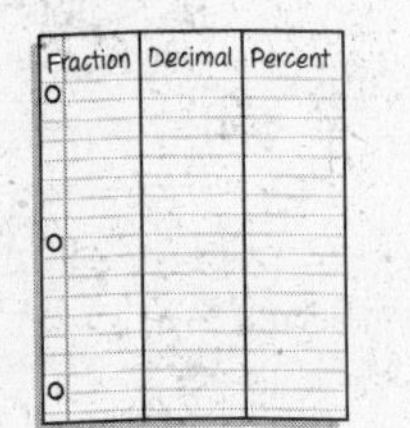

EXAMPLE Find Percent of Decrease

3 CLOTHING **A $110 sweater is on sale for $88. What is the percent of change?**

Step 1 Subtract to find the amount of change.

☐ – ☐ = ☐ sale price – original price

Step 2 percent of change = $\frac{\text{amount of change}}{\text{original price}}$

= ☐ = ☐ or ☐

The percent of change is ☐. In this case, the percent of change is a percent of ☐.

Your Turn A $145 pair of tennis shoes is on sale for $105. What is the percent of change?

HOMEWORK ASSIGNMENT

Page(s):

Exercises:

6–9 Probability and Predictions

What You'll Learn

- Find the probability of simple events.
- Use a sample to predict the actions of a larger group.

Key Concept

Probability The probability of an event is a ratio that compares the number of favorable outcomes to the number of possible outcomes.

EXAMPLE Find Probability

1 **Suppose a number cube is rolled. What is the probability of rolling a 4 or a 5?**

There are ______ numbers that are a 4 or a 5.

There are 6 possible outcomes: 1, 2, 3, 4, 5, and 6.

$P(4 \text{ or } 5) = \dfrac{\text{number of favorable outcomes}}{\text{number of possible outcomes}}$

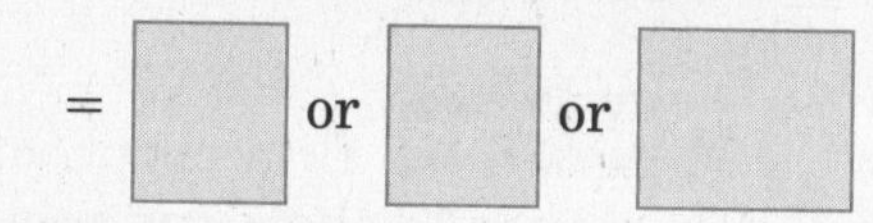

EXAMPLE Find Probability

2 **Suppose that two number cubes are rolled. Find the probability of rolling two identical numbers.**

Make a table showing the sample space when rolling two number cubes.

	1	2	3	4	5	6
1	(1, 1)	(1, 2)	(1, 3)	(1, 4)	(1, 5)	(1, 6)
2	(2, 1)	(2, 2)	(2, 3)	(2, 4)	(2, 5)	(2, 6)
3	(3, 1)	(3, 2)	(3, 3)	(3, 4)	(3, 5)	(3, 6)
4	(4, 1)	(4, 2)	(4, 3)	(4, 4)	(4, 5)	(4, 6)
5	(5, 1)	(5, 2)	(5, 3)	(5, 4)	(5, 5)	(5, 6)
6	(6, 1)	(6, 2)	(6, 3)	(6, 4)	(6, 5)	(6, 6)

P (two identical numbers) = 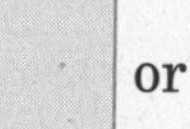 or 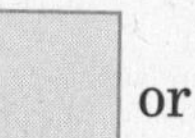 or 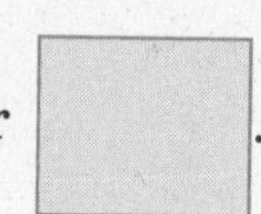.

Your Turn

a. Suppose a number cube is rolled. What is the probability of rolling a number that is divisible by 3?

b. Suppose that two number cubes are rolled. Find the probability of rolling two numbers whose sum is 8.

BUILD YOUR VOCABULARY (pages 130–131)

Experimental probability is what actually occurs when conducting a probability experiment. **Theoretical probability** is what should occur.

EXAMPLE Find Experimental Probability

3 Use the table to determine the experimental probability of landing on heads for this experiment.

Outcome	Tally	Frequency
Heads	卌 卌 ||||	14
Tails	卌 卌 |	11

$\frac{\text{number of times head occur}}{\text{number of possible outcomes}} =$ ☐ or ☐ or ☐.

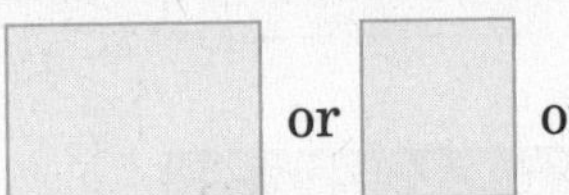

Your Turn Brian is shooting baskets with a basketball. He makes 13 shots and misses 9 shots. Determine the experimental probability of Brian making a shot.

EXAMPLE Make a Prediction

4 SPORTS Miss Newman surveyed her class to see which sport they preferred watching. 44% preferred football, 28% basketball, 20% soccer, and 8% tennis. Out of 560 students in the entire school, how many would you expect to say they prefer watching basketball?

part → $\frac{a}{560} = \frac{28}{100}$ ← percent
base →

$100 \cdot a = 560 \cdot 28$

☐ = ☐

$a =$ ☐

About ☐ students to say they prefer watching basketball.

Your Turn The students in an art class were surveyed about their favorite color. 32% preferred blue, 29% preferred red, 23% preferred yellow, and 16% preferred green. Out of 450 students in the entire school, how many would you expect to say they prefer red?

HOMEWORK ASSIGNMENT

Page(s):

Exercises:

CHAPTER 6

BRINGING IT ALL TOGETHER

STUDY GUIDE

FOLDABLES	VOCABULARY PUZZLEMAKER	BUILD YOUR VOCABULARY
Use your **Chapter 6 Foldable** to help you study for your chapter test.	To make a crossword puzzle, word search, or jumble puzzle of the vocabulary words in Chapter 6, go to: www.glencoe.com/sec/math/t_resources/free/index.php	You can use your completed **Vocabulary Builder** (pages 130–131) to help you solve the puzzle.

6-1 Ratios and Rates

Underline the correct term or phrase to fill the blank in each sentence.

1. A ____________ is a ratio of two measurements having different kinds of units. (fraction, unit, rate)
2. A unit rate has a ____________ of 1. (numerator, denominator, simplest form)
3. A ratio is a comparison of two numbers by ____________. (addition, multiplication, division)
4. Express the ratio *16 novels out of 40 books* as a fraction in simplest form.

6-2 Using Proportions

Determine whether each pair of ratios forms a proportion.

5. $\frac{8}{45}, \frac{1.6}{9}$
6. $\frac{4}{12}, \frac{1}{4}$
7. $\frac{5}{24}, \frac{15}{72}$

8. A restaurant's fruit salad recipe calls for 15 ounces of coconut for 6 quarts of salad. How many ounces of coconut are needed for 8 quarts of salad?

6-3 Scale Drawings and Models

9. A swimming pool is 36 feet long and 15 feet wide. Make a scale drawing of the pool that has a scale of $\frac{1}{4}$ in. = 3 ft.

10. The right arm of the Statue of Liberty is 42 feet long. A scale model of the statue has a 3-inch long right arm. What is the scale of the model?

6-4 Fractions, Decimals, and Percents

Underline the greatest number in each set.

11. $\{\frac{4}{7}$, 45%, 0.42, 5 out of 8$\}$

12. $\{\frac{2}{11}$, 11%, 0.17, 1 out of 12$\}$

Write each list of numbers in order from least to greatest.

13. 59%, 0.53, $\frac{5}{9}$

14. 81%, 0.8, $\frac{5}{6}$

6-5 Using the Percent Proportion

15. 11 is 20% of what number?

16. What is 36% of 75?

17. 18 is what percent of 60?

18. Of the pieces of fruit in a crate, 28% are apples. If the crate contains 50 pieces of fruit, how many are apples?

6-6 Finding Percents Mentally

Estimate.

19. $\frac{2}{3}$% of 155

20. 147% of 78

21. 84% of 31

6-7

Using Percent Equations

Solve each problem using the percent equation.

22. 7 is what percent of 25?

23. What is 40.4% of 50?

24. 32 is 5% of what number?

25. Find 140% of 75.

26. A CD player is on sale at a 20% discount. If it normally sells for $49.95, what is the sale price?

27. What is the annual interest rate if $2800 is invested for 4 years and $364 in interest is earned?

6-8

Percent of Change

28. A $775 computer is marked down to $620. Find the percent of change.

29. Refer to the table shown. Which school had the smallest percent of increase in the number of students from 1994 to 2004?

School	1994	2004
Oakwood	672	702
Jefferson	433	459
Marshall	764	780

6-9

Probability and Predictions

30. A box contains 7 black, 10 blue, 5 green and 8 red pens. One pen is selected at random. Find the probability that it is *not* green.

31. Refer to the graph. Out of a group of 3500 people, how many would you expect to say that family time is their favorite leisure-time activity?

Leisure-time favorites

Reading 29%

TV watching 19%

Family time 12%

Source: *USA Today*

Checklist

ARE YOU READY FOR THE CHAPTER TEST?

Visit **pre-alg.com** to access your textbook, more examples, self-check quizzes, and practice tests to help you study the concepts in Chapter 6.

Check the one that applies. Suggestions to help you study are given with each item.

☐ **I completed the review of all or most lessons without using my notes or asking for help.**

- You are probably ready for the Chapter Test.
- You may want take the Chapter 6 Practice Test on page 321 of your textbook as a final check.

☐ **I used my Foldable or Study Notebook to complete the review of all or most lessons.**

- You should complete the Chapter 6 Study Guide and Review on pages 316–320 of your textbook.
- If you are unsure of any concepts or skills, refer back to the specific lesson(s).
- You may also want to take the Chapter 6 Practice Test on page 321.

☐ **I asked for help from someone else to complete the review of all or most lessons.**

- You should review the examples and concepts in your Study Notebook and Chapter 6 Foldable.
- Then complete the Chapter 6 Study Guide and Review on pages 316–320 of your textbook.
- If you are unsure of any concepts or skills, refer back to the specific lesson(s).
- You may also want to take the Chapter 6 Practice Test on page 321.

Student Signature

Parent/Guardian Signature

Teacher Signature

Equations and Inequalities

Use the instructions below to make a Foldable to help you organize your notes as you study the chapter. You will see Foldable reminders in the margin of this Interactive Study Notebook to help you in taking notes.

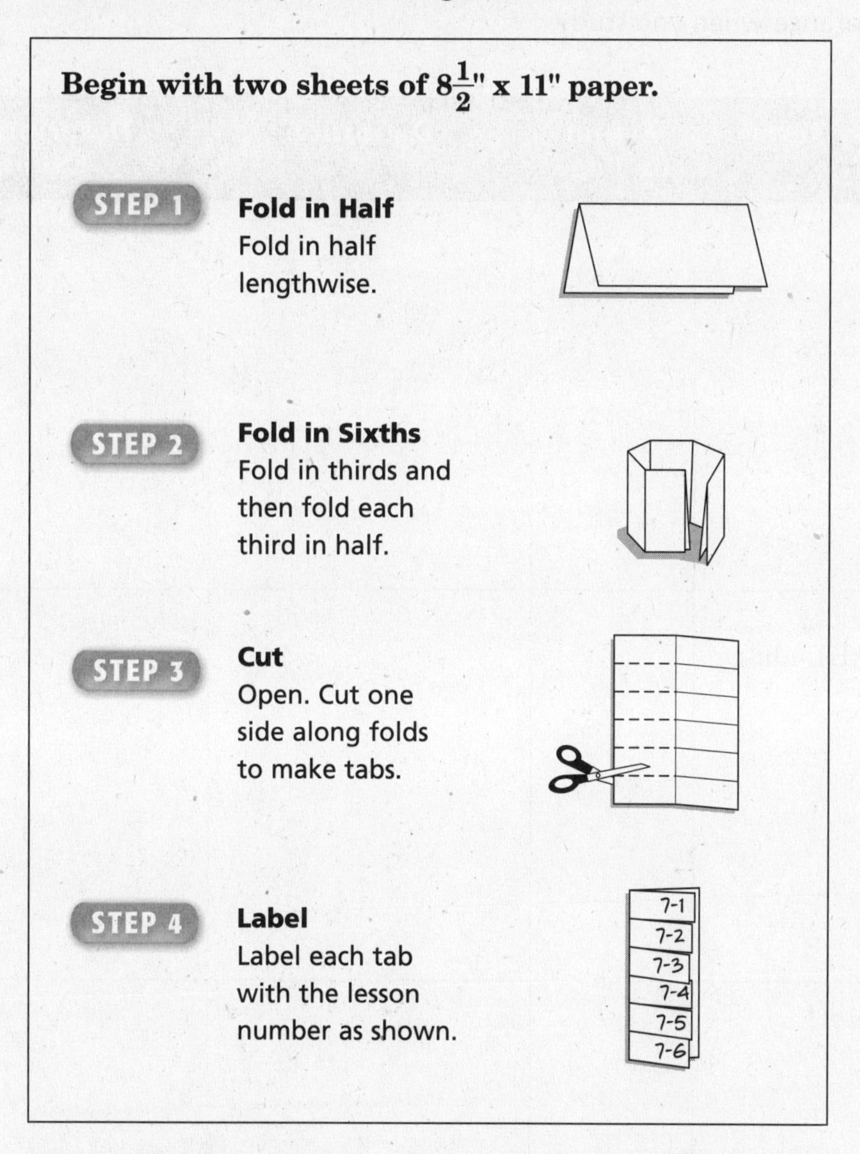

Begin with two sheets of $8\frac{1}{2}$" x 11" paper.

STEP 1 **Fold in Half**
Fold in half lengthwise.

STEP 2 **Fold in Sixths**
Fold in thirds and then fold each third in half.

STEP 3 **Cut**
Open. Cut one side along folds to make tabs.

STEP 4 **Label**
Label each tab with the lesson number as shown.

NOTE-TAKING TIP: Write down questions that you have about what you are reading in the lesson. Then record the answer to each question as you study the lesson.

BUILD YOUR VOCABULARY

This is an alphabetical list of new vocabulary terms you will learn in Chapter 7. As you complete the study notes for the chapter, you will see Build Your Vocabulary reminders to complete each term's definition or description on these pages. Remember to add the textbook page number in the second column for reference when you study.

Vocabulary Term	Found on Page	Definition	Description or Example
identity			
inequality [IHN-ih-KWAHL-uht-ee]			
null or empty set [NUHL]			

7-1 Solving Equations with Variables on Each Side

WHAT YOU'LL LEARN

- Solve equations with variables on each side.

EXAMPLE Equations with Variables on Each Side

1 **Solve $5x + 12 = 2x$.**

$5x + 12 = 2x$	Write the equation.
$5x - \square + 12 = 2x - \square$	Subtract $\square$ from each side.
$\square = \square$	Simplify.
$\square = x$	Mentally divide each side by $\square$.

Your Turn Solve $7x = 5x + 6$.

EXAMPLE Equations with Variables on Each Side

2 **a. Solve $7x + 3 = 2x + 23$.**

$7x + 3 = 2x + 23$	Write the equation.
$7x - \square + 3 = 2x - \square + 23$	Subtract $\square$ from each side.
$\square = 23$	Simplify.
$\square - \square = 23 - \square$	Subtract $\square$ from each side.
$\square = \square$	Simplify.
$x = \square$	Mentally divide.

FOLDABLES

ORGANIZE IT

As you read through Lesson 7-1, write down one or more questions you have behind the 7-1 tab of your Foldable. As you study the lesson, take notes, and record information that answers your questions.

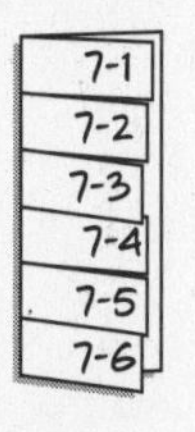

b. Solve $1.7 + a = 2.5a - 4.9$.

$1.7 + a = 2.5a - 4.9$	Write the equation.
$1.7 + a - \square = 2.5a - \square - 4.9$	Subtract $\square$ from each side.
$1.7 = \square - 4.9$	Simplify.
$1.7 + \square = 1.5a - \square + \square$	Add $\square$ to each side.
$\square = 1.5a$	Simplify.
$\square = a$	Divide.

Your Turn **Solve each equation.**

a. $4x + 15 = 2x - 7$

b. $2.4 - 3m = 6.4m - 8.88$

HOMEWORK ASSIGNMENT

Page(s):

Exercises:

7–2 Solving Equations with Grouping Symbols

WHAT YOU'LL LEARN

- Solve equations that involve grouping symbols.
- Identify equations that have no solution or an infinite number of solutions.

EXAMPLE Solve Equations with Parentheses

1 Solve $3h = 5(h - 2)$.

$3h = 5(h - 2)$ Write the equation.

$3h = \square - \square$  Property

$3h = \square$ Simplify.

$3h - \square = \square$ Subtract $\square$ from each side.

$\square = \square$ Simplify.

$h = \square$ Simplify.

Ths solution is $\square$.

EXAMPLE No Solution

2 Solve $4x - 0.3 = 4x + 0.9$.

$4x - 0.3 = 4x + 0.9$ Write the equation.

$4x - \square - 0.3 = 4x - \square + 0.9$ Subtract $\square$ from each side.

$\square = \square$ Simplify.

The sentence is $\square$ true. So, the solution set is .

FOLDABLES

ORGANIZE IT

As you read through Lesson 7-2, write down one or more questions you have behind the 7-2 tab of your Foldable. As you study the lesson, take notes, and record information that answers your questions.

Your Turn **Solve each equation.**

a. $4t = 7(t - 3)$

b. $16 + 1.3m = -12 + 1.3m$

BUILD YOUR VOCABULARY (page 158)

An equation that is ______ for every value of the ______ is called an **identity**.

EXAMPLE **All Numbers as Solutions**

3 **Solve $3(4x - 2) + 15 = 12x + 9$.**

$3(4x - 2) + 15 = 12x + 9$ Write the equation.

______ $+ 15 = 12x + 9$ Distributive Property

______ $= 12x + 9$ Simplify.

______ = ______ Subtract ______ from each side.

______ = ______ Mentally divide each side by ______.

The sentence is ______ true. The solution set is ______.

Your Turn Solve $10a - 9 = 5(2a - 3) + 6$.

HOMEWORK ASSIGNMENT

Pages(s):

Exercises:

7–3 Inequalities

What You'll Learn

- Write inequalities.
- Graph inequalities.

Build Your Vocabulary (page 158)

A mathematical sentence that contains ______ or ______ is called an **inequality**.

Foldables

Organize It

As you read through Lesson 7-3, write down one or more questions you have behind the 7-3 tab of your Foldable. As you study the lesson, take notes, and record information that answers your questions.

7-1
7-2
7-3
7-4
7-5
7-6

EXAMPLE Write Inequalities with $<$ or $>$

1 **Write an inequality for each sentence.**

a. Your age is less than 19 years.

Variable Let a represent ______.

Inequality ______

b. Your height is greater than 52 inches.

Variable Let h represent ______.

Inequality ______

EXAMPLE Write Inequalities with $\leq$ or $\geq$

2 **Write an inequality for each sentence.**

a. Your speed is less than or equal to 62 miles per hour.

Variable Let s represent ______.

Inequality ______

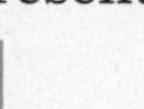

b. Your speed is greater than or equal to 42.

Variable Let s represent ______.

Inequality ______

Your Turn **Write an inequality for each sentence.**

a. Your height is less than 48 inches.

______

b. Your age is greater than 12 years.

c. Your weight is less than or equal to 120 pounds.

d. Your speed is greater than or equal to 35.

EXAMPLE Determine Truth of an Inequality

WRITE IT

Describe one way to remember the difference between the $>$ symbol and the $\geq$ symbol.

____

____

____

____

3 For the given value, state whether the inequality is *true* or *false*.

a. $s - 9 < 4, s = 6$

$\square - 9 \overset{?}{<} 4$ Replace s with $\square$.

$\square < 4$ Simplify.

The sentence is $\square$.

b. $14 \leq \frac{a}{3} + 1, a = 36$

$14 \leq \frac{\square}{3} + 1$ Replace a with $\square$.

$14 \overset{?}{\leq} \square + 1$ Simplify.

$14 \not\leq \square$ Simplify.

The sentence is $\square$.

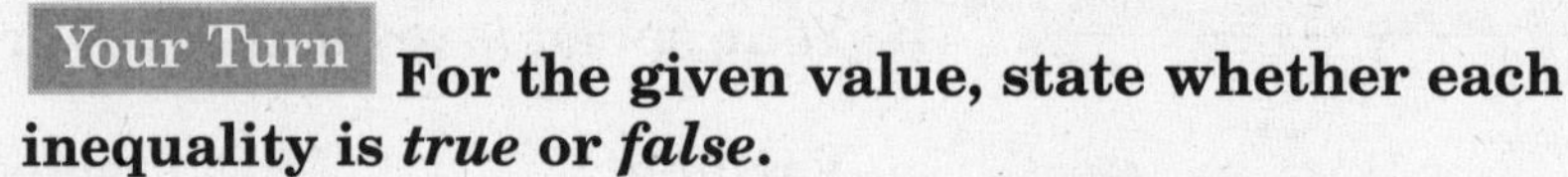

Your Turn **For the given value, state whether each inequality is *true* or *false*.**

a. $12 - m > 7, m = 5$

b. $\frac{20}{x} + 3 \leq 6, x = 10$

EXAMPLE Graph Inequalities

4 **a. Graph $x > 10$.**

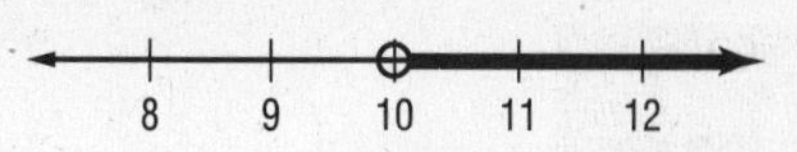

The open circle means the number 10 is ______.

b. Graph $x \geq 10$.

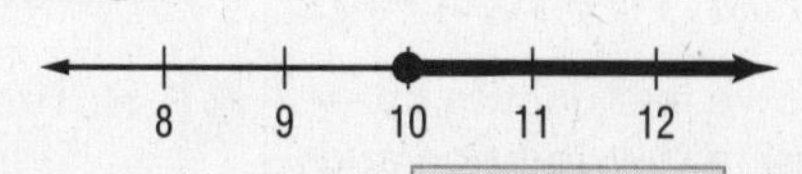

The closed circle means the number 10 is ______.

c. Graph $x < 10$.

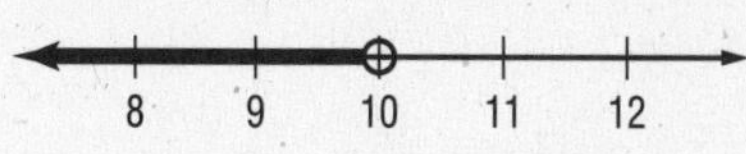

The open circle means the number 10 is .

Your Turn **Graph each inequality.**

a. $x < 3$

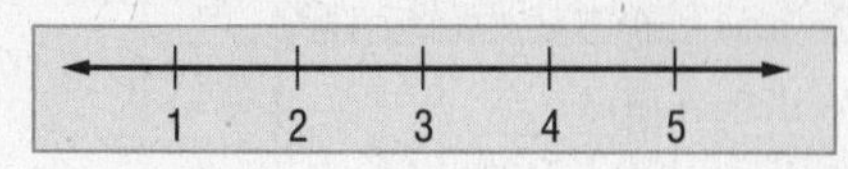

b. $x > 3$

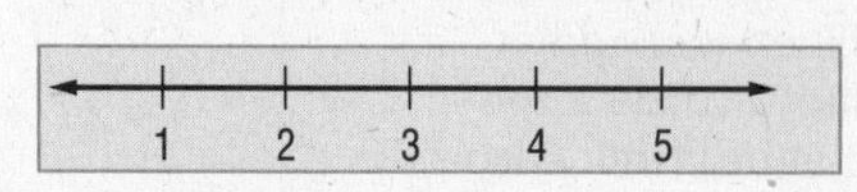

c. $x \geq 3$

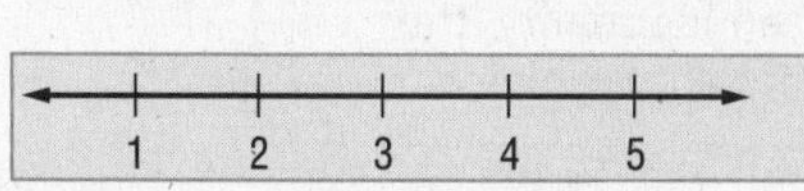

EXAMPLE Write an Inequality

5 **Write the inequality for the graph.**

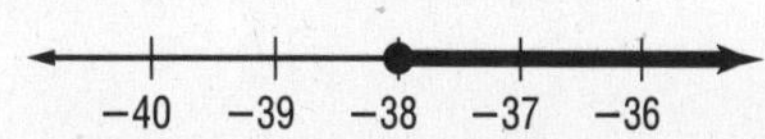

A closed circle is on -38, so the point -38 is ______ in the graph. The arrow points to the ______, so the graph includes all numbers ______ than or ______ -38. That is, ______.

Your Turn Write the inequality for the graph.

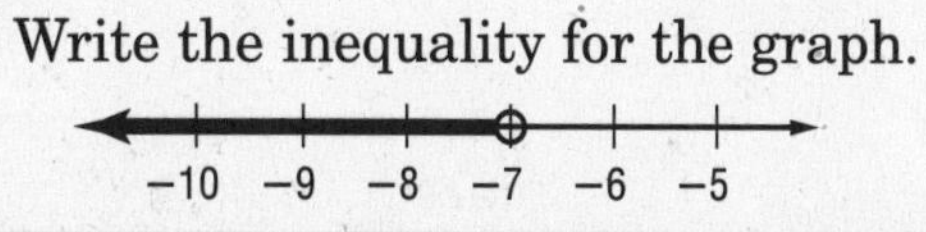

HOMEWORK ASSIGNMENT

Page(s):

Exercises:

7-4 Solving Inequalities by Adding or Subtracting

WHAT YOU'LL LEARN

- Solve inequalities by using the Addition and Subtraction Properties of Inequality.

EXAMPLE Solve an Inequality Using Subtraction

1 **Solve $y + 5 > 11$.**

$y + 5 > 11$ Write the inequality.

$y + 5 -$ ______ $> 11 -$ ______ Subtract 5 from each side.

______ $>$ ______ Simplify.

EXAMPLE Solve an Inequality Using Addition

KEY CONCEPT

Addition and Subtraction Properties When you add or subtract the same number from each side of an inequality, the inequality remains true.

2 **Solve $-21 \geq d - 8$.**

$-21 \geq d - 8$ Write the inequality.

$-21 +$ ______ $\geq d - 8 +$ ______ Add ______ to each side.

______ $\geq$ ______ Simplify.

EXAMPLE Graph Solutions of Inequalities

3 **Solve $h - 1\frac{1}{2} < 5$. Graph the solution on a number line.**

$h - 1\frac{1}{2} < 5$ Write the inequality.

$h - 1\frac{1}{2} +$ ______ $< 5 +$ ______ Add ______ to each side.

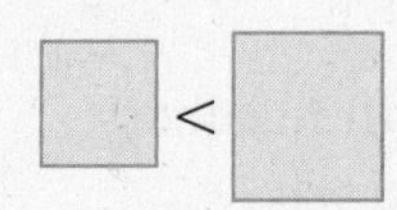

______ $<$ ______ Simplify.

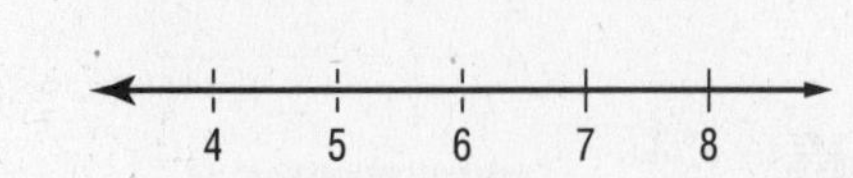

Place ______ at ______. Draw a line and arrow to the ______.

FOLDABLES™

ORGANIZE IT

As you read through Lesson 7-4, write down questions you have behind the 7-4 tab of your Foldable. As you study the lesson, take notes, and record information that answers your questions.

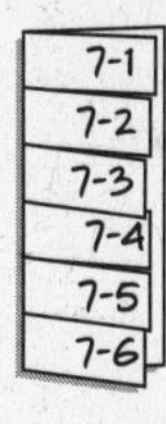

Your Turn **Solve each equation.**

a. $x + 9 < 13$

b. $m + 8 < -2$

c. Solve $x - \frac{3}{4} \geq \frac{1}{2}$. Graph the solution on a number line.

HOMEWORK ASSIGNMENT

Page(s):

Exercises:

7–5 Solving Inequalities by Multiplying or Dividing

WHAT YOU'LL LEARN

- Solve inequalities by multiplying or dividing by a positive number.
- Solve inequalities by multiplying or dividing by a negative number.

KEY CONCEPT

Multiplication and Division Properties
When you multiply or divide each side of an inequality by the same or positive number, the inequality remains true.

FOLDABLES

ORGANIZE IT

As you study the lesson, take notes, and record information about solving inequalities.

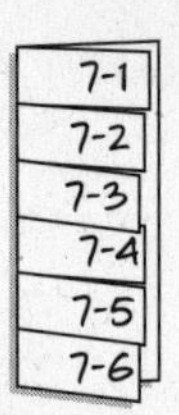

EXAMPLE Multiply or divide by a Positive Number

1 **a. Solve $9x \le 54$.**

$9x \le 54$ — Write the inequality.

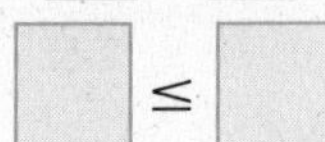

$\frac{9x}{\square} \le \frac{54}{\square}$ — Divide each side by ☐

$\square \le \square$ — Simplify.

b. Solve $\frac{d}{9} > 4$.

$\frac{d}{9} > 4$ — Write the inequality.

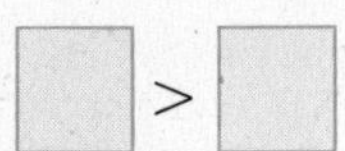

$9\left(\frac{d}{9}\right) > \square(4)$ — Multiply each side by ☐.

$\square > \square$ — Simplify.

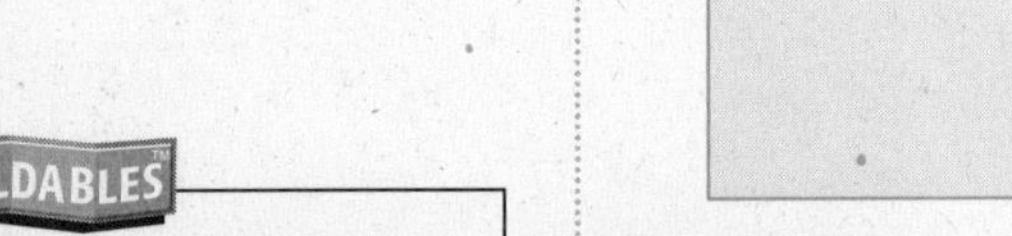

Your Turn **Solve each inequality.**

a. $3x > 21$ **b.** $6 \le \frac{p}{3}$

EXAMPLE Write an Inequality

2 **Martha earns \$9 per hour working for a fast-food restaurant. Which inequality can be used to find how many hours she must work in a week to earn at least \$243?**

Let x represent the number of hours worked.

Amount earned per hour	times	number of hours	is at least	amount earned each week
☐	☐	☐	☐	☐

The inequality is ☐.

Your Turn Ed earns $6 per hour working at the library. Write an inequality that can be used to find how many hours he must work in a week to earn more than $100?

KEY CONCEPT

Multiplication and Division Properties When you multiply or divide each side of an inequality by the same negative number, the inequality symbol must be reversed for the inequality to remain true.

EXAMPLE Multiply or Divide by a Negative Number

3 **Solve each inequality and check your solution. Then graph the solution on a number line.**

a. $\frac{x}{-5} \geq 7$

$\frac{x}{-5} \geq 7$ Write the inequality.

$\square\left(\frac{x}{-5}\right) \leq \square(7)$ Multiply each side by $\square$ and reverse the symbol.

$x \leq \square$

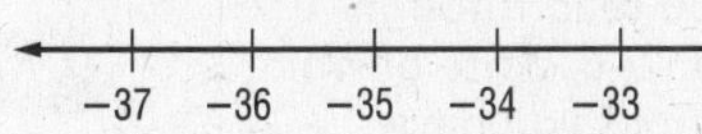

b. $-9x < -27$

$-9x < -27$ Write the inequality.

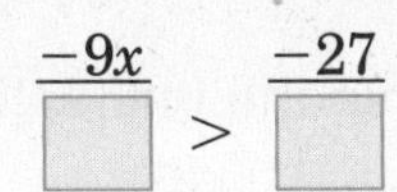

Divide each side by $\square$ and reverse the symbol.

$x >$ 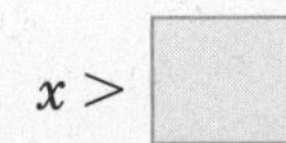

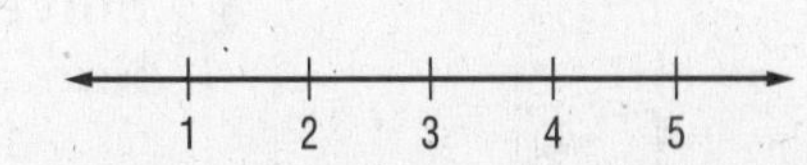

Your Turn **Solve each inequality and check your solution. Then graph the solution on a number line.**

a. $\frac{x}{-3} > 6$

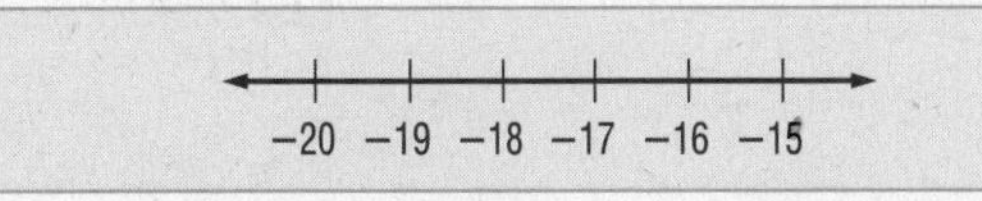

b. $-5x \leq -40$

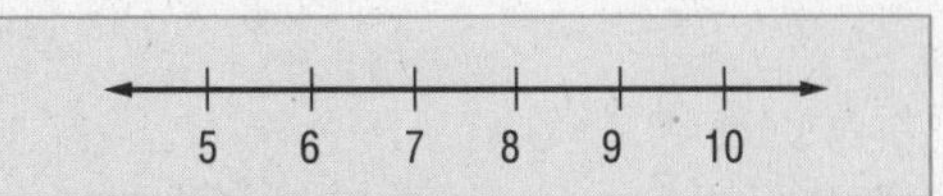

Page(s):

Exercises:

7–6 Solving Multi-Step Inequalities

WHAT YOU'LL LEARN

- Solve inequalities that involve more than one operation.

FOLDABLES

ORGANIZE IT

As you read through Lesson 7-6, write down questions you have behind the 7-6 tab of your Foldable. As you study the lesson, take notes, and record information that answers your questions.

7-1
7-2
7-3
7-4
7-5
7-6

EXAMPLE Solve a Two-Step Inequality

1 **Solve $5x + 13 > 83$. Graph the solution on a number line.**

$5x + 13 > 83$ Write the inequality.

$5x + 13 - \_\_\_ > 83 - \_\_\_$ Subtract ___ from each side.

$\_\_\_ > \_\_\_$ Simplify.

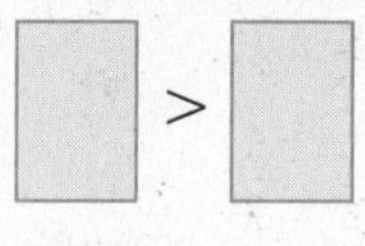

Divide each side by ___.

$\_\_\_ > \_\_\_$

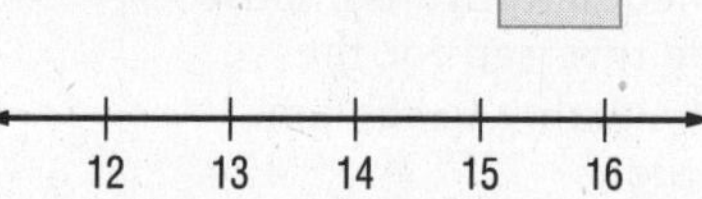

Your Turn Solve $3x - 9 < 18$. Graph the solution on a number line.

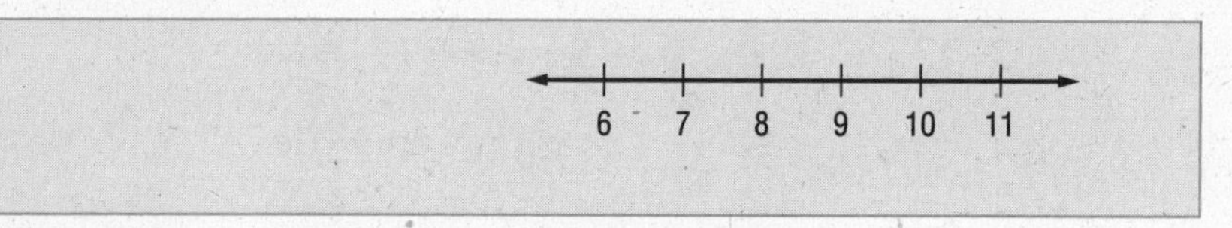

EXAMPLE Reverse the Inequality Symbol

2 **Solve $7 - 4a \leq 23 - 2a$. Graph the solution on a number line.**

$7 - 4a \leq 23 - 2a$ Write the inequality.

$7 - 4a + \_\_\_ \leq 23 - 2a + \_\_\_$ Add ___ to each side.

$\_\_\_ \leq \_\_\_$ Simplify.

$\_\_\_ \leq \_\_\_$ Subtract ___ from each side.

$\_\_\_ \leq \_\_\_$ Simplify.

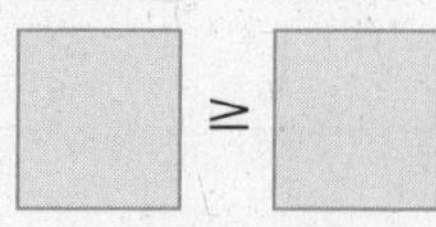

Divide each side by ___ and change $\leq$ to $\geq$.

$a \geq \_\_\_$

−10 −9 −8 −7 −6

Your Turn Solve $8 + 2x < 5x - 7$. Graph the solution on a number line.

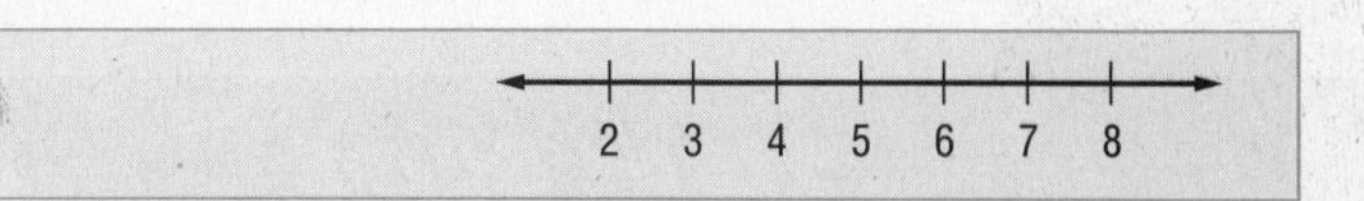

EXAMPLE Inequalities with Grouping Symbols

3 BACKPACKING **A person weighing 213 pounds has a 10-pound backpack. If three times the weight of your backpack and its contents should be less than you body weight, what is the maximum weight for the contents of the pack?**

Let c represent the weight of the contents of the pack.

Words	3	times	weight of pack and contents	should be less than	body weight.
Inequality					

$\_\_\_ < \_\_\_$ Write the inequality.

$\_\_\_ < \_\_\_$ Distributive Property

$\_\_\_ < \_\_\_$ Subtract ___ from each side.

$\_\_\_ < \_\_\_$ Simplify.

$c < \_\_\_$ Divide each side by ___.

The weight of the contents should be less than ___.

Your Turn A person weighing 168 pounds has a 7-pound backpack. If three times the weight of your backpack and its contents should be less than your body weight, what is the maximum weight for the contents of the pack?

FOLDABLES™

ORGANIZE IT

Under the tab for Lesson 7-6, write an example of an inequality that requires two steps to solve. Label each step with the operation being undone.

7-1
7-2
7-3
7-4
7-5
7-6

HOMEWORK ASSIGNMENT

Page(s):

Exercises:

CHAPTER 7

BRINGING IT ALL TOGETHER

STUDY GUIDE

FOLDABLES™	VOCABULARY PUZZLEMAKER	BUILD YOUR VOCABULARY
Use your **Chapter 7 Foldable** to help you study for your chapter test.	To make a crossword puzzle, word search, or jumble puzzle of the vocabulary words in Chapter 7, go to: www.glencoe.com/sec/math/t_resources/free/index.php	You can use your completed **Vocabulary Builder** (page 158) to help you solve the puzzle.

7-1 Solving Equations with Variables on Each Side

Number the steps in the correct order for solving the equation $2x + 4 = 4x - 8$. Some steps may be used more than once.

1. ☐ Simplify.
2. ☐ Subtract $2x$ from each side.
3. ☐ Write the equation.
4. ☐ Add 8 to each side.
5. ☐ Divide each side by 2

7-2 Solving Equations with Grouping Symbols

6. The perimeter of a rectangle is 74 inches. Find the dimensions and the area if the length is 5 inches shorter than twice the widths.

7-3 Inequalities

For each of the following phrases, write the corresponding inequality symbol in the blank. Use $<$, $>$, $\leq$, or $\geq$.

7. is greater than ☐
8. is less than or equal to ☐
9. Write an inequality for the sentence: *Seven less than a number is at least 15.* ☐

7-4 Solving Inequalities by Adding or Subtracting

10. Is 6 a solution for the inequality $17 + x > 23$? Explain.

Solve each inequality. Then graph the solution on a number line.

11. $b + 6 < 19$

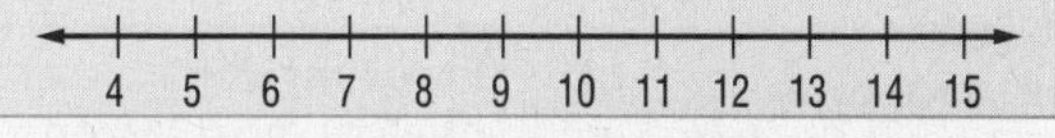

12. $21 > n + 27$

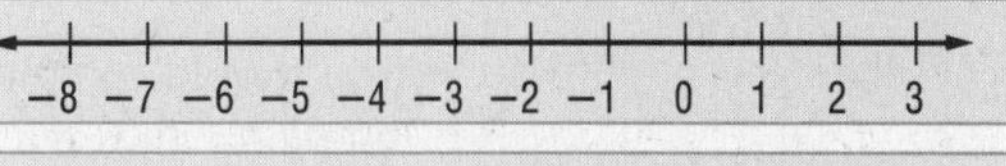

13. $-8 \leq -15 + x$

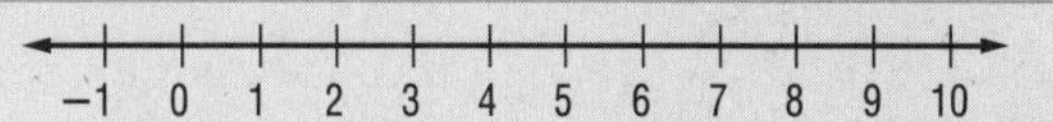

7-5 Solving Inequalities by Multiplying or Dividing

Match each inequality with its graph.

14. $2x \geq 6$

15. $\frac{x}{-3} > -1$

16. $12x < -36$

17. $-3x < -9$

a. −4 −3 −2 −1 0 1 2 3 4

b. −4 −3 −2 −1 0 1 2 3 4

c. −4 −3 −2 −1 0 1 2 3 4

d. −4 −3 −2 −1 0 1 2 3 4

e. −4 −3 −2 −1 0 1 2 3 4

7-6 Solving Multi-Step Inequalities

Underline the correct term or phrase to complete each statement.

18. Remember to (reverse, delete) the inequality symbol when multiplying or dividing both sides of the inequality by a negative number.

19. To check the solution $x > 14$, you should try a number (smaller, greater) than 14 in the original inequality.

Solve each inequality. Graph the solution on a number line.

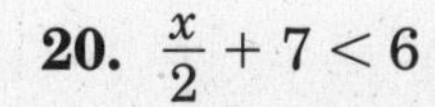

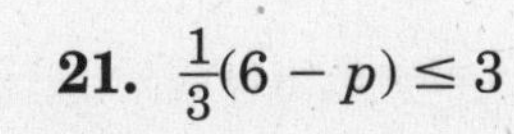

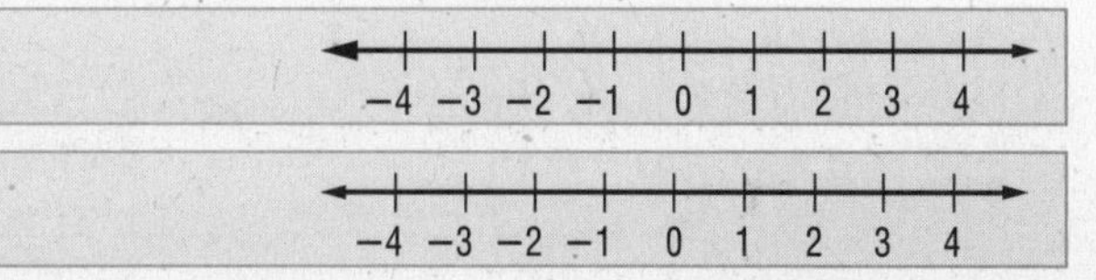

Checklist

ARE YOU READY FOR THE CHAPTER TEST?

Visit **pre-alg.com** to access your textbook, more examples, self-check quizzes, and practice tests to help you study the concepts in Chapter 7.

Check the one that applies. Suggestions to help you study are given with each item.

☐ **I completed the review of all or most lessons without using my notes or asking for help.**

- You are probably ready for the Chapter Test.
- You may want take the Chapter 7 Practice Test on page 363 of your textbook as a final check.

☐ **I used my Foldable or Study Notebook to complete the review of all or most lessons.**

- You should complete the Chapter 7 Study Guide and Review on pages 360–362 of your textbook.
- If you are unsure of any concepts or skills, refer back to the specific lesson(s).
- You may also want to take the Chapter 7 Practice Test on page 363.

☐ **I asked for help from someone else to complete the review of all or most lessons.**

- You should review the examples and concepts in your Study Notebook and Chapter 7 Foldable.
- Then complete the Chapter 7 Study Guide and Review on pages 360–362 of your textbook.
- If you are unsure of any concepts or skills, refer back to the specific lesson(s).
- You may also want to take the Chapter 7 Practice Test on page 363.

Student Signature

Parent/Guardian Signature

Teacher Signature

Functions and Graphing

Use the instructions below to make a Foldable to help you organize your notes as you study the chapter. You will see Foldable reminders in the margin of this Interactive Study Notebook to help you in taking notes.

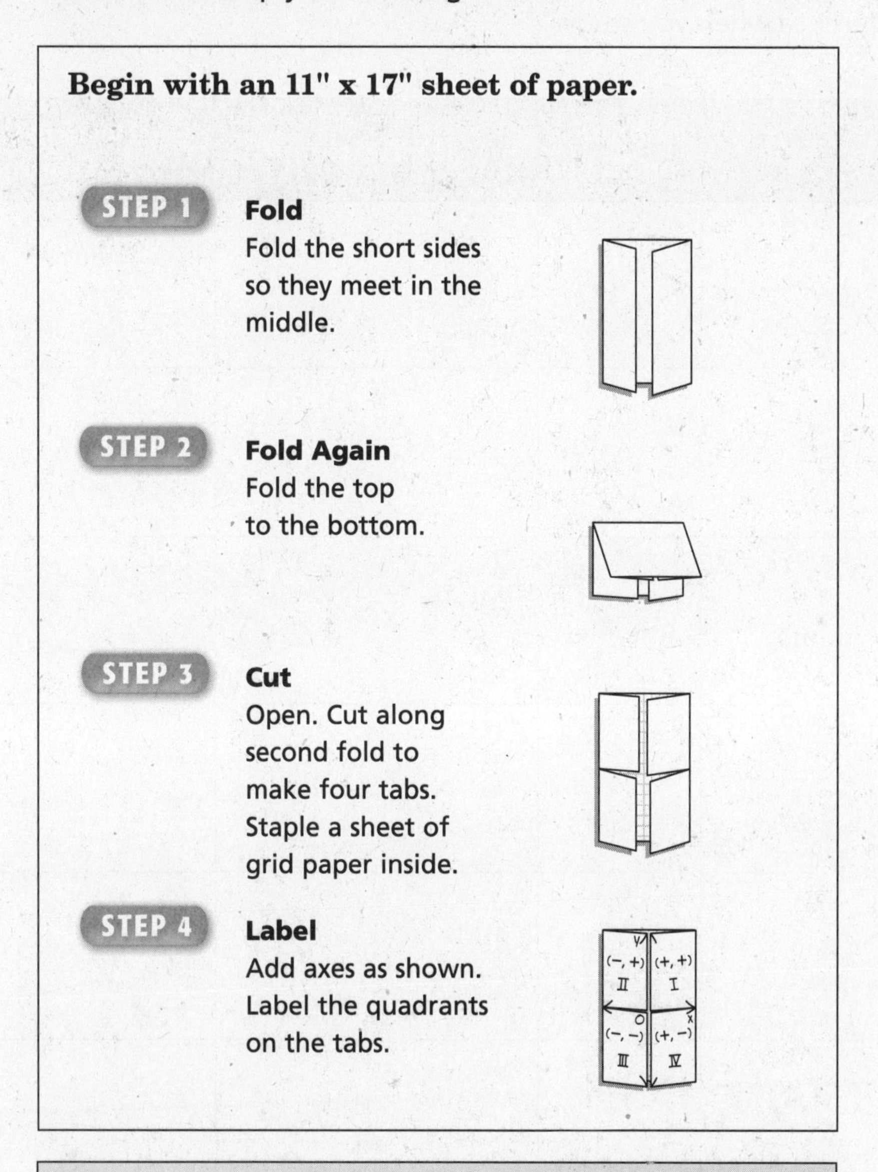

Begin with an 11" x 17" sheet of paper.

STEP 1 **Fold**
Fold the short sides so they meet in the middle.

STEP 2 **Fold Again**
Fold the top to the bottom.

STEP 3 **Cut**
Open. Cut along second fold to make four tabs. Staple a sheet of grid paper inside.

STEP 4 **Label**
Add axes as shown. Label the quadrants on the tabs.

NOTE-TAKING TIP: When you take notes, listen or read for main ideas. Then record those ideas for future reference.

BUILD YOUR VOCABULARY

This is an alphabetical list of new vocabulary terms you will learn in Chapter 8. As you complete the study notes for the chapter, you will see Build Your Vocabulary reminders to complete each term's definition or description on these pages. Remember to add the textbook page number in the second column for reference when you study.

Vocabulary Term	Found on Page	Definition	Description or Example
best-fit line			
boundary			
constant of variation [VEHR-ee-Ay-shuhn]			
direct variation			
function			
half plane			
linear equation [LINH-ee-uhr]			

Vocabulary Term	Found on Page	Definition	Description or Example
rate of change			
slope			
slope-intercept form [IHNT-uhr-SEHPT]			
substitution			
system of equations			
vertical line test			
x-intercept			
y-intercept			

8–1 Functions

BUILD YOUR VOCABULARY (page 176)

A **function** is a special relation in which each member of the domain is paired with *exactly* one member in the range.

WHAT YOU'LL LEARN

- Determine whether relations are functions.
- Use functions to describe relationships between two quantities.

EXAMPLE Ordered Pairs and Tables as Functions

1 **Determine whether each relation is a function. Explain.**

a. $\{(-3, -3), (-1, -1), (0, 0), (-1, 1), (3, 3)\}$

_____; -1 in the domain is paired with both _____ and _____ in the _____.

b.

x	7	6	5	2	−3	−6
y	2	4	6	4	2	−2

_____, each x value is paired with _____ y value.

Your Turn **Determine whether each relation is a function. Explain.**

a. $\{(2, 5), (4, -1), (3, 1), (6, 0), (-2, -2)\}$

_____

b.

x	3	1	−1	−3	1	−5
y	5	4	3	−4	2	1

_____

REMEMBER IT

If any vertical line drawn on the graph of a relation passes through no more than one point on the graph, then the relation is a function. This is a **vertical line test**.

EXAMPLE Use a Graph to Identify Functions

2 **Determine whether the graph is a function. Explain.**

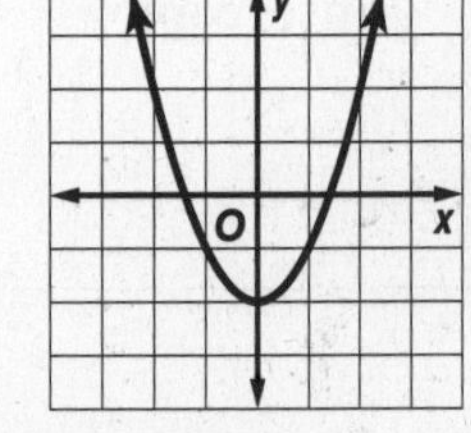

_____, it passes the _____.

FOLDABLES

ORGANIZE IT

In your notes, draw a graph of a relation that is a function and a graph of a relation that is not a function. Explain why the second relation is not a function.

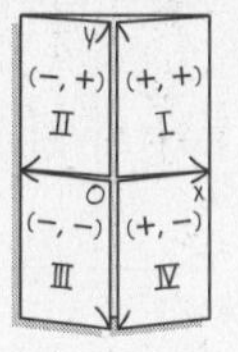

Your Turn Determine whether the graph is a function. Explain.

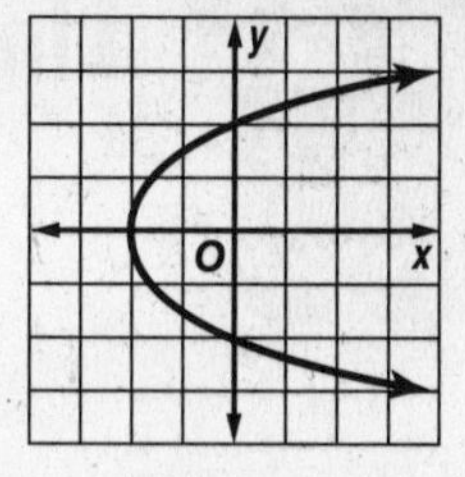

EXAMPLE Use a Function to Describe Data

3 BUSINESS The table shows the number of boxes made.

Number of Hours	Number of Boxes
0	0
10	3000
20	6000
30	9000

a. Do these data represent a function? Explain.

______; for each 10 hours, only ______ of boxes is made.

b. Describe how box production is related to hours of operation.

As the number of hours ______, the number of boxes produced ______.

Your Turn **The table shows the number of chairs made.**

Number of Hours	Number of Boxes
5	120
10	240
15	360
20	480

a. Do these data represent a function? Explain.

b. Describe how chair production is related to hours of operation.

HOMEWORK ASSIGNMENT

Page(s):

Exercises:

8–2 Linear Equations in Two Variables

WHAT YOU'LL LEARN

- Solve linear equations with two variables.
- Graph linear equations using ordered pairs.

BUILD YOUR VOCABULARY (page 176)

A **linear equation** in two variables is an equation in which the ______ appear in ______ terms and neither variable contains an ______ other than 1.

EXAMPLE Find Solutions

1 Find four solutions of $y = 4x + 3$.

Choose four values for x. Then substitute each value into the equation to solve for y. There are many possible solutions. The solutions you find depend on which x values you choose.

x	$y = 4x + 3$	y	(x, y)
0	$y = 4$ ___ $+ 3$		
1	$y = 4$ ___ $+ 3$		
2	$y = 4$ ___ $+ 3$		
3	$y = 4$ ___ $+ 3$		

Four solutions are ______, ______, ______, and ______.

Your Turn Find four solutions of $y = 2x - 4$.

FOLDABLES

ORGANIZE IT

In your notes, write a linear equation, then explain how to solve it using the four steps for finding solutions of equations.

(−, +) II | (+, +) I | (−, −) III | (+, −) IV

EXAMPLE Solve an Equation for y

2 BUSINESS **At a local software company, Level 1 employees x earn \$48,000 and Level 2 employees y earn \$24,000. Find four solutions of $48{,}000x + 24{,}000y = 216{,}000$ to determine how many employees at each level the company can hire for \$216,000.**

$48{,}000x + 24{,}000y = 216{,}000$ Write the equation.

$24{,}000y = 216{,}000 -$ ______ Subtract ______ from each side.

$\frac{24{,}000y}{\_\_\_} = \frac{216{,}000}{\_\_\_} - \frac{48{,}000x}{\_\_\_}$ Divide each side by ______.

$y =$ ______ Simplify.

Choose four x values and substitute them into ______.

x	$y = 9 - 2x$	y	(x, y)
0	$y = 9 - 2$ ___		
1	$y = 9 - 2$ ___		
2	$y = 9 - 2$ ___		
3	$y = 9 - 2$ ___		

(0, ___) 0 Level 1, ___ Level 2

(1, ___) 1 Level 1, ___ Level 2

(2, ___) 2 Level 1, ___ Level 2

(3, ___) 3 Level 1, ___ Level 2

The company can hire 0 Level 1 and ___ Level 2 employees, 1 Level 1 and ___ Level 2 employees, 2 Level 1 and ___ Level 2 employees, 3 Level 1 and ___ Level 2 employees.

Your Turn At a local bookstore, hardbacks are on sale for \$6 and paperbacks are on sale for \$3. Bob has \$42 to spend on books. Find four solutions to determine how many books of each type Bob can buy with his \$42.

EXAMPLE Graph a Linear Equation

REVIEW IT

What are the signs of the *x*- and *y*- coordinates in the four quadrants of the coordinate plane? *(Lesson 2–6)*

____

____

____

____

____

3 **Graph $y = x - 3$ by plotting ordered pairs.**

First, find ordered pair solutions.

x	$y = x - 3$	*y*	(*x*, *y*)
−1	$y =$ ☐ -3		
0	$y =$ ☐ -3		
1	$y =$ ☐ -3		
2	$y =$ ☐ -3		

Plot these ordered pairs and draw a line through them. The line is a complete graph of the function.

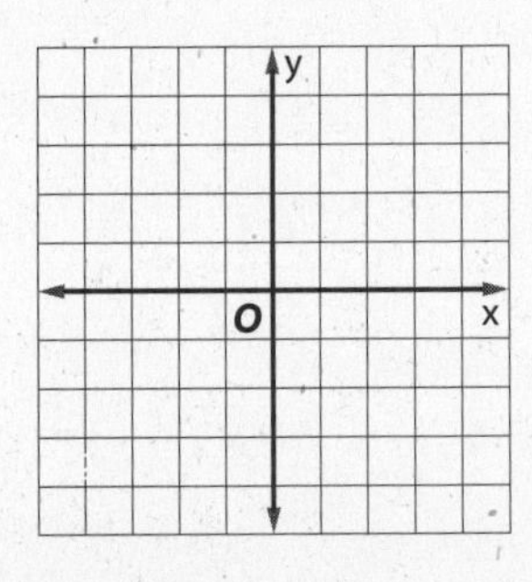

Your Turn Graph $y = 5 - x$ by plotting ordered pairs.

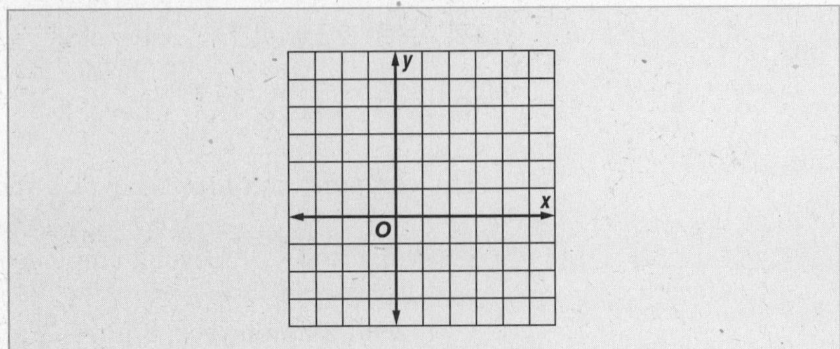

HOMEWORK ASSIGNMENT

Page(s): ______________

Exercises: ______________

____

8–3 Graphing Linear Equations Using Intercepts

What You'll Learn

- Find the x- and y-intercepts of graphs.
- Graph linear equations using the x- and y-intercepts.

Build Your Vocabulary (page 177)

The **x-intercept** is the x-coordinate of a point where a graph crosses the x-axis. The y-coordinate of this point is 0.

The **y-intercept** is the y-coordinate of a point where a graph crosses the y-axis. The x-coordinate of this point is 0.

Key Concept

Intercepts of Lines
To find the x-intercept, let $y = 0$ in the equation and solve for x.

To find the y-intercept, let $x = 0$ in the equation and solve for y.

EXAMPLE Find Intercepts From Graphs

1 State the x-intercept and the y-intercept of the line.

The graph crosses the x-axis at

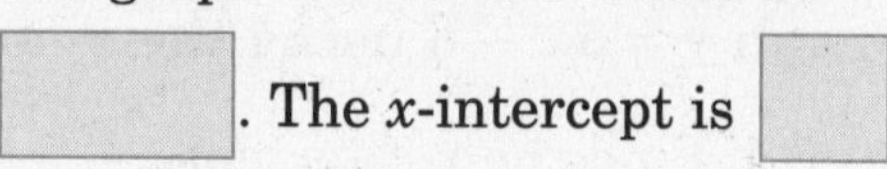

_____. The x-intercept is _____.

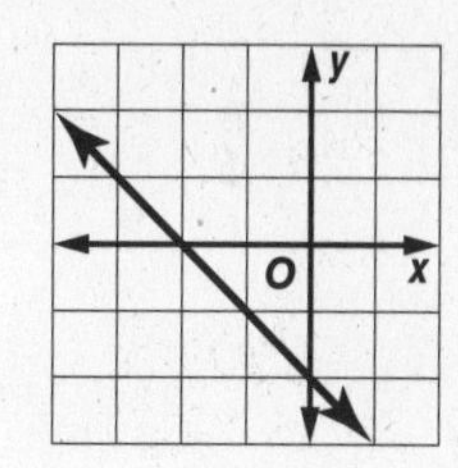

The graph crosses the y-axis at

_____. The y-intercept is _____.

Your Turn State the x-intercept and the y-intercept of the line.

_____

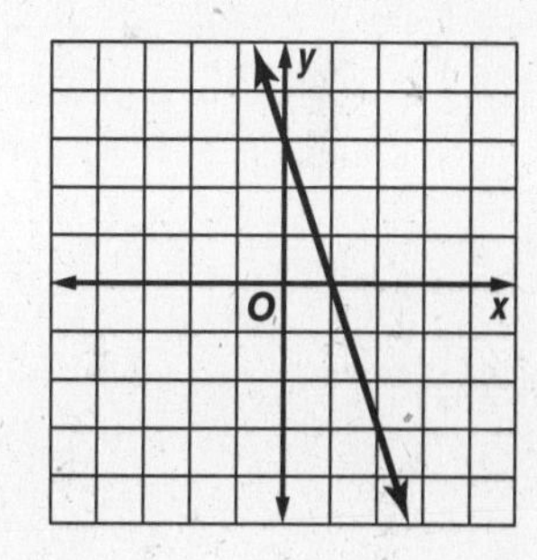

EXAMPLE Find Intercepts from Equations

2 Find the x-intercept and the y-intercept for the graph of $x + 2y = 4$.

To find the x-intercept, let _____.

$x + 2y = 4$ Write the equation.

$x + 2(\_\_\_) = 4$ Replace y with _____.

$x = \_\_\_$ The x-intercept is _____.

FOLDABLES

ORGANIZE IT

In your notes, draw a linear graph that crosses both the x and y axes. Then explain how to find the x-intercept and y-intercept.

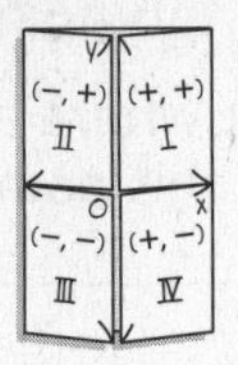

To find the y-intercept, let ______.

$x + 2y = 4$ Write the equation.

______ $+ 2y = 4$ Replace x with 0.

______ $= 4$ Simplify.

$y = 2$ The y-intercept is 2.

Your Turn Find the x-intercept and the y-intercept for the graph of $x + 3y = 9$.

EXAMPLE Use Intercepts to Graph Equations

3 **Graph $y = 3x - 6$ using the x- and y-intercepts.**

Step 1 Find the x-intercept.

$0 = 3x - 6$ Replace y with 0.

______ $= 3x$ Add ______ to each side.

______ $= x$ The x-intercept is ______.

Step 2 Find the y-intercept.

$y = 3(0) - 6$ Replace x with 0.

$y =$ ______ The y-intercept is ______.

Step 3 Graph the points ______ and ______ and draw a line through them.

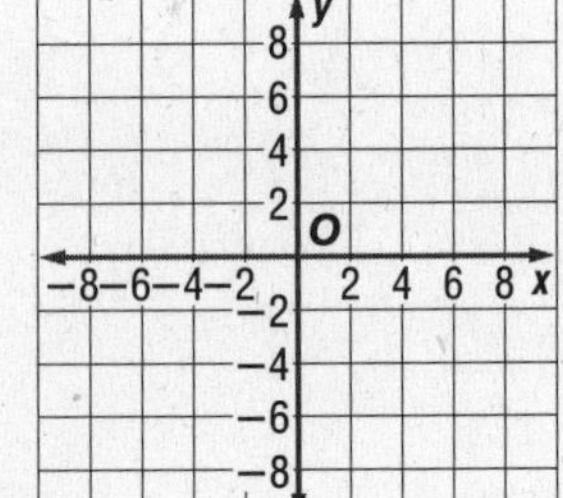

Your Turn Graph $y = 2x + 4$ using the x- and y-intercepts.

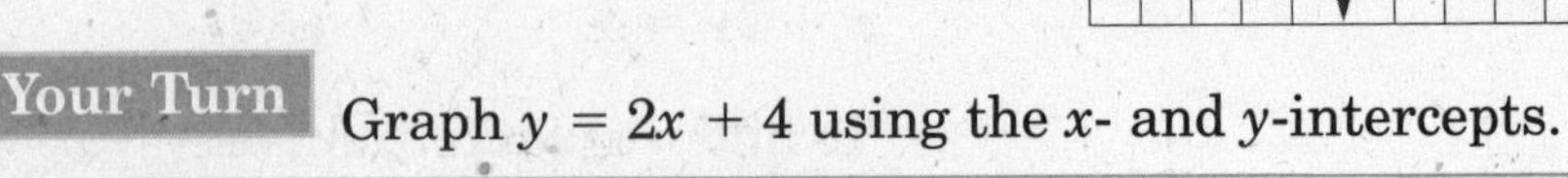

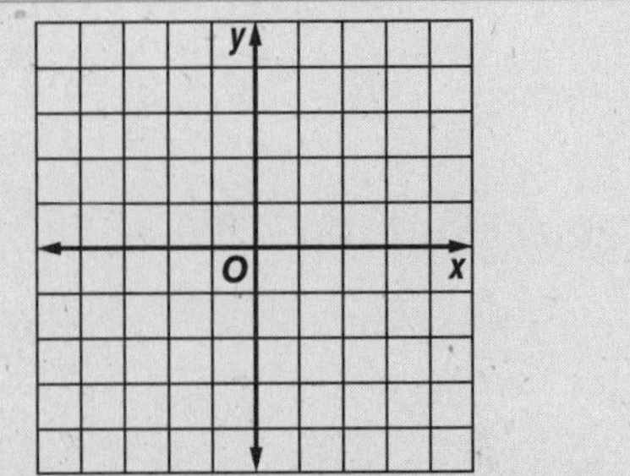

HOMEWORK ASSIGNMENT

Page(s):

Exercises:

8–4 Slope

BUILD YOUR VOCABULARY (page 177)

Slope describes the ________ of a line. It is the ratio of the *rise*, or the ________ change, to the *run*, or the ________ change.

WHAT YOU'LL LEARN

- Find the slope of a line.

KEY CONCEPT

Slope The slope m of a line passing through points (x_1, y_1) and (x_2, y_2) is the ratio of the difference in y-coordinates to the corresponding difference in x-coordinates.

EXAMPLE Positive Slope

1 **Find the slope of the line.**

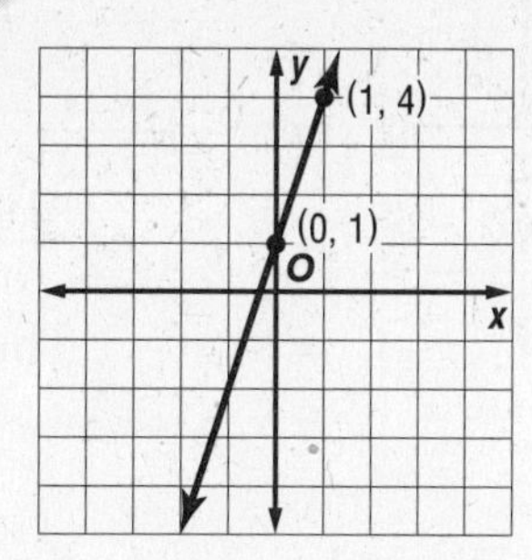

$m = \dfrac{y_2 - y_1}{x_2 - x_1}$ Definition of slope

$m = \dfrac{\qquad}{\qquad}$ $(x_1, y_1) = (0, 1)$
$(x_2, y_2) = (1, 4)$

$m =$ ____ or ____

The slope is ____.

EXAMPLE Negative Slope

2 **Find the slope of the line.**

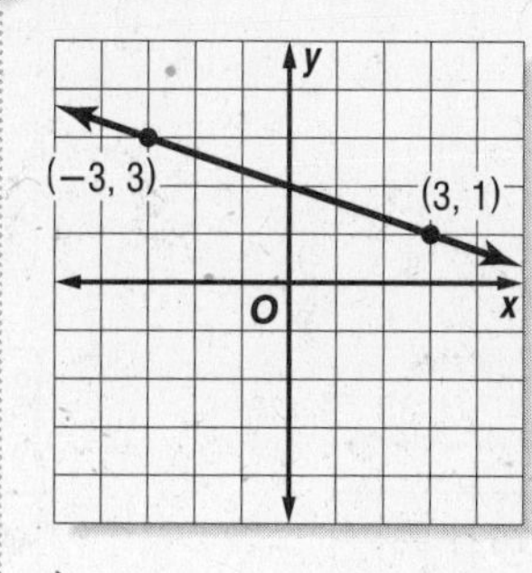

$m = \dfrac{y_2 - y_1}{x_2 - x_1}$ Definition of slope

$m = \dfrac{\qquad}{\qquad}$ $(x_1, y_1) = (3, 1)$
$(x_2, y_2) = (-3, 3)$

$m =$ ____ or ____

The slope is .

EXAMPLE Zero Slope

3 Find the slope of the line.

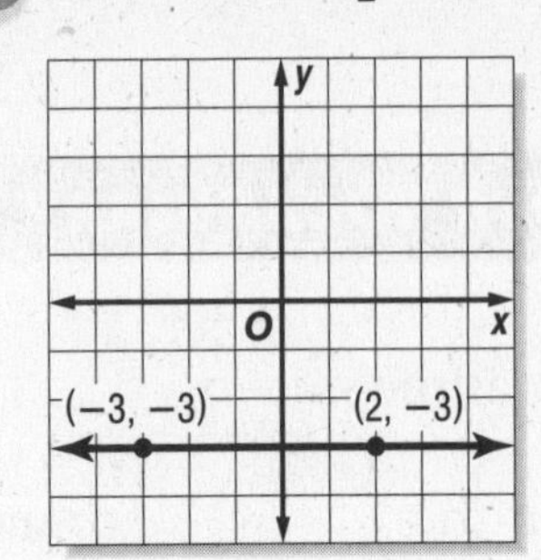

$m = \dfrac{y_2 - y_1}{x_2 - x_1}$ Definition of slope

$m = \dfrac{\square}{\square}$ $(x_1, y_1) = (-3, -3)$; $(x_2, y_2) = (2, -3)$

$m = \square$ or $\square$

FOLDABLES™

ORGANIZE IT

In your notes, write a sample equation for each slope: positive, negative, zero, and undefined. Then graph each equation and write its slope.

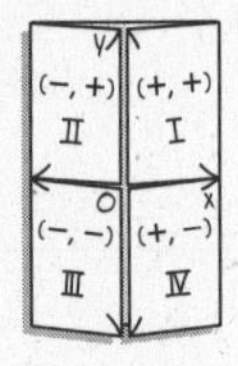

EXAMPLE Undefined Slope

4 Find the slope of the line.

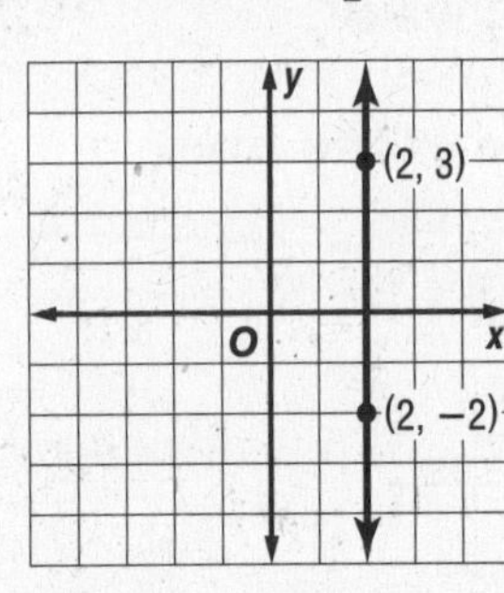

$m = \dfrac{y_2 - y_1}{x_2 - x_1}$ Definition of slope

$m = \dfrac{\square}{\square}$ $(x_1, y_1) = (2, -2)$; $(x_2, y_2) = (2, 3)$

$m = \square$ The slope is $\square$.

Your Turn **Find the slope of each line.**

a.

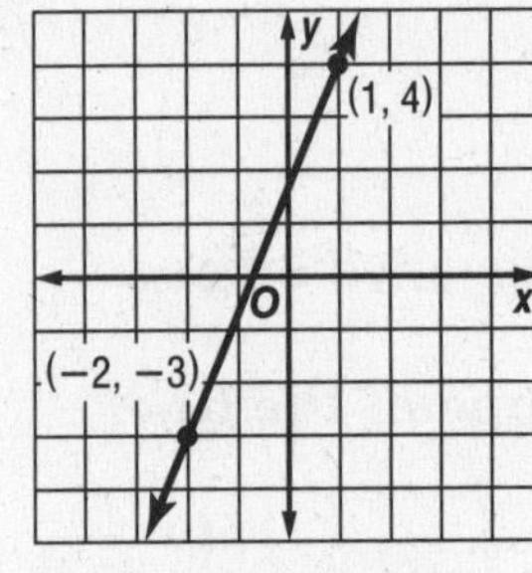

b.

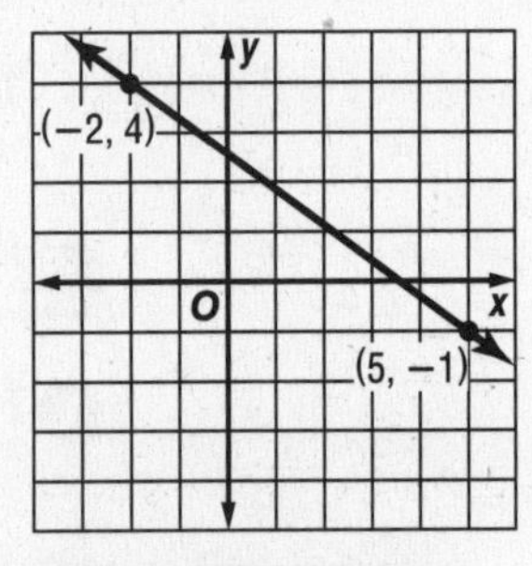

c.

(−1, 3) (4, 3)

d.

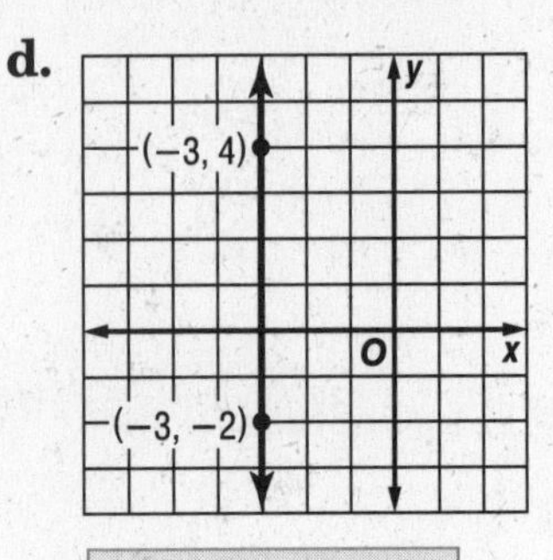

HOMEWORK ASSIGNMENT

Page(s):

Exercises:

8–5 Rate of Change

WHAT YOU'LL LEARN

- Find rates of change.
- Solve problems involving direct variation.

BUILD YOUR VOCABULARY (page 177)

A ______ in one ______ with respect to another quantity is called the **rate of change.**

EXAMPLE Find a Rate of Change

1 SCHOOL **The graph shows Jadon's quiz scores for the first five weeks after he joined a study group. Find the rate of change from Week 2 to Week 5.**

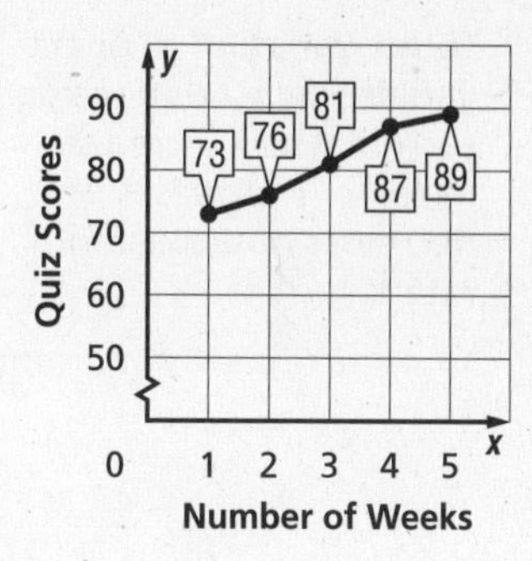

rate of change = slope

$= \frac{\square}{\square}$ Definition of slope

$= \frac{\square}{\square}$ ← change in quiz score / ← change in time

≈ 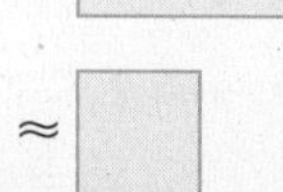

So, the expected rate of change in quiz scores is an increase of about ______ per week.

Your Turn The graph shows the number of campers enrolled at a summer camp during its first five years of operation. Find the rate of change from Year 2 to Year 5.

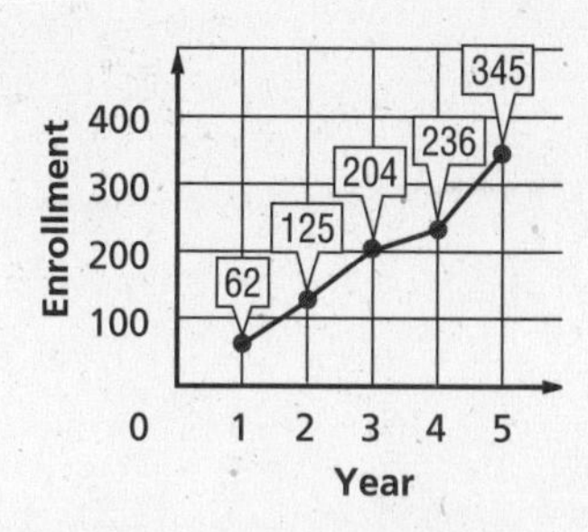

BUILD YOUR VOCABULARY (page 176)

A special type of linear equation that describes ________ is called a **direct variation**.

In the equation $y = kx$, k is called the **constant of variation**.

It is the ________, or ________.

KEY CONCEPT

Direct Variation A direct variation is a relationship such that as x increases in value, y increases or decreases at a constant rate k.

EXAMPLE Write a Direct Variation Equation

2 **Suppose y varies directly with x and $y = 6$ when $x = 54$. Write an equation relating x and y.**

Step 1 Find the value of k.

$y = kx$ Direct variation

$\square = k\square$ 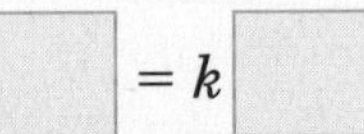 Replace y with $\square$ and x with $\square$.

$\square = k$ 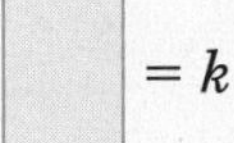 Simplify.

Step 2 Use k to write an equation.

$y = kx$ Direct variation

$y = \square x$ Replace k with $\square$. 

Your Turn Suppose y varies directly with x and $y = 75$ when $x = 15$. Write an equation relating x and y.

EXAMPLE Use Direct Variation to Solve Problems

FOLDABLES

ORGANIZE IT

In your notes, write an example of a direct variation equation. Then explain how its slope, rate of change, and constant of variation are related.

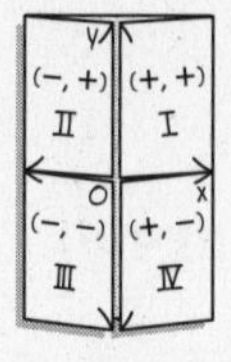

3 LANDSCAPING **The depth of a wide hole for a backyard pond as it is being dug is recorded in the table.**

a. Write an equation that relates time and hole depth.

Step 1 Find the ratio of y to x for each recorded time. The ratios are approximately equal to 0.8.

Hole Depth (in.)	Time (min)	$k = \frac{y}{x}$
y	x	
8	10	
15	20	
24	30	
31	40	

Step 2 Write an equation.

$y = kx$ Direct variation

$y = \_\_\_ x$ Replace k with ___.

b. Predict how long it will take to dig a depth of 36 inches.

$y = \_\_\_ x$ Write the direct variation equation.

$\_\_\_ = \_\_\_ x$ Replace y with ___.

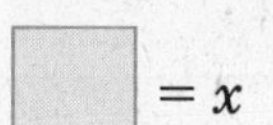

$\_\_\_ = x$ It will take about ___ minutes.

Your Turn The number of units produced on a factory assembly line is recorded in the table. Write an equation that relates time and units produced. Then predict how long it will take to produce 315 units.

Time (min) x	Units produced y
5	35
10	68
15	105
20	140

HOMEWORK ASSIGNMENT

Page(s):

Exercises:

8–6 Slope-Intercept Form

BUILD YOUR VOCABULARY (page 177)

An equation written in the form $y = mx + b$, where m is the slope and b is the y-intercept, is in **slope-intercept form**.

WHAT YOU'LL LEARN

- Determine slopes and y-intercepts of lines.
- Graph linear equations using the slope and y-intercept.

EXAMPLE Find the Slope and y-Intercept

1 **State the slope and the y-intercept of the graph of $y = \frac{1}{2}x + 3$.**

$y = \frac{1}{2}x + 3$ Write the equation in the form $y = mx + b$.

$y = mx + b$ The slope is ____. The y-intercept is .

EXAMPLE Write the Equation in Slope-Intercept Form

2 **State the slope and the y-intercept of the graph of $-4x + 5y = -10$.**

$-4x + 5y = -10$ Write the equation.

$-4x + 5y + 4x = -10 + 4x$ Add $4x$ to each side.

____ = ____ Simplify.

$y =$ ____ Divide each side by ____.

$y =$ ____ Slope-intercept form

The slope is ____, and the y-intercept is ____.

Your Turn **State the slope and the y-intercept of the graph of each line.**

a. $y = 2x - 7$

b. $-5x + y = 1$

FOLDABLES™

ORGANIZE IT

In your notes, write an example of a linear equation in slope-intercept form. Graph the equation using its slope and *y*-intercept and list the steps involved.

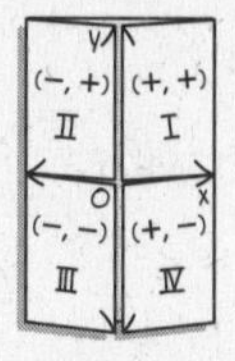

WRITE IT

What are the two ways to interpret a negative slope when graphing an equation?

______

______

______

______

______

HOMEWORK ASSIGNMENT

Page(s):

Exercises:

EXAMPLE Graph an Equation

3 Graph $3x + y = 9$ using the slope and y-intercept.

Step 1 Find the slope and y-intercept.

$3x + y = 9$ Write the equation.

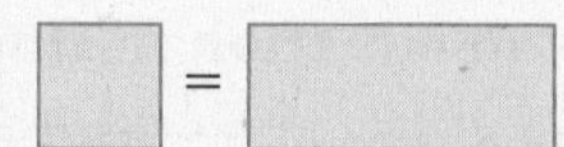

Subtract ______ from each side to write in slope-intercept form.

slope = ______ y-intercept = ______

Step 2 Graph the y-intercept point at ______.

Step 3 Write the slope ______ as ______.

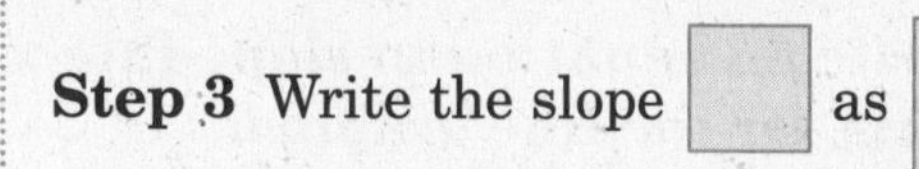

Use it to locate a second point on the line.

$m =$ ______ / ______ change in y: ______ / change in x: ______

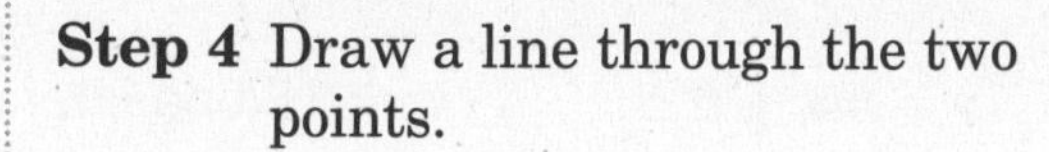

Another point on the line is at ______.

Step 4 Draw a line through the two points.

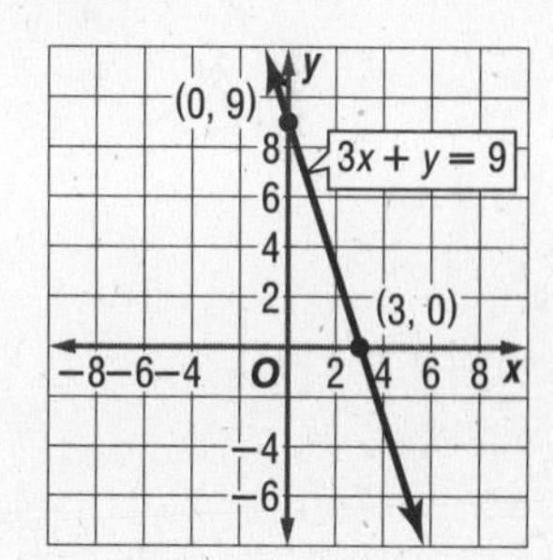

Your Turn Graph $-2x + 3y = 12$ using the slope and y-intercept.

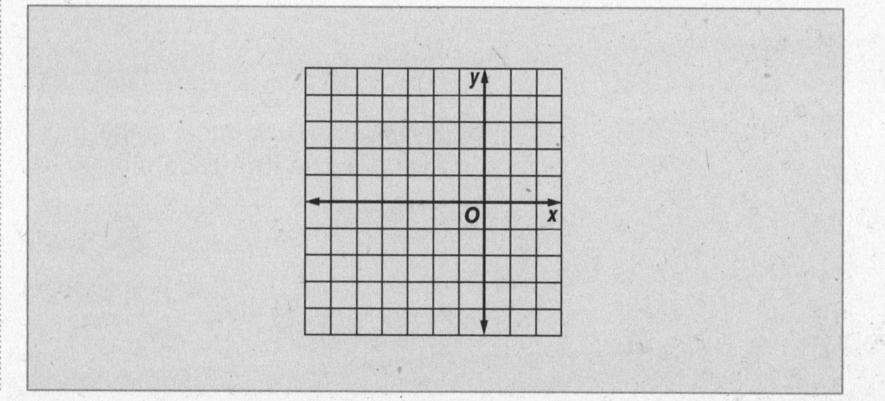

8–7 Writing Linear Equations

WHAT YOU'LL LEARN

- Write equations given the slope and y-intercept, a graph, a table, or two points.

EXAMPLE Write Equations From Slope and y-Intercept

1 **Write an equation in slope-intercept form for the line having slope of $-\frac{1}{4}$ and a y-intercept of 7.**

$y = mx + b$ — Slope-intercept form

$y = -\frac{1}{4}x + 7$ — Replace m with ___ and b with ___.

EXAMPLE Write an Equation From a Graph

2 **Write an equation in slope-intercept form for the line graphed.**

The y-intercept is ___. From ___,

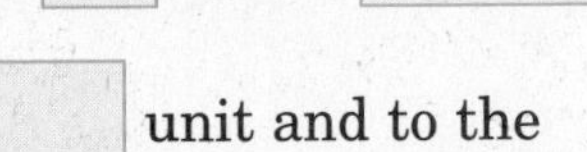

you can go up ___ unit and to the ___ one unit to another point on the line. So, the slope is ___.

$y = mx + b$ — Slope-intercept form

$y =$ ___ $x +$ ___ — Replace m with ___ and b with ___.

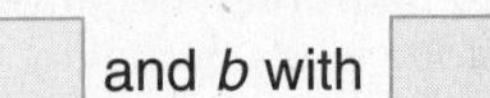

$y =$ ___ — Simplify.

Your Turn

a. Write an equation in slope-intercept form for the line having slope of -3 and a y-intercept of -5.

b. Write an equation in slope-intercept form for the line graphed.

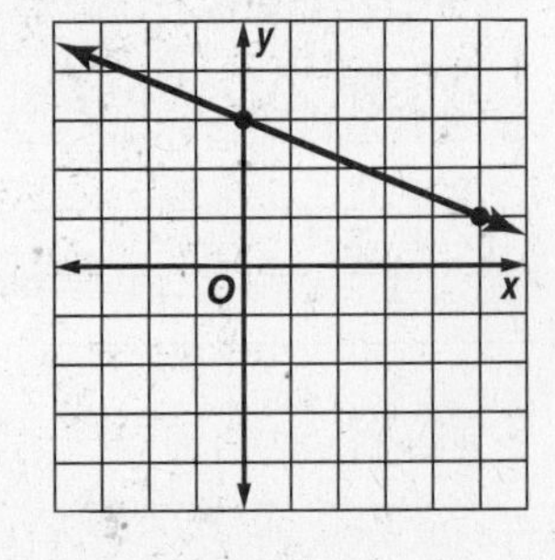

FOLDABLES™

ORGANIZE IT

In your notes, write two points, find the equation of the line that passes through them, and graph the line.

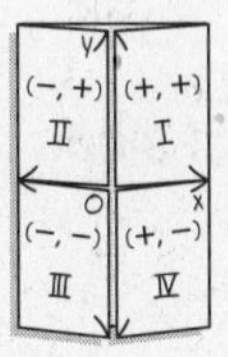

EXAMPLE Write an Equation Given Two Points

3 **Write an equation for the line that passes through (7, 0) and (6, 3).**

Step 1 Find the slope m.

$m = \frac{y_2 - y_1}{x_2 - x_1}$ — Definition of slope

$m = \frac{\_\_\_\_}{\_\_\_\_}$ or ____ — $(x_1, y_1) = (7, 0)$; $(x_2, y_2) = (6, 3)$

Step 2 Find the y-intercept b. Use the slope and the coordinates of either point.

$y = mx + b$ — Slope-intercept form

____ = ____ — Replace m with ____, x with

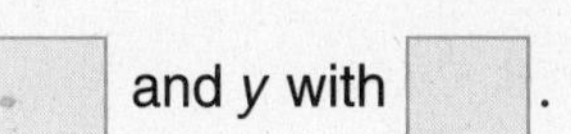

____ and y with ____.

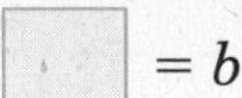

____ $= b$ — Simplify.

Step 3 Substitute the slope and y-intercept.

$y = mx + b$ — Slope-intercept form

$y =$ ____ — Replace m with ____, and b with ____.

Your Turn Write an equation for the line that passes through (4, −2) and (−2, −14).

EXAMPLE Write an Equation From a Table

4 **Use the table of values to write an equation is slope-intercept form.**

x	y
−2	16
−1	10
0	4
1	−2

Step 1 Find the slope m. Use the coordinates of any two points.

$m = \frac{y_2 - y_1}{x_2 - x_1}$ Definition of slope

$m = \frac{☐}{☐}$ or ☐ $(x_1, y_1) = (-2, 16)$

$(x_2, y_2) = (-1, 10)$

Step 2 Find the y-intercept b. Use the slope and the coordinates of either point.

$y = mx + b$ Slope-intercept form

☐ = ☐ $+ b$ Replace m with 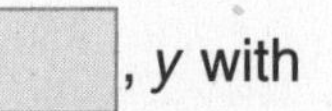 ☐, y with

☐, and x with .

☐ $= b$ Simplify.

Step 3 Substitute the slope and y-intercept.

$y = mx + b$ Slope-intercept form

$y =$ ☐ + ☐ Replace m with ☐, and

b with .

Your Turn

Use the table of values to write an equation in slope-intercept form.

x	y
−6	4
−3	2
3	−2
6	−4

HOMEWORK ASSIGNMENT

Page(s):

Exercises:

8–8 Best-Fit Lines

WHAT YOU'LL LEARN

- Draw best-fit lines for sets of data.
- Use best-fit lines to make predictions about data.

BUILD YOUR VOCABULARY (page 176)

A **best-fit line** is a line that is very ______ to most of the data points on a points.

EXAMPLE Make Predictions from a Best-Fit Line

1 AGRICULTURE **The table shows the amount of land in the U.S. farms from 1980 to 2000.**

Year	Land (million acres)
1980	1039
1985	1012
1990	986
1995	963
2000	943

a. Make a scatter plot and draw the best-fit line for the data.

Draw a line that best fits the data.

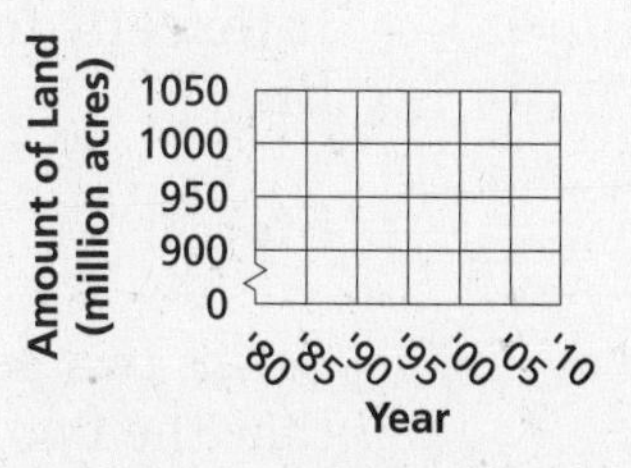

REMEMBER IT

A best-fit line is only an estimation. Different lines with different slopes can be drawn to approximate the data.

b. Use the best-fit line to predict the amount of land in the year 2010.

Extend the line so that you can find the y value for an x value of ______. The y value for ______ is about ______.

So, a prediction for the amount of farm land in 2010 is approximately ______ million acres.

Your Turn **The table shows the number of laptop computers sold at a local computer store from 1998 to 2001.**

Year	Land (million acres)
1998	215
1999	298
2000	395
2001	430

a. Make a scatter plot and draw a best-fit line for the data.

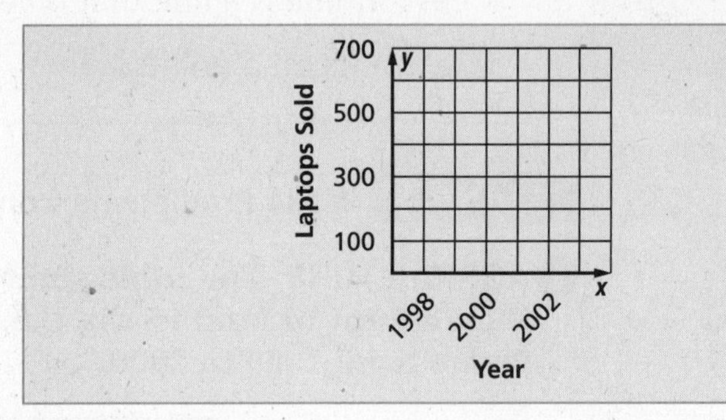

b. Use the best-fit line to predict the number of laptops sold in the year 2003.

Write It

Does a best-fit line always give a good prediction? Explain.

EXAMPLE Make Predictions from an Equation

2 AGRICULTURE **The scatter plot shows the number of milk cow operations in New Mexico from 1975 to 1995.**

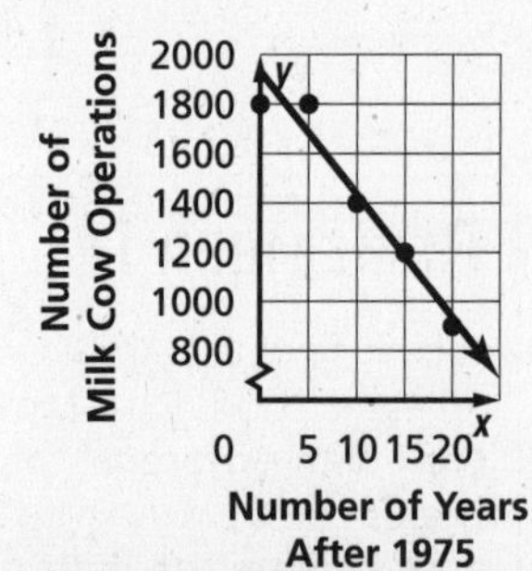

a. Write an equation in slope-intercept form for the best-fit line.

Step 1

Select two points on the line and find the slope. Notice that the two points on the best-fit line are not original data points.

$m = \dfrac{y_2 - y_1}{x_2 - x_1}$ Definition of slope

$(x_1, y_1) = (15, 1200)$

$m = \dfrac{\square}{\square}$ $(x_2, y_2) = (20, 900)$

$m = \square$

Step 2 Next, find the y-intercept.

$y = mx + b$ Slope-intercept form

[ ] = [ ] $+ b$ $(x, y) = (15, 1200)$, and $m =$ [ ].

[ ] $= b$ Simplify.

Step 3 Write the equation.

$y = mx + b$ Slope-intercept form

$y =$ [ ] $m =$ [ ], $b =$ [ ].

b. Predict the number of milk cow operations in the year 2005.

$y =$ [ ] Write the equation for the best-fit line.

$y =$ [ ] Replace x with [ ].

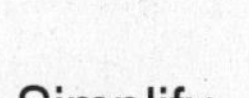

$y =$ [ ] Simplify.

A prediction for the number of milk cow operations is about [ ].

Your Turn **The scatter plot shows the heating bill for the month of January for different size houses.**

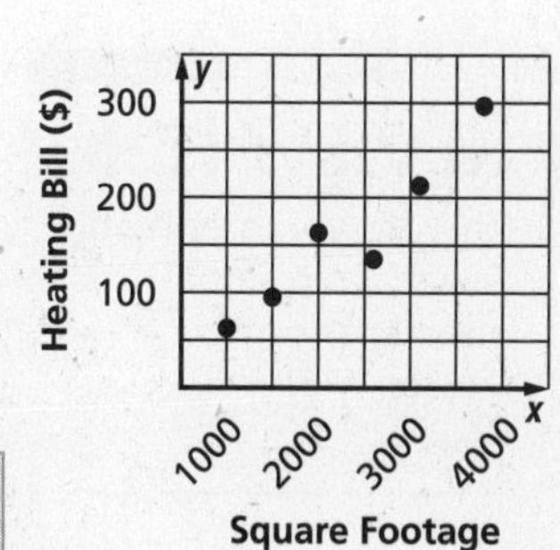

a. Write an equation in slope-intercept form for the best-fit line.

[ ]

b. Predict the heating bill for a house that is 4100 square feet in size.

HOMEWORK ASSIGNMENT

Page(s):

Exercises:

8–9 Solving Systems of Equations

What You'll Learn

- Solve systems of linear equations by graphing.
- Solve systems of linear equations by substitution.

A **system of equations** is a set of equations with the same ________. The solution of the system is the ________ that is a solution for all the equations, and is where the graphs of the equations ________.

EXAMPLE Solve by Graphing

1 Solve the system of equations by graphing.

$y = x - 1$
$y = 2x - 2$

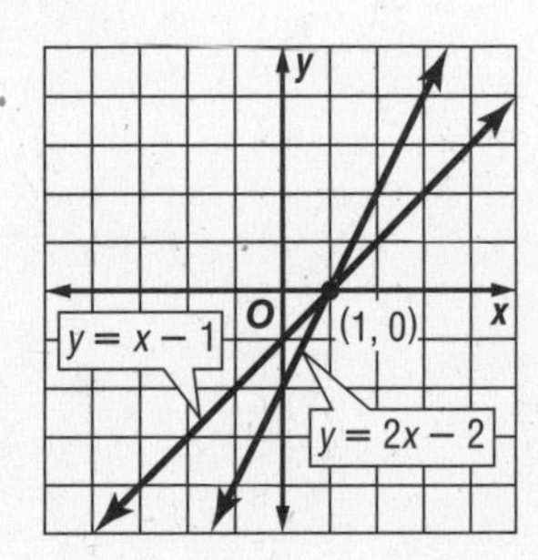

The graphs appear to intersect at ________. Check this estimate by ________ the ________ into each ________.

Check $y = x - 1$ $\quad y = 2x - 2$

$0 \stackrel{?}{=} 1 - 1$ $\quad 0 = 2(1) - 2$

$0 = 0$ ✓ $\quad 0 = 0$ ✓

Review It

How do you graph an equation in slope-intercept form? *(Lesson 8-6)*

____

____

____

____

____

____

Your Turn Solve the system of equations by graphing.

$y = 2x + 3$
$y = x + 4$

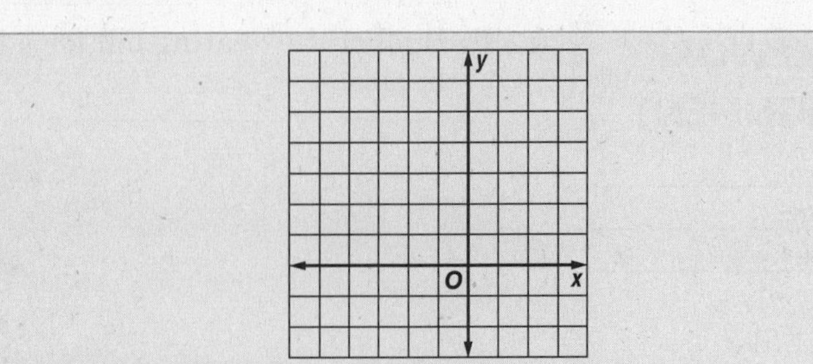

EXAMPLE One Solution

2 ENTERTAINMENT **One CD club, A, charges members a $25.00 annual fee. After that, CDs cost $5.00 each. Another CD club, B, has no annual fee and sells its members CDs for $7.50 each.**

a. How many CDs would you need to buy in a year for the clubs to cost the same?

Write an equation to represent each plan, and then graph the equations to find the solution.

Let x = number of CDs purchased and let y = the total cost.

	total cost		cost of a CD		annual fee
Club A	___	=	___	+	___
Club B	___	=	___	+	___

The graph of the system shows the solution is ___. This means that if ___ CDs are bought, both clubs have the same cost of ___.

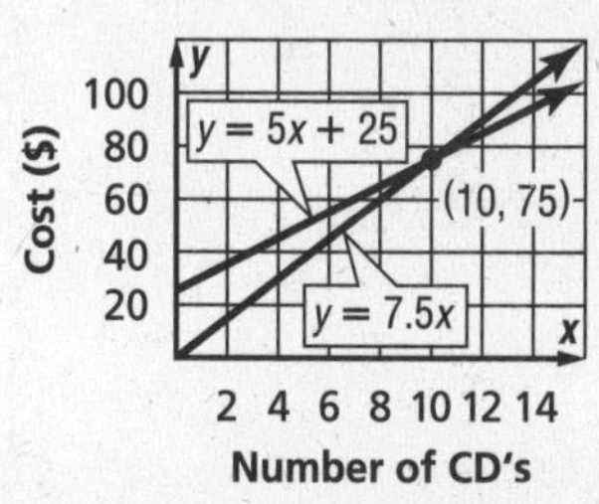

b. Which club would be a better deal for a customer buying 20 CDs in a year?

For ___, the line representing ___ has a smaller y value. So, ___ would be a better deal.

Your Turn **One film store charges an annual fee of $25.00 and then charges $5 per roll for film possessing. A second store offers an annual fee of $10 and then charges its customers $6.50 per roll for film processing.**

a. How many rolls of film would you need to have processed for the cost to be the same at both stores?

b. Which store would be a better deal for a customer developing 8 rolls of film?

EXAMPLE No Solution

3 **Solve the system of equations by graphing.**

$y = \frac{1}{3}x - 5$

$y = \frac{1}{3}x + 2$

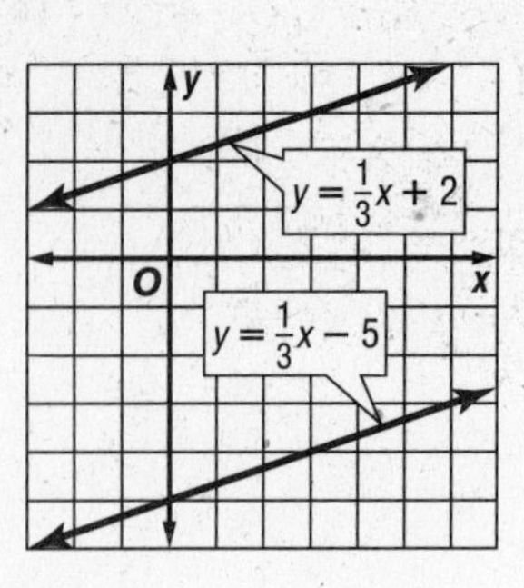

The graphs appear to be ______ lines. Since there is ______ coordinate pair that is a solution to both equations, there is ______ of this system of equations.

Your Turn Solve the system of equations by graphing.

$y = 3x + 7$

$y = 3x - 2$

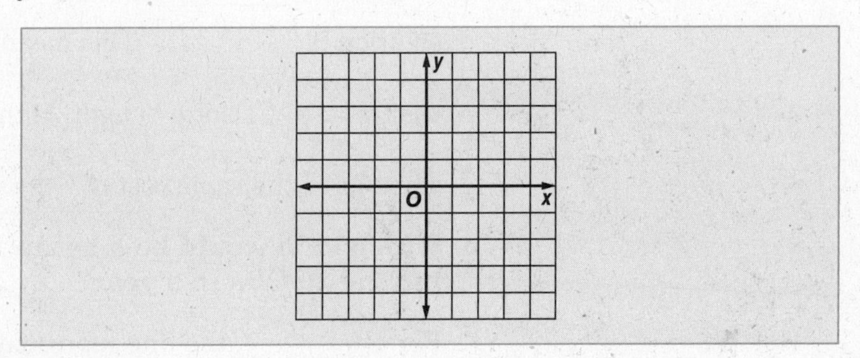

FOLDABLES

ORGANIZE IT

In your notes, write examples of systems of two equations with one solution, no solution, and infinitely many solutions. Graph each example.

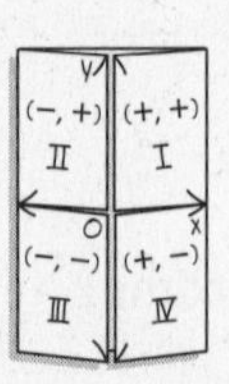

EXAMPLE Infinitely Many Solutions

4 **Solve the system of equations by graphing.**

$y = \frac{3}{4}x + 6$

$4y - 3x = 24$

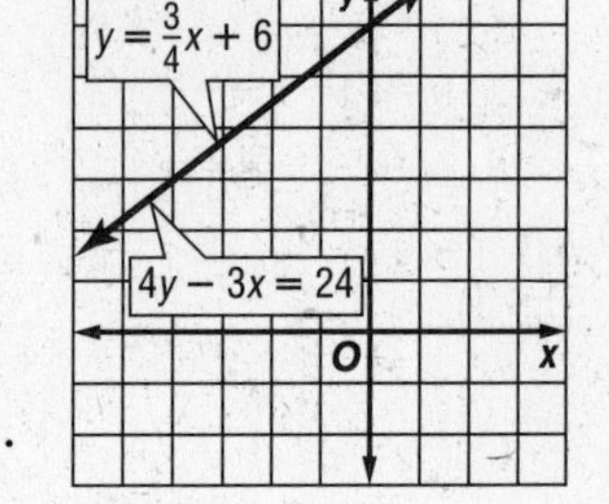

Both equations have the same ______.

Any ordered pair on the graph will satisfy ______ equations.

Therefore, there are ______ solutions to this system of equations.

Your Turn Solve the system of equations by graphing.

$y = 2x - 3$

$-6x + 3y = 29$

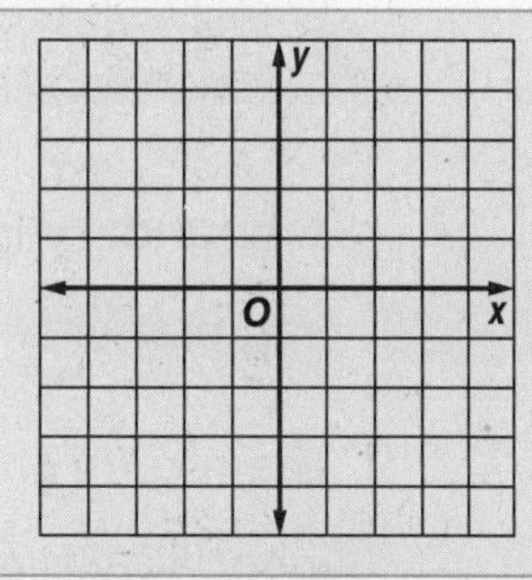

BUILD YOUR VOCABULARY (page 177)

When you solve a system of equations using **substitution**, you find the value of one variable from one equation and substitute that value into the other equation.

EXAMPLE Solve by Substitution

5 Solve the system of equations by substitution.

$y = 3x - 4$

$y = 2$

Since y must have the same value in both equations, you can replace y with ______ in the first equation.

$y = 3x - 4$	Write the first equation.
______ $= 3x - 4$	Replace y with ______.
______ = ______	Add ______ to each side.
______ = ______	Divide each side by ______.

The solution of this system of equations is ______. You can check the solution by ______.

Your Turn Solve the system $y = 4x + 1$ and $x = -1$ by substitution.

HOMEWORK ASSIGNMENT

Page(s):

Exercises:

8–10 Graphing Inequalities

WHAT YOU'LL LEARN

- Graph linear inequalities.
- Describe solutions of linear inequalities.

BUILD YOUR VOCABULARY (page 176)

A **boundary** is a line that ______ a plane into half planes.

The region that contains the ______ of an inequality is called a **half plane**.

REMEMBER IT

When the boundary is part of the solution (the inequality contains the $\leq$ or the $\geq$ symbol), use a solid line.

When the boundary is *not* part of the solution (the inequality contains the $<$ or the $>$ symbol), use a dashed line.

EXAMPLE Graph Inequalities

1 **Graph $y > 3x - 3$.**

Graph $y = 3x - 3$. Draw a ______ line since the boundary is not part of the graph.

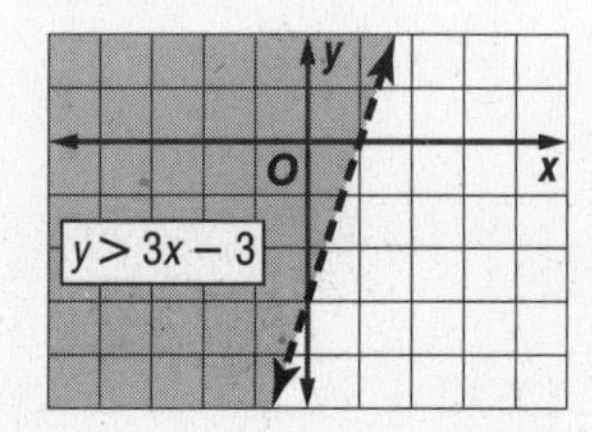

Test (0, 0): $y > 3x - 3$

______ $\overset{?}{>}$ ______ $- 3$ Replace (x, y) with ______.

______ $>$ ______

Thus, the graph is all points in the region ______ the boundary.

Your Turn **Graph each inequality.**

a. $y < 2x + 1$

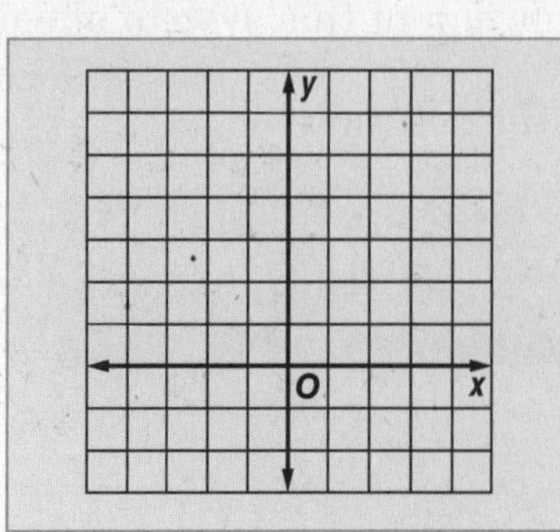

b. $y \geq \frac{3}{4}x - 1$

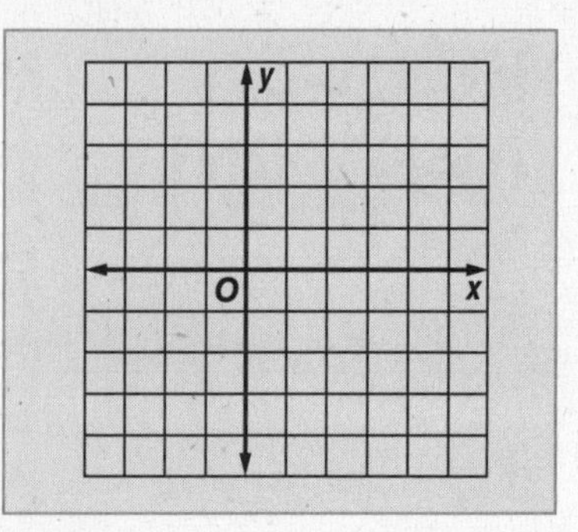

FOLDABLES

ORGANIZE IT

In your notes, write and graph an example of an inequality. List the steps you took to graph the inequality.

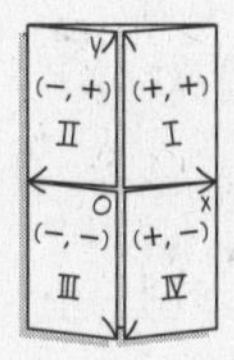

EXAMPLE Write and Graph an Inequality to Solve a Problem

2 BAKING **Terrence needs 3 cups of flour to make a batch of cookies and 5 cups to make a loaf of bread. He has 30 cups of flour. How many batches of cookies and loaves of bread can he bake?**

Step 1 Write an inequality.

Let x represent the number of batches of cookies and let y represent the number of loaves of bread. The inequality that represents the situation is ____ + ____ $\leq$ ____.

Step 2 Graph the inequality.

To graph the inequality, first solve for ____.

$3x + 5y \leq 30$ — Write the inequality.

$5y \leq -3x + 30$ — Subtract ____ from each side.

$y \leq -\frac{3}{5}x + 6$ — Divide each side by ____.

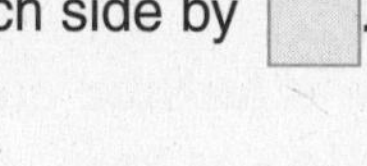

Graph $y \leq -\frac{3}{5}x + 6$ as a ____ line since the boundary is part of the graph. The ____ is part of the graph since $0 \leq -\frac{3}{5}(0) + 6$ is true. Thus, the coordinates of all points in the shaded region are possible solutions.

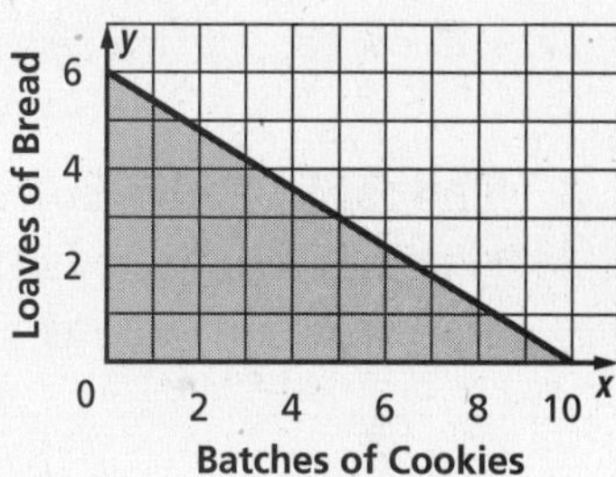

(0, 6) = ____ batches of cookies, ____ loaves

(4, 3) = ____ batches of cookies, ____ loaves

(5, 3) = ____ batches of cookies, ____ loaves

(7, 1) = ____ batches of cookies, ____ loaf

Your Turn Sue has at most $50 to spend on shirts and jeans. Each shirt costs $5 and each pair of jeans costs $10. How many shirts and pairs of jeans can Sue buy?

HOMEWORK ASSIGNMENT

Page(s): ____

Exercises: ____

CHAPTER 8

BRINGING IT ALL TOGETHER

STUDY GUIDE

FOLDABLES	VOCABULARY PUZZLEMAKER	BUILD YOUR VOCABULARY
Use your **Chapter 8 Foldable** to help you study for your chapter test.	To make a crossword puzzle, word search, or jumble puzzle of the vocabulary words in Chapter 8, go to: www.glencoe.com/sec/math/t_resources/free/index.php	You can use your completed **Vocabulary Builder** (pages 176–177) to help you solve the puzzle.

8-1

Functions

Determine whether each relation is a function.

1. $\{(2, 5), (3, 7), (-2, 5), (1, 8)\}$

2. $\{(-1, 1), (3, 4), (2, 2), (-1, 5)\}$

3. The table shows how age affects the value of one type of computer. Is the relation a function? Describe how age is related to value.

Age (years)	Value
0	$1,500
1	$1,200
2	$800
3	$300

8-2

Linear Equations in Two Variables

Determine whether each equation is linear or nonlinear.

4. $y = x + 1$

5. $y = x^2 + 1$

6. $xy = 4$

The equation $y = 3.28x$ describes the approximate number of feet y in x meters.

7. Describe what the solution (5, 16.4) means.

8. About how many feet is a 200 meter dash?

9. Graph the equation $y = 4x - 3$ by plotting ordered pairs.

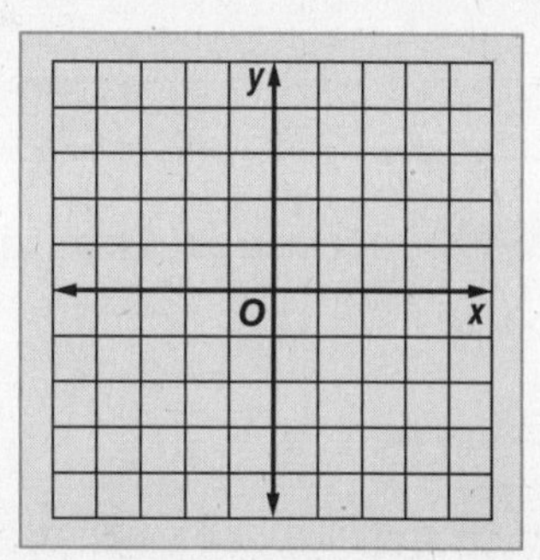

8-3

Graphing Linear Equations Using Intercepts

Find the x-intercept and the y-intercept for the graph of each equation.

10. $y = x - 2$

11. $y = -5$

12. $x + 2y = 6$

Graph each equation using the x- and y-intercepts.

13. $y = 2x - 3$

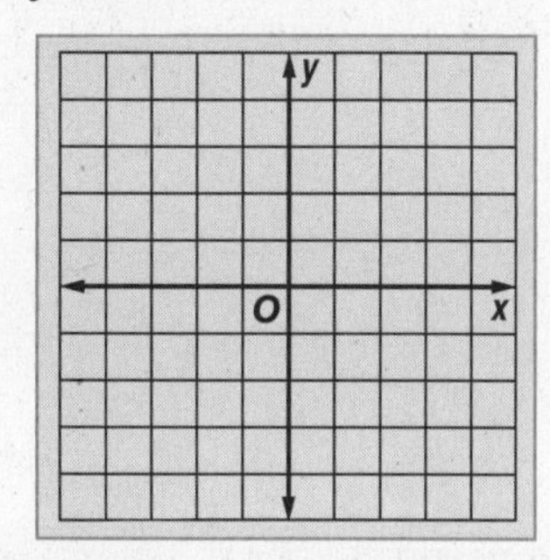

14. $3x - y = 1$

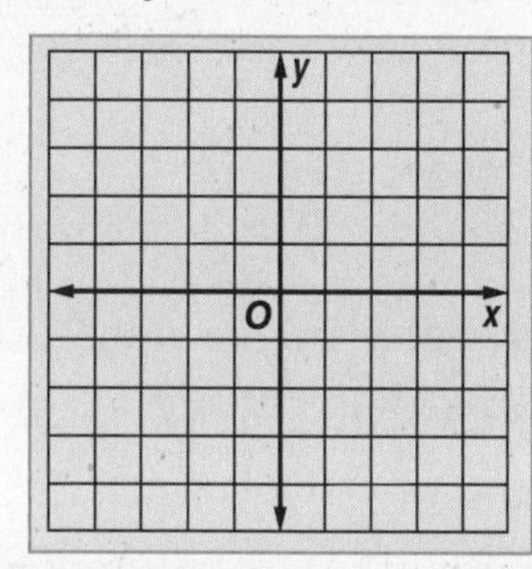

8-4

Slope

Find the slope of the line that passes through each pair of points.

15. $A(2, 3)$ and $B(1, 1)$

16. $S(6, -5)$ and $T(4, 1)$

8-5

Rate of Change

Suppose y varies directly with x. Write an equation relating x and y.

17. $y = 3$ when $x = 6$

18. $y = -12$ when $x = 9$

19. The cost of walnuts varies directly with the number of pounds bought. If 3 pounds cost $9.75, find the cost of 1.4 pounds.

8-6

Slope-Intercept Form

Find the slope and y-intercept, then graph each equation.

20. $y = 2x - 1$

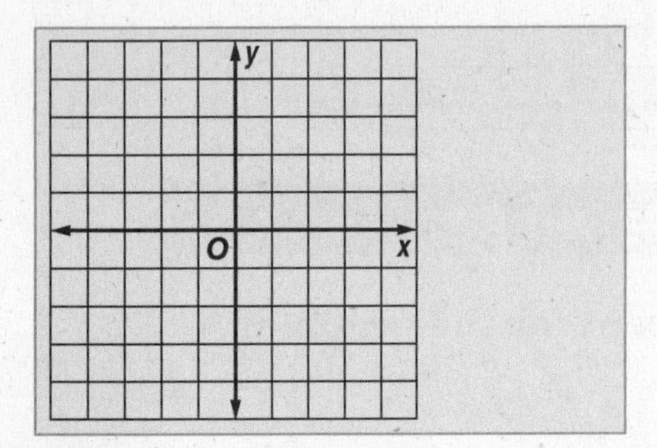

21. $4x + 2y = 5$

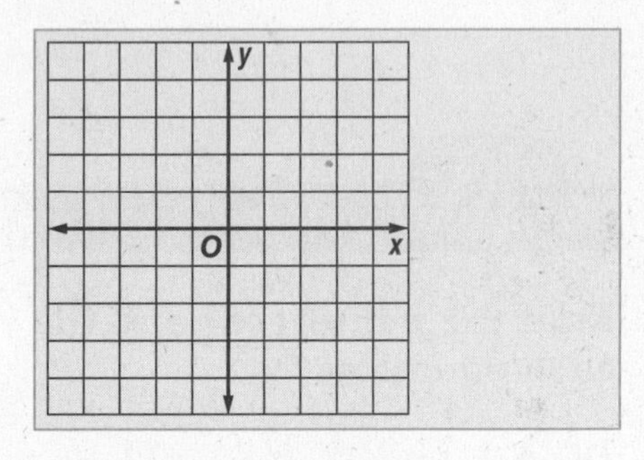

8-7

Writing Linear Equations

Write an equation in slope-intercept form for each line.

22. slope $= -3$, y-intercept $= 7$

23. slope $= \frac{5}{8}$, y-intercept $= 0$

24. Write an equation in slope-intercept form for the line passing through $(-3, 4)$ and $(1, 2)$.

8-8

Best-Fit Lines

The table shows the number of digital still cameras sold in Japan.

Year	Sales (millions)
1999	1.8
2000	3.6
2001	5.9
2002	6.7
2003	9.2*

*Projected in Nov. 2003
Digital Photography Review

25. Make a scatter plot and draw a best-fit line. Then predict how many digital still cameras will be sold in Japan in 2008.

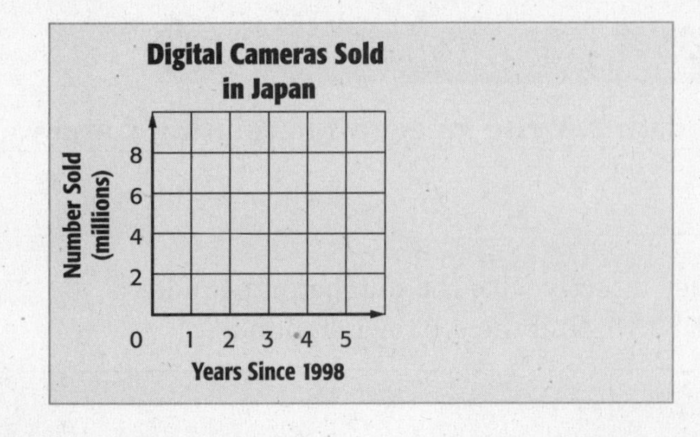

8-9 Solving Systems of Equations

State whether each sentence is *true* or *false*. If false, replace the underlined word to make a true sentence.

26. The solution to a system of equations is the ordered pair that solves both equations.

27. If two equations have the same graph, there are no solutions to that system of equations.

28. Solve the system of equations $2x + y = 3$ and $y = x - 3$ by graphing.

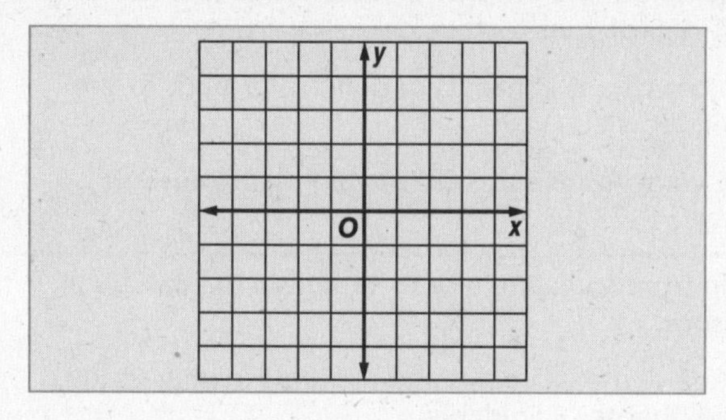

29. Solve the system of equations $x - y = 8$ and $y = 3x$ by substitution.

8-10 Graphing Inequalities

Pablo can paint a large canvas in 3 hours and a small canvas in 2 hours. This month, he has no more than 40 hours to paint canvasses for an upcoming art show.

30. Write an inequality to represent this situation.

31. Graph the inequality.

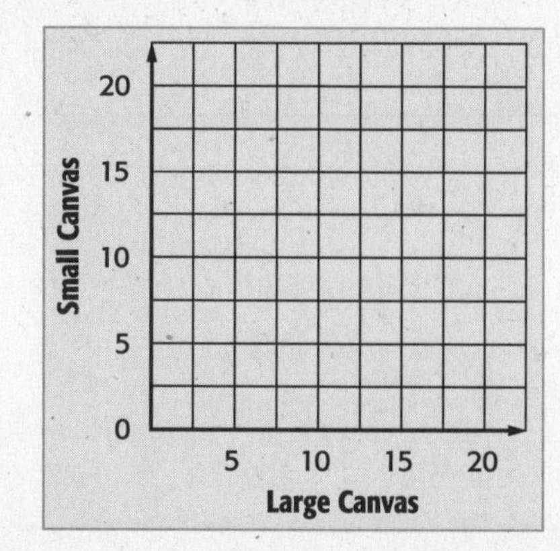

32. Use the graph to determine how many or each type of canvas he could paint this month. List three possibilities.

Checklist

ARE YOU READY FOR THE CHAPTER TEST?

Visit **pre-alg.com** to access your textbook, more examples, self-check quizzes, and practice tests to help you study the concepts in Chapter 8.

Check the one that applies. Suggestions to help you study are given with each item.

☐ **I completed the review of all or most lessons without using my notes or asking for help.**

- You are probably ready for the Chapter Test.
- You may want take the Chapter 8 Practice Test on page 429 of your textbook as a final check.

☐ **I used my Foldable or Study Notebook to complete the review of all or most lessons.**

- You should complete the Chapter 8 Study Guide and Review on pages 424–428 of your textbook.
- If you are unsure of any concepts or skills, refer back to the specific lesson(s).
- You may also want to take the Chapter 8 Practice Test on page 429.

☐ **I asked for help from someone else to complete the review of all or most lessons.**

- You should review the examples and concepts in your Study Notebook and Chapter 8 Foldable.
- Then complete the Chapter 8 Study Guide and Review on pages 424–428 of your textbook.
- If you are unsure of any concepts or skills, refer back to the specific lesson(s).
- You may also want to take the Chapter 8 Practice Test on page 429.

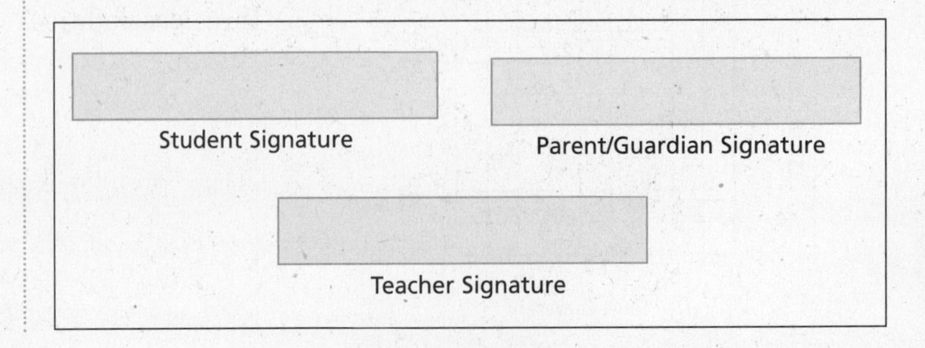

Real Numbers and Right Triangles

Use the instructions below to make a Foldable to help you organize your notes as you study the chapter. You will see Foldable reminders in the margin of this Interactive Study Notebook to help you in taking notes.

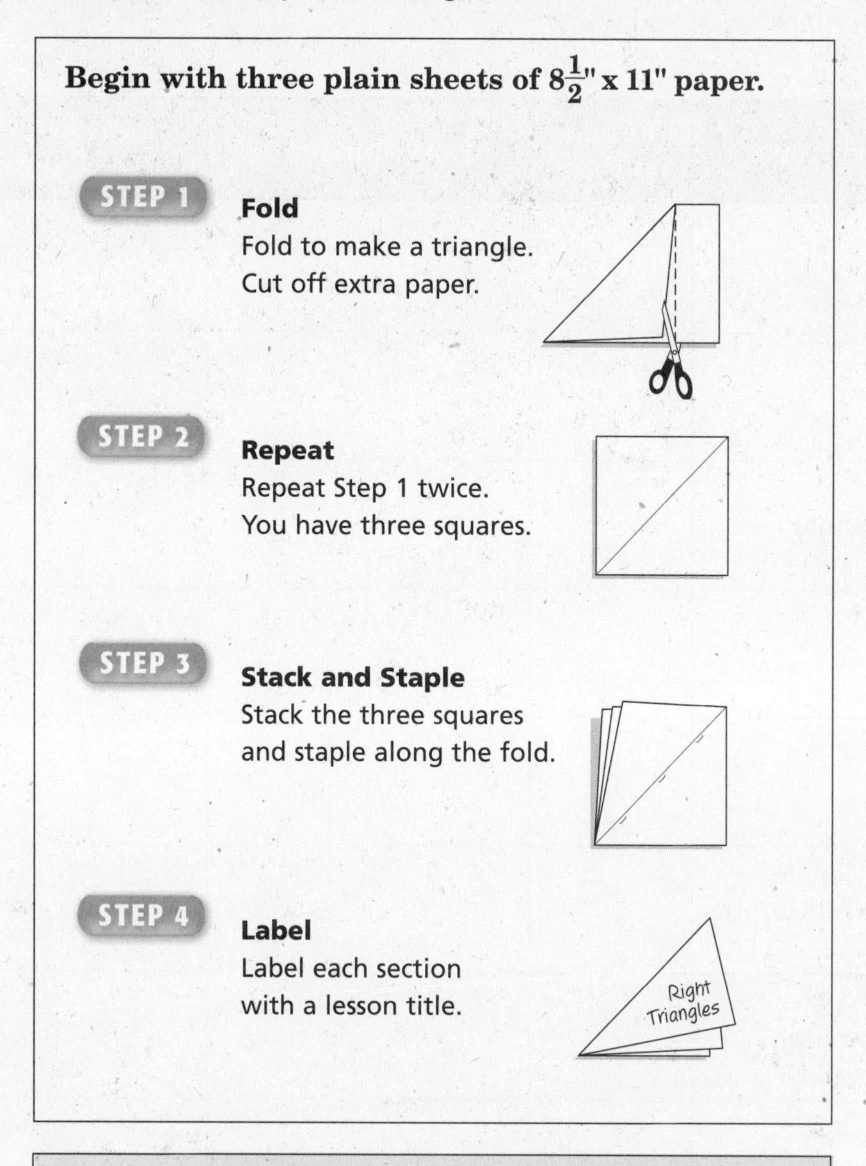

NOTE-TAKING TIP: A visual (graph, diagram, picture, chart) can present information in a concise, easy-to-study format. Clearly label your visuals and write captions when needed.

BUILD YOUR VOCABULARY

This is an alphabetical list of new vocabulary terms you will learn in Chapter 9. As you complete the study notes for the chapter, you will see Build Your Vocabulary reminders to complete each term's definition or description on these pages. Remember to add the textbook page number in the second column for reference when you study.

Vocabulary Term	Found on Page	Definition	Description or Example
acute angle			
angle			
congruent [kuhn-GROO-uhnt]			
degree			
Distance Formula			
equilateral triangle [EE-kwuh-LAT-uh-ruhl]			
hypotenuse [hy-PAHT-uhn-noos]			
isosceles triangle [eye-SAHS-uh-LEEZ]			
Midpoint Formula			

Vocabulary Term	Found on Page	Definition	Description or Example
obtuse angle [ahb-TOOS]			
protractor			
Pythagorean Theorem [puh-THAG-uh-REE-uhn]			
real numbers			
right angle			
scalene triangle [SKAY-LEEN]			
similar triangles			
square root			
straight angle			
trigonometric ratio [TRIHG-uh-nuh-MEH-trihk]			
vertex			

9–1 Squares and Square Roots

WHAT YOU'LL LEARN

- Find squares and square roots.
- Estimate square roots.

KEY CONCEPT

Square Root A square root of a number is one of its two equal factors.

BUILD YOUR VOCABULARY (page 211)

A **square root** of a number is one of two ________ of the number.

A **radical sign,** $\sqrt{\ }$, is used to ________ the square root.

EXAMPLE Find Square Roots

1 **Find each square root.**

a. $\sqrt{64}$ indicates the ________ square root of 64.

Since ________ = 64, $\sqrt{64}$ = ________.

b. $-\sqrt{121}$ indicates the ________ square root of 121.

Since ________ = 121, $-\sqrt{121}$ = ________.

c. $\pm\sqrt{4}$ indicates both square roots of 4. Since 2^2 = ________,

$\sqrt{4}$ = ________ and $-\sqrt{4}$ = ________.

EXAMPLE Calculate Square Roots

2 **Use a calculator to find each square root to the nearest tenth.**

a. $\sqrt{23}$

2nd [$\sqrt{\ }$] 23 ENTER ________ Use a calculator.

$\sqrt{23} \approx$ ________ Round to the nearest tenth.

b. $-\sqrt{46}$

2nd [$\sqrt{\ }$] 46 ENTER ________ Use a calculator.

$\sqrt{46} \approx$ ________ Round to the nearest tenth.

Your Turn **Find each square root.**

a. $\sqrt{25}$

b. $-\sqrt{144}$

c. $\pm\sqrt{16}$

Use a calculator to find each square root to the nearest tenth.

d. $\sqrt{71}$ 

e. $^{-}\sqrt{38}$

FOLDABLES

ORGANIZE IT

Under the tab for Lesson 9-1, list and then estimate three square roots to the nearest whole number.

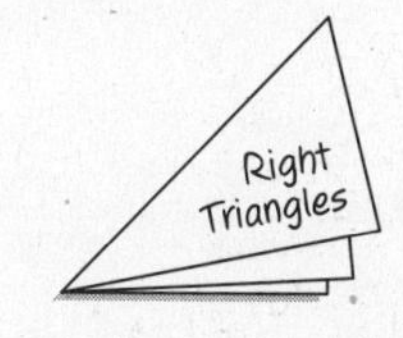

EXAMPLE Estimate Square Roots

3 **Estimate $\sqrt{22}$ to the nearest whole number.**

Find the two ______ closest to 22. To do this, list some perfect squares.

___, ___, ___, ___, ___, . . .

___ and ___ are closest to 22.

___ < 22 < ___ 22 is between ___ and ___.

___ $< \sqrt{22} <$ ___ $\sqrt{22}$ is between ___ and ___. 

___ $< \sqrt{22} <$ ___

Since 22 is closer to ___ than ___, the best whole 

number estimate for $\sqrt{22}$ is ___.

HOMEWORK ASSIGNMENT

Page(s):

Exercises:

Your Turn **Estimate each square root to the nearest whole number.**

a. $\sqrt{54}$ 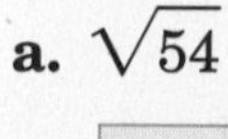

b. $-\sqrt{152}$

9–2 The Real Number System

BUILD YOUR VOCABULARY (page 211)

The set of ______ numbers and the set of ______ numbers together make up the set of **real numbers.**

WHAT YOU'LL LEARN

- Identify and compare numbers in the real number system.
- Solve equations by finding square roots.

KEY CONCEPT

Irrational Number An irrational number is a number that cannot be expressed as $\frac{a}{b}$, where a and b are integers and b does not equal 0.

EXAMPLE Classify Real Numbers

1 **Name all of the sets of numbers to which each real number belongs.**

a. 17 This number is a natural number, a whole number, an ______, and a ______ number.

b. $-\frac{72}{6}$ Since $-\frac{72}{6} =$ ______, this number is an ______ and a ______.

c. $\sqrt{225}$ Since $\sqrt{225} =$ ______, this number is a ______

d. $0.2\overline{46}$ This repeating decimal is a ______ number because it is equivalent to ______.

Your Turn **Name all of the sets of numbers to which each real number belongs.**

a. $-\sqrt{81}$ ______

b. 4.375 ______

c. $\frac{45}{9}$ ______

d. $\sqrt{83}$ ______

FOLDABLES

ORGANIZE IT

Under the tab for Lesson 9-2, explain how to compare real numbers on a number line. Be sure to include an example.

EXAMPLE Compare Real Numbers on a Number Line

2 Replace • with <, >, or = to make $\sqrt{125}$ • $11\frac{7}{8}$ a true statement.

Express each number as a ______. Then ______ the numbers.

$\sqrt{125}$ = ______

$11\frac{7}{8}$ = ______

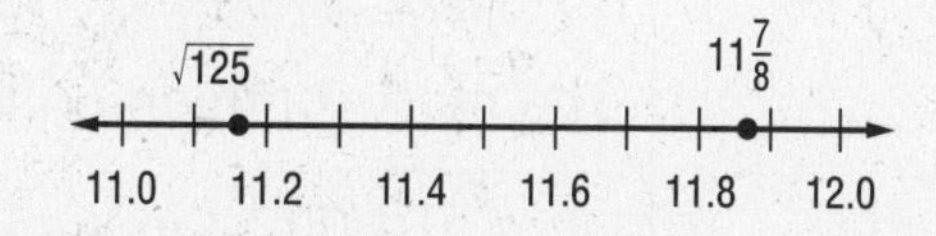

Since $\sqrt{125}$ is to the ______ of $11\frac{7}{8}$, $\sqrt{125}$ ______ $11\frac{7}{8}$.

Your Turn Replace • with <, >, or = to make $\sqrt{61}$ • $7\frac{3}{4}$ a true statement.

EXAMPLE Solve Equations

3 Solve w^2 = 169. Round to the nearest tenth, if necessary.

$w^2 = 169$	Write the equation.
______ = ______	Take the square root of each side.
w = ______ or w = ______	Find the positive and negative square root.
w = ______ or w = ______	

HOMEWORK ASSIGNMENT

Page(s):

Exercises:

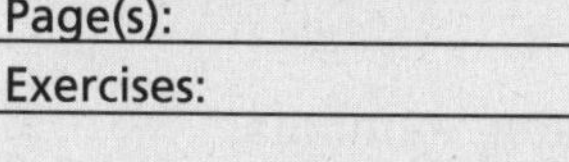

Your Turn Solve $m^2 = 81$. Round to the nearest tenth, if necessary.

9–3 Angles

BUILD YOUR VOCABULARY (pages 210–211)

Two ______ that have the same ______ form an **angle**. The common endpoint is called the **vertex**.

The most common unit of ______ for ______ is called a **degree**.

A **protractor** is used to measure ______.

WHAT YOU'LL LEARN

- Measure and draw angles.
- Classify angles as acute, right, obtuse, or straight.

Measure Angles

1 **Use a protractor to measure $\angle RSW$.**

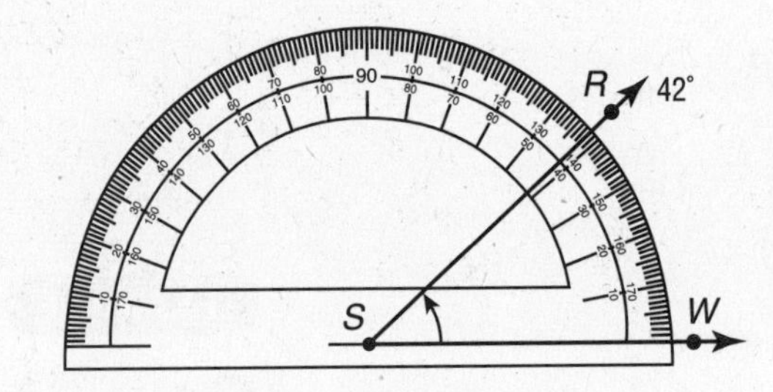

Step 1 Place the ______ of the protractor base on vertex ______. Align the ______ with side ______ so that the marker for ______ is on the ______.

Step 2 Use the scale that begins with ______ at ______.

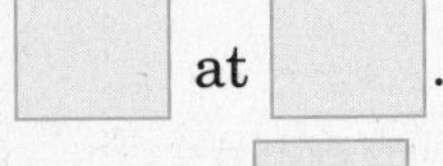

Read where the other side of the angle, ______, crosses this scale.

The measure of angle *RSW* is ______. Using symbols, $m\angle RSW =$ ______.

ORGANIZE IT

Under the tab for this lesson, explain how to draw an angle having a given measure.

Your Turn Use a protractor to measure $\angle ABC$.

EXAMPLE Draw Angles

2 **Draw $\angle R$ having a measurement of 145°.**

REMEMBER IT

The numbers on a protractor go in two directions. When measuring or drawing an angle, make sure you are looking at the correct scale.

Step 1 Draw a ______ with endpoint ______.

Step 2 Place the center point of the protractor on ______. Align the mark labeled ______ with the ______.

Step 3 Use the scale that begins with ______. Locate the mark labeled ______. Then draw the other side of the angle.

Your Turn Draw $\angle M$ having a measurement of 47°.

KEY CONCEPT

Types of Angles

Acute Angle

$0^\circ < m\angle A < 90^\circ$

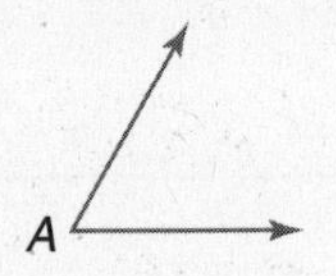

Right Angle

$m\angle A = 90^\circ$

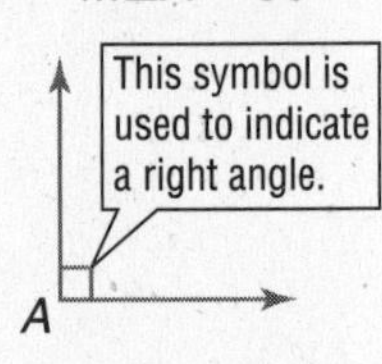

Obtuse Angle

$90^\circ < m\angle A < 180^\circ$

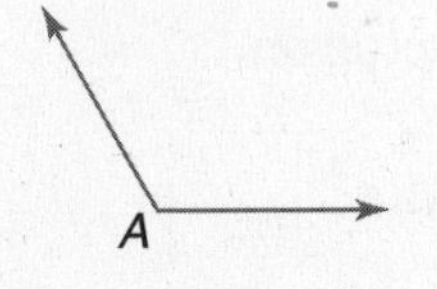

Straight Angle

$m\angle A = 180^\circ$

A

EXAMPLE Classify Angles

3 Classify each angle as *acute*, *obtuse*, *right*, or *straight*.

a.

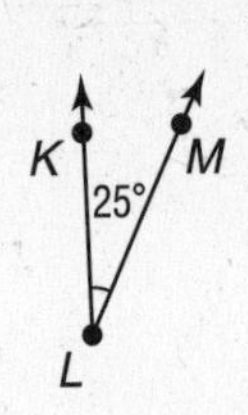

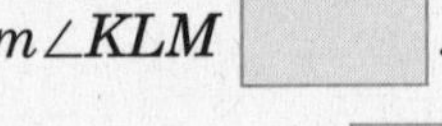

$m\angle KLM$ ______.

So, $\angle KLM$ is ______.

b.

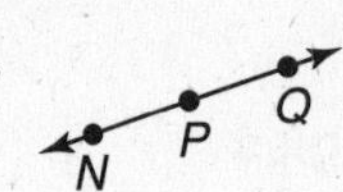

$m\angle NPQ$ ______.

So, $\angle NPQ$ is ______.

c.

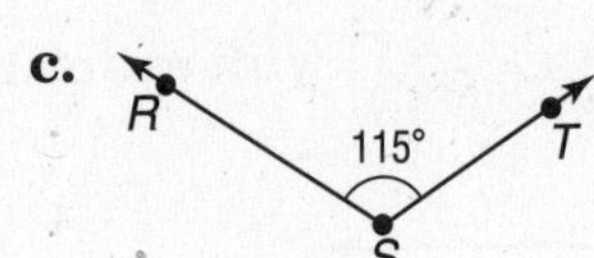

$m\angle RST$ ______.

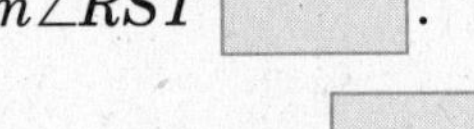

So, $\angle RST$ is ______.

Your Turn **Classify each angle as *acute*, *obtuse*, *right*, or *straight*.**

a.

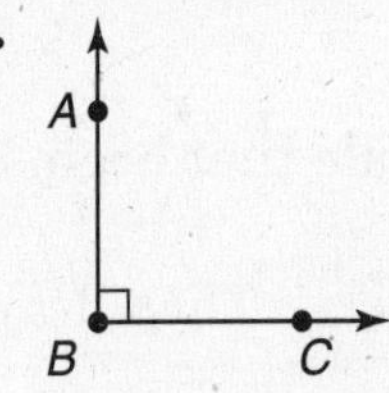

b.

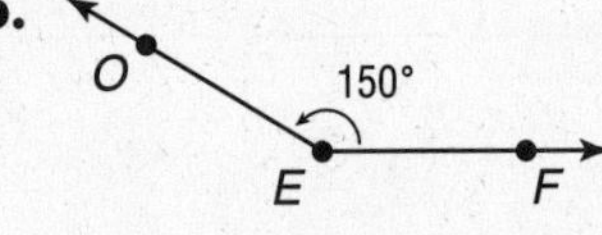

c.

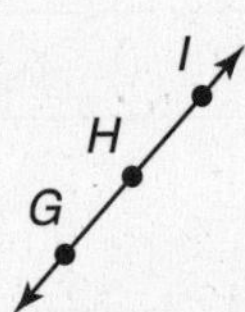

HOMEWORK ASSIGNMENT

Page(s):

Exercises:

9-4 Triangles

Reinforcement of Standard 6MG2.2 Use the properties of complementary and supplementary angles and the sum of the angles of a triangle to solve problems involving an unknown angle. (Key)

WHAT YOU'LL LEARN

- Find the missing angle measure of a triangle.
- Classify triangles by angles and by sides.

KEY CONCEPT

Angles of a Triangle The sum of the measures of the angles of a triangle is 180°.

EXAMPLE Find Angle Measures

1 Find the value of x in $\triangle DEF$.

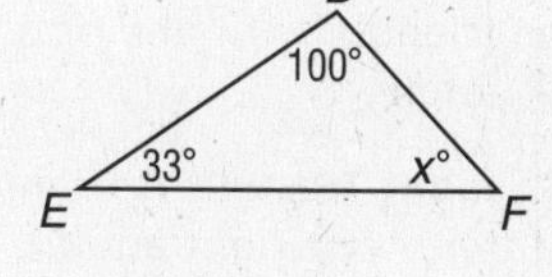

The sum of the measures in a triangle is 180°.

$m\angle D + m\angle E + m\angle F = 180$

___ + ___ + ___ = 180 $m\angle D =$ ___

$m\angle E =$ ___

___ = 180 Simplify.

___ = 180 − 133 Subtract.

$x =$ ___ The measure of $\angle F$ is ___.

EXAMPLE Use Ratios to Find Angle Measures

2 ALGEBRA The measures of the angles of a certain triangle are the ratio 2:3:5. What are the measures of the angles?

Let ___ represent the measure of one angle, ___ the measure of a second angle, and ___ the measure of the third angle.

___ = 180 The sum of the measures is 180.

___ = 180 Combine like terms.

___ = ___ Divide each side by ___.

$x =$ ___ Simplify.

$2x = 2($___$)$ or ___, $3x = 3($___$)$ or ___, and

$5x = 5($___$)$ or ___.

The measures of the angles are ___, ___, and ___.

KEY CONCEPT

Classify Triangles by their Angles and by their Sides

Acute Triangle A triangle with all acute angles.

Obtuse Triangle A triangle with one obtuse angle.

Right Triangle A triangle with one right angle.

Scalene Triangle A triangle with no congruent sides.

Isosceles Triangle A triangle with at least two sides congruent.

Equilateral Triangle A triangle with all sides congruent.

Your Turn

a. Find the value of x in $\triangle MNO$.

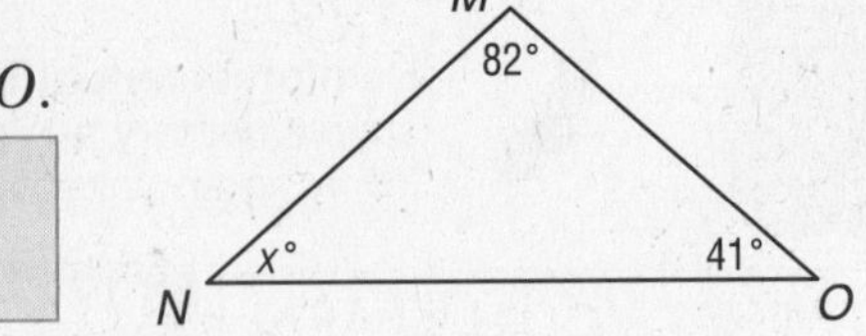

b. The measures of the angles of a certain triangle are in the ratio 3:5:7. What are the measures of the angles?

Triangles can be classified by their ______.

Congruent sides have the same ______.

EXAMPLE Classify Triangles

3 Classify each triangle by its angles and by its sides.

a.

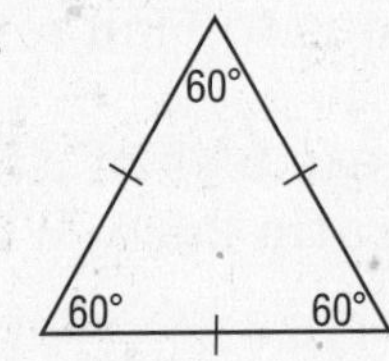

Angles All angles are ______.

Sides All sides are ______.

The triangle is an ______ ______ triangle.

b.

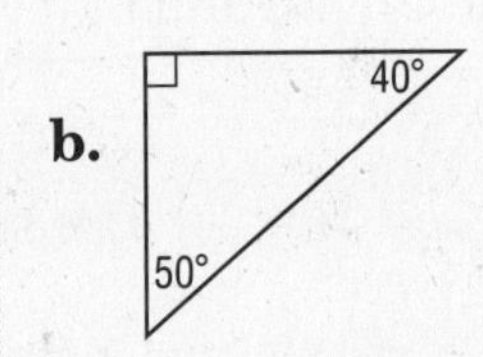

Angles The triangle has a 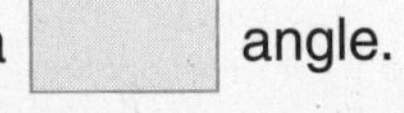 angle.

Sides The triangle has no ______.

The triangle is a ______ triangle.

HOMEWORK ASSIGNMENT

Page(s):

Exercises:

Your Turn

Classify each triangle by its angles and by its sides.

a.

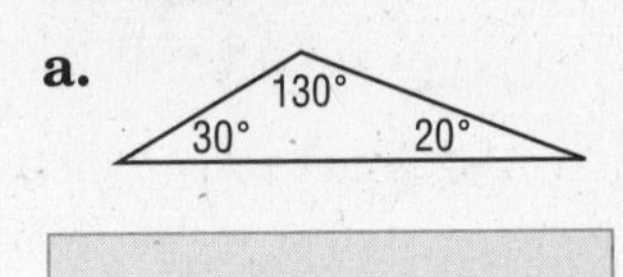

b.

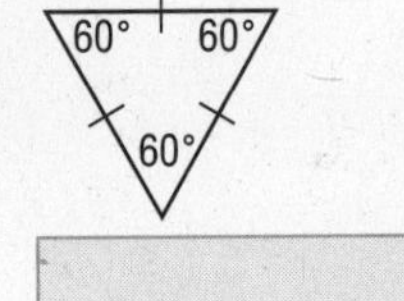

9–5 The Pythagorean Theorem

WHAT YOU'LL LEARN

- Use the Pythagorean Theorem to find the length of a side of a right triangle.
- Use the converse of the Pythagorean Theorem to determine whether a triangle is a right triangle.

BUILD YOUR VOCABULARY (pages 210–211)

In a right triangle, the side opposite the ______ angle is the **hypotenuse**.

If you know the lengths of two ______ of a right triangle, you can use the **Pythagorean Theorem** to find the length of the ______ side. This is called **solving a right triangle**.

KEY CONCEPT

Pythagorean Theorem
If a triangle is a right triangle, then the square of the length of the hypotenuse is equal to the sum of the squares of the lengths of the legs.

EXAMPLE Find the Length of the Hypotenuse

1 **Find the length of the hypotenuse of the right triangle.**

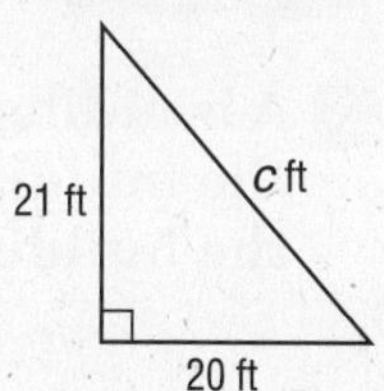

$c^2 = a^2 + b^2$	Pythagorean Theorem
$c^2 = \square^2 + \square^2$	Replace a with ______ and b with ______.
$c^2 = \square + \square$	Evaluate ______ and ______.
$c^2 = \square$	Add ______ and ______.
$c^2 = \square$	Take the ______ of each side.
$c = \square$	The length is ______.

Your Turn Find the length of the hypotenuse of the right triangle.

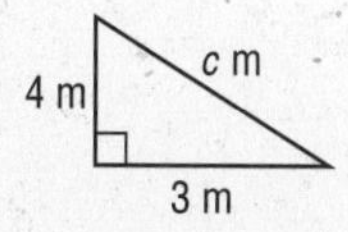

EXAMPLE Solve a Right Triangle

2 Find the length of the leg of the right triangle.

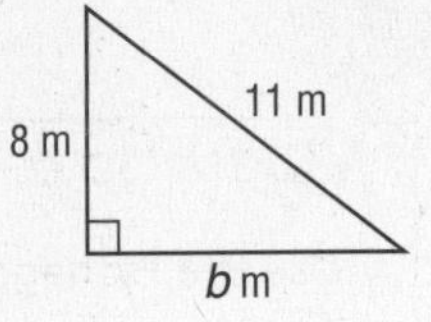

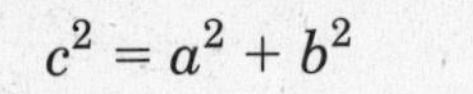

$c^2 = a^2 + b^2$ Pythagorean Theorem

$\square^2 = \square^2 + b^2$ Replace c with $\square$ and a with $\square$.

$\square = \square + b^2$ Evaluate $\square$ and $\square$.

$\square = b^2$ Subtract 64 from each side.

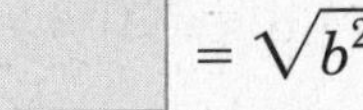

$\square = \sqrt{b^2}$ Take the $\square$ of each side.

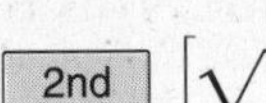  $\square$  $\square$

2nd [√] $\square$ ENTER $\square$

The length of the leg is about $\square$.

FOLDABLES

ORGANIZE IT

Under the tab for Lesson 9–5, write the Pythagorean Theorem. Then draw a right triangle and label the sides *a*, *b*, and *c* as used in the theorem.

EXAMPLE Use the Pythagorean Theorem

3 A building is 10 feet tall. A ladder is positioned against the building so that the base of the ladder is 3 feet from the building. How long is the ladder?

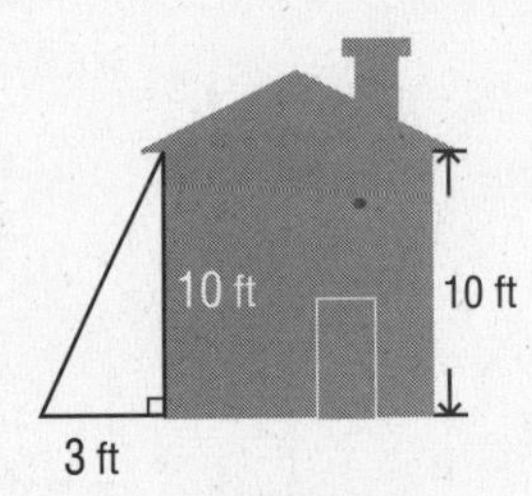

$c^2 = a^2 + b^2$ Pythagorean Theorem

$c^2 = \square^2 + \square^2$ Replace a with $\square$ and b with $\square$.

$c^2 = \square + \square$ Evaluate $\square$ and $\square$.

$c^2 = \square$ Simplify.

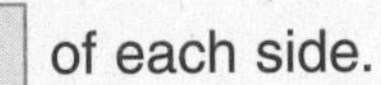

$\sqrt{c^2} = \sqrt{109}$ Take the $\square$ of each side.

$c \approx \square$ Round to the nearest tenth.

The ladder is about $\square$ tall.

Your Turn

a. Find the length of the leg of the right triangle.

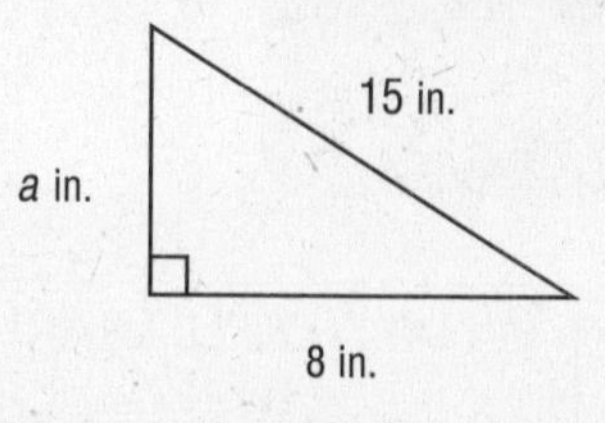

b. An 18-foot ladder is placed against a building which is 14 feet tall. About how far is the base of the ladder from the building?

EXAMPLE Identify a Right Triangle

4 **The measures of three sides of a triangle are 48 feet, 60 feet, and 78 feet. Determine whether the triangle is a right triangle.**

$c^2 = a^2 + b^2$ — Pythagorean Theorem

$\square^2 \stackrel{?}{=} \square^2 + \square^2$ — Replace a with ☐, b with ☐, and c with ☐.

☐ $\stackrel{?}{=}$ ☐ + ☐ — Evaluate.

☐ — Simplify.

The triangle ☐ a right triangle.

Your Turn The measures of three sides of a triangle are 42 inches, 61 inches, 84 inches. Determine whether the triangle is a right triangle.

HOMEWORK ASSIGNMENT

Page(s):

Exercises:

9–6 The Distance and Midpoint Formulas

What You'll Learn

- Use the Distance Formula to determine lengths on a coordinate plane.
- Use the Midpoint Formula to find the midpoint of a line segment on the coordinate plane.

Key Concept

Distance Formula The distance d between two points with coordinates (x_1, y_1) and (x_2, y_2), is given by $d = \sqrt{(x_2 - x_1)^2 + (y_2 - y_1)^2}$.

EXAMPLE Use the Distance Formula

1 Find the distance between $M(8, 4)$ and $N(-6, -2)$. Round to the nearest tenth, if necessary.

$d = \sqrt{(x_2 - x_1)^2 + (y_2 - y_1)^2}$ Distance Formula

$MN =$ ______ $(x_1, y_1) = (8, 4)$, $(x_2, y_2) = (-6, -2)$

$MN =$ ______ Simplify.

$MN =$ ______ Evaluate ______ and ______.

$MN =$ ______ Add ______ and ______.

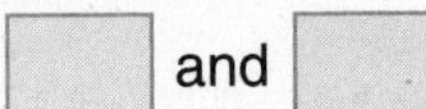

$MN \approx$ ______

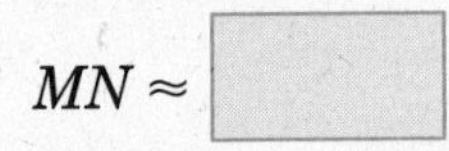

The distance between points M and N is about ______.

Your Turn Find the distance between $A(-4, 5)$ and $B(3, -9)$. Round to the nearest tenth, if necessary.

EXAMPLE Use the Distance Formula to Solve a Problem

2 GEOMETRY Find the perimeter of $\triangle XYZ$ to the nearest tenth.

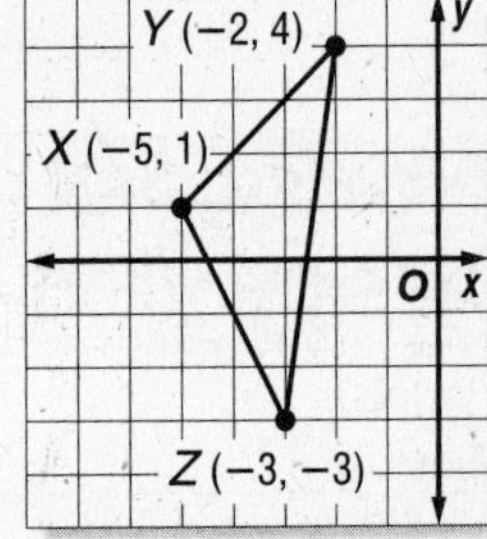

First, use the Distance Formula to find the length of each side of the triangle.

Distance Formula:

WRITE IT

Which point should be used for $(x_1 - y_1)$ in the distance formula? Explain.

____

____

____

____

Side $\overline{XY}$: $X(-5, 1)$, $Y(-2, 4)$

$XY =$ ______ $(x_1, y_1) = (-5, 1)$, $(x_2, y_2) = (-2, 4)$

$XY =$ ______ Simplify.

$XY =$ ______ Simplify.

Side $\overline{YZ}$: $Y(-2, 4)$, $Z(-3, -3)$

$YZ =$ ______ $(x_1, y_1) = (-2, 4)$, $(x_2, y_2) = (-3, -3)$

$YZ =$ ______ Simplify.

$YZ =$ ______ Simplify.

Side $\overline{ZX}$: $Z(-3, -3)$, $X(-5, 1)$

$ZX =$ ______ $(x_1, y_1) = (-3, -3)$, $(x_2, y_2) = (-5, 1)$

$ZX =$ ______ Simplify.

$ZX =$ ______ Simplify.

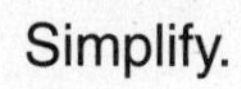

The perimeter is ______ + ______ + ______ or about

.

Your Turn Find the perimeter of △ABC to the nearest tenth.

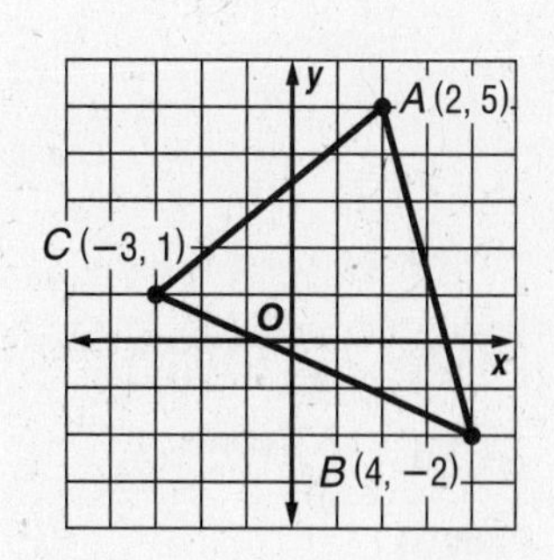

KEY CONCEPT

Midpoint Formula On a coordinate plane, the coordinates of the midpoint of a segment whose endpoints have coordinates at (x_1, y_1) and (x_2, y_2) are given by $\left(\frac{x_1+x_2}{2}, \frac{y_1+y_2}{2}\right)$.

FOLDABLES Write the Distance Formula and the Midpoint Formula under the tab for this lesson. Illustrate each formula.

BUILD YOUR VOCABULARY (page 210)

On a ______, the point that is ______ between the ______ is called the **midpoint**.

EXAMPLE Use the Midpoint Formula

3 Find the coordinates of the midpoint of $\overline{RS}$.

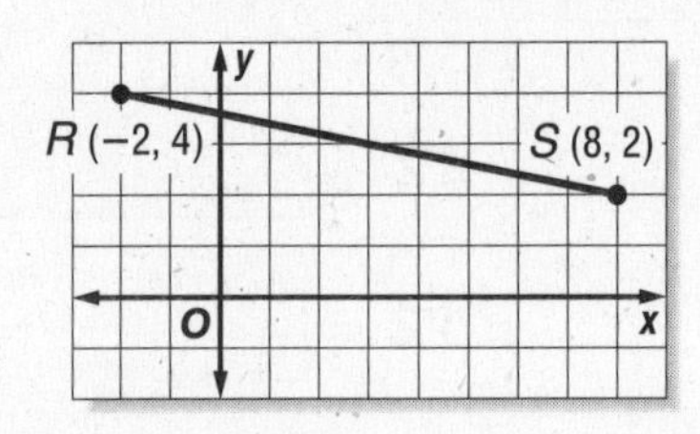

midpoint = ______ Midpoint Formula

= ______ Substitution

= ______ The coordinates of the midpoint of $\overline{RS}$ are ______.

Your Turn Find the coordinates of the midpoint of $\overline{GH}$.

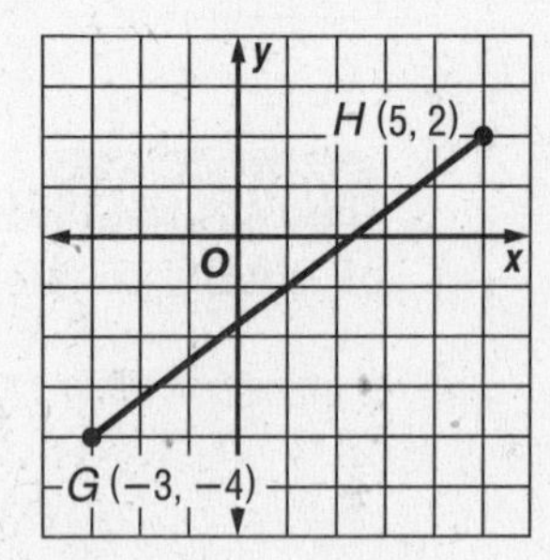

HOMEWORK ASSIGNMENT

Page(s):

Exercises:

9–7 Similar Triangles and Indirect Measurement

BUILD YOUR VOCABULARY (page 211)

Triangles that have the same  but not necessarily the same size are called **similar triangles**.

WHAT YOU'LL LEARN

- Identify corresponding parts and find missing measures of similar triangles.
- Solve problems involving indirect measurement using similar triangles.

EXAMPLE Find Measures of Similar Triangles

1 If $\triangle RUN \sim \triangle CAB$, what is the value of x?

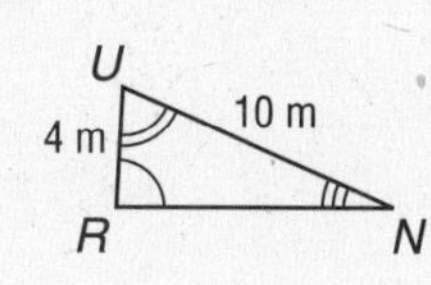

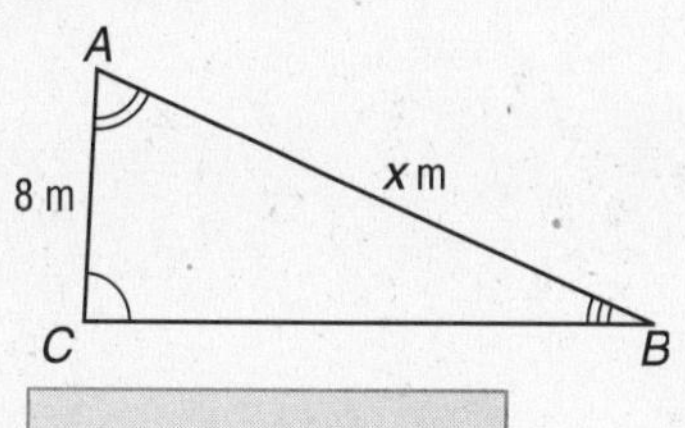

The corresponding sides are ______.

$\frac{UR}{AC} = \frac{UN}{AB}$ Write a proposition.

____ = ____ $UR =$ ____, $AC =$ ____, $UN =$ ____, $AB =$ ____

____ = ____ Find the cross products.

____ = ____ Simplify.

$x =$ ____ Divide each side by ____.

The value of x is ____.

KEY CONCEPT

Corresponding Parts of Similar Triangles If two triangles are similar, then the corresponding angles have the same measure, and the corresponding sides are proportional.

FOLDABLES Draw an example of similar triangles in your notes. Label the corresponding sides and angles.

Your Turn If $\triangle ABC \sim \triangle DEF$, what is the value of x?

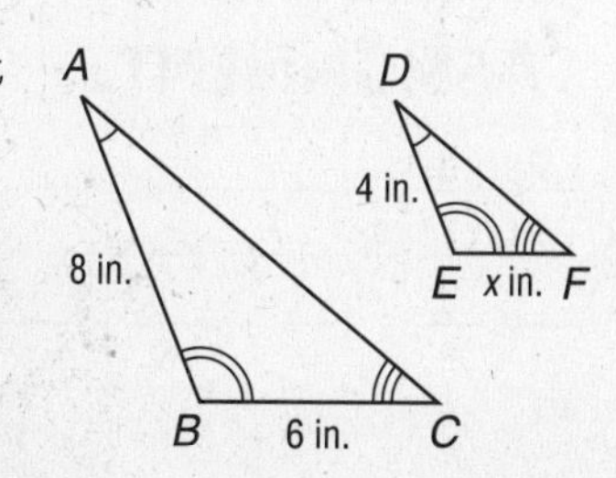

EXAMPLE Use Indirect Measurement

2 A surveyor wants to find the distance RS across the lake. He constructs $\triangle PQT$ similar to $\triangle PRS$ and measures the distances as shown. What is the distance across the lake?

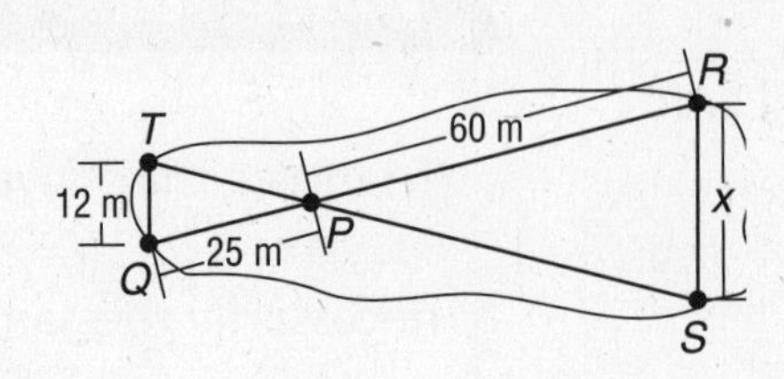

___ = ___ Write a ______.

___ = ___ Substitution

___ = ___ Find the ______.

___ = ___ Simplify.

___ $= x$ Divide each side by ___.

The distance across the lake is ______.

Your Turn In the figure, $\triangle MNO$ is similar to $\triangle OPQ$. Find the distance across the park.

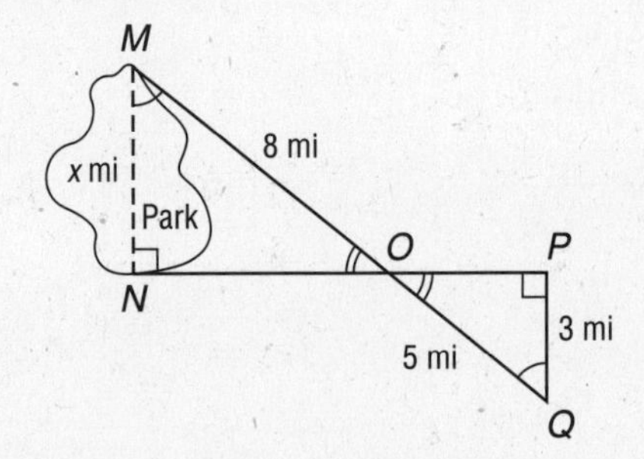

HOMEWORK ASSIGNMENT

Page(s):

Exercises:

9–8 Sine, Cosine, and Tangent Ratios

What You'll Learn

- Find sine, cosine, and tangent ratios.
- Solve problems by using the trigonometric ratios.

Key Concept

Trigonometric Ratios

$\sin A = \frac{a}{c}$

$\cos A = \frac{b}{c}$

$\tan A = \frac{a}{b}$

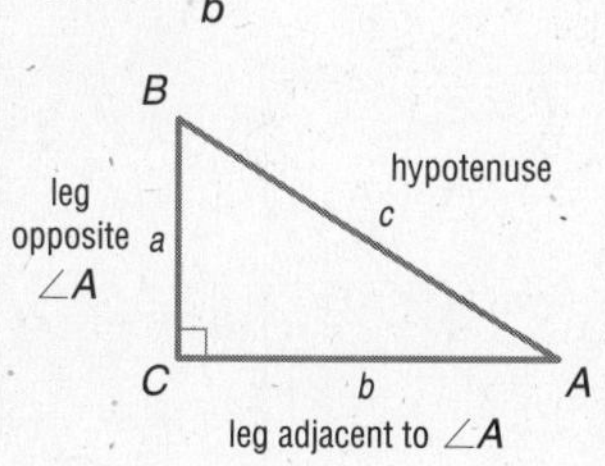

Build Your Vocabulary (page 211)

A **trigonometric ratio** is a ratio of the ______ of two sides of a ______ triangle.

EXAMPLE Find Trigonometric Ratios

1 Find sin *A*, cos *A*, and tan *A*.

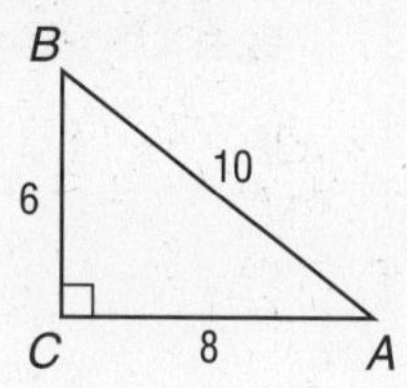

$\sin A = \frac{\text{measure of leg opposite } \angle A}{\text{measure of hypotenuse}} =$ ______ or ______

$\cos A = \frac{\text{measure of leg adjacent to } \angle A}{\text{measure of hypotenuse}} =$ ______ or ______

$\tan A = \frac{\text{measure of leg opposite } \angle A}{\text{measure of leg adjacent to } \angle A} =$ ______ or ______

Your Turn Find sin *B*, cos *B*, and tan *B*.

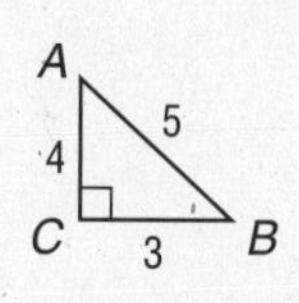

EXAMPLE Use a Calculator to Find trigonometric Ratios

2 Find each value to the nearest ten thousandth.

a. sin 19°

SIN  ENTER 

So, sin 19° is about ______.

b. cos 51°

So, cos 51° is about ____.

c. tan 24°

So, tan 24° is about ____.

FOLDABLES

ORGANIZE IT

Under the tab for Lesson 9-8, tell how to find the value of a trigonometric ratio using a calculator.

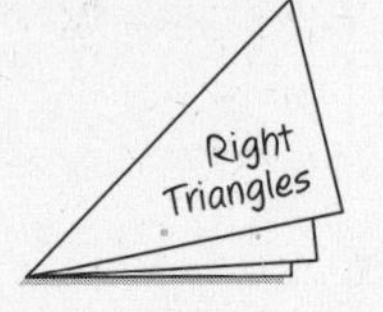

Your Turn **Find each value to the nearest ten thousandth.**

a.  63° **b.**  cos 14° **c.** tan 41°

EXAMPLE Use Trigonometric Ratios

3 **Find the missing measure. Round the nearest tenth.**

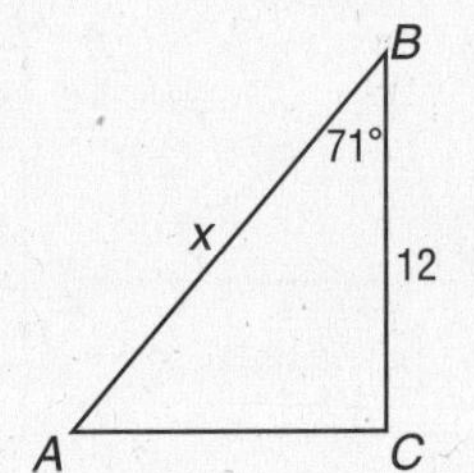

Use the cosine ratio.

$$\cos B = \frac{\text{measure of leg adjacent to } \angle B}{\text{hypotenuse}}$$

cos 71° = ____ Substitution

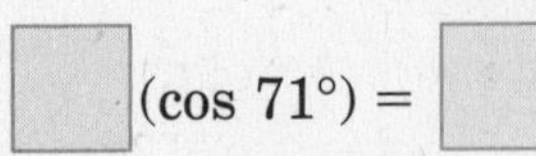

____ (cos 71°) = ____ Multiply each side by .

  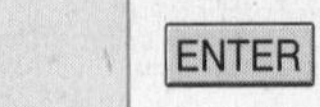 

≈ ____

The measure of the hypotenus is about 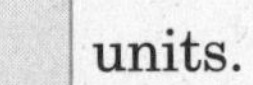 units.

Your Turn Find the missing measure. Round to the nearest tenth.

B, A, C, 35°, x, 15

EXAMPLE Use Trigonometric Ratios to Solve a Problem

4 ARCHITECTURE **A tourist visiting the Petronas Towers in Kuala Lumpur, Malaysia, stands 261 feet away from their base. She looks at the top at an angle of 80° with the ground. How tall are the Towers?**

Use the tangent ratio.

$\tan A =$ ______

$\tan 80° =$ ______ Substitution

______ $(\tan 80°) =$ ______ Multiply each side by ______.

______ × TAN ______ ENTER ______

______ $\approx x$

The height of the Towers is about ______ feet.

Your Turn Jenna stands 142 feet away from the base of a building. She looks at the top at an angle of 62° with the ground. How tall is the building?

HOMEWORK ASSIGNMENT

Page(s):

Exercises:

CHAPTER 9

BRINGING IT ALL TOGETHER

STUDY GUIDE

FOLDABLES	VOCABULARY PUZZLEMAKER	BUILD YOUR VOCABULARY
Use your **Chapter 9 Foldable** to help you study for your chapter test.	To make a crossword puzzle, word search, or jumble puzzle of the vocabulary words in Chapter 9, go to: www.glencoe.com/sec/math/t_resources/free/index.php	You can use your completed **Vocabulary Builder** (pages 210–211) to help you solve the puzzle.

9-1 Squares and Square Roots

Find each square root, if possible.

1. $\sqrt{361}$ 2. $\sqrt{-196}$ 3. $-\sqrt{441}$

Estimate each square root to the nearest whole number. Do not use a calculator.

4. $\sqrt{120}$ 5. $\sqrt{150}$  6. $-\sqrt{70}$

9-2 The Real Number System

Underline the correct term to complete each sentence.

7. Numbers with decimals that (are, are not) repeating or terminating are irrational numbers.
8. All square roots (are, are not) irrational numbers.
9. Irrational numbers (are, are not) real numbers.

Name all of the sets of numbers to which each real number belongs. Let N = natural numbers, W = whole numbers, Z = integers, Q = rational numbers, and I = irrational numbers.

10. -49 11. $\sqrt{48}$ 12. 11 

Solve each equation. Round to the nearest tenth, if necessary.

13. $b^2 = 225$ 14. $z^2 = 44$

9-3 Angles

Classify each as *acute*, *right*, *obtuse*, or *straight*.

15. 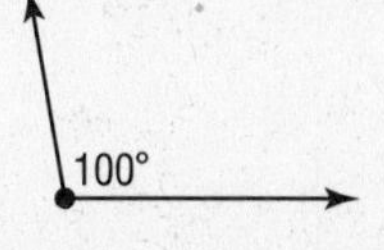

16. 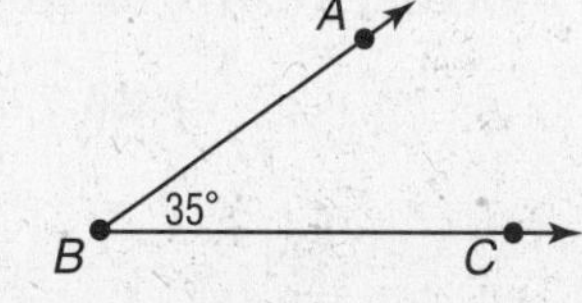

Use a protractor to find the measure of each angle. Then classify each angle as *acute*, *obtuse*, *right*, or *straight*.

17. $\angle RNS$

18. $\angle MNQ$

19. $\angle SNM$

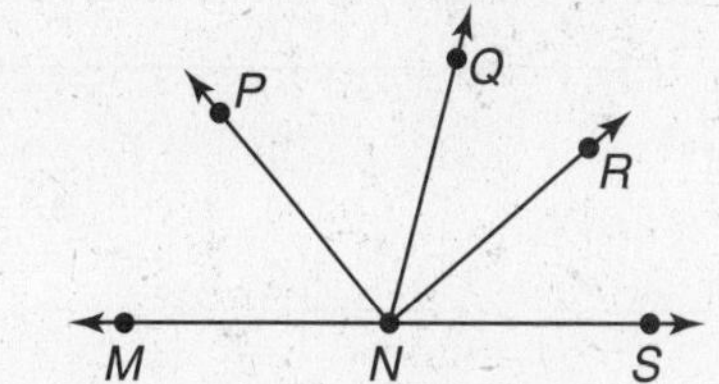

9-4 Triangles

Find the value of *x* in the triangle. Then classify the triangle as *acute*, *right*, or *obtuse*.

20. 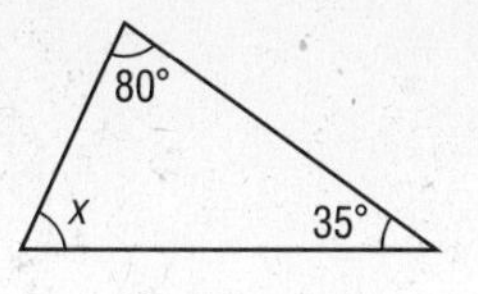

21. 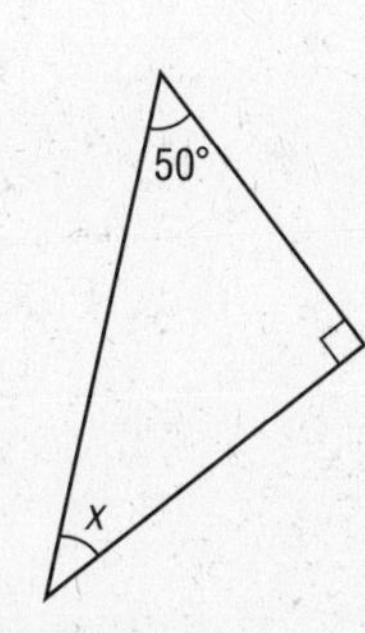

22. The measures of the angles of a triangle are in the ratio 3:4:5. What is the measure of each angle?

9-5 The Pythagorean Theorem

If c is the measure of the hypotenuse, find each missing measure. Round to the nearest tenth, if necessary.

23. $a = 12, b = ?, c = 37$

24. $a = ?, b = 6, c = 16$

25. The length of the sides of a triangle are 10, 24, and 26. Determine whether the triangle is a right triangle.

9-6 The Distance and Midpoint Formulas

Complete.

26. The ______ is used to find the length of a segment on a coordinate plane.

27. One a line segment, the point that is halfway between the endpoints is called the ______.

Find the distance between each pair of points. Round to the nearest tenth, if necessary.

28. $J(8, -3)$, $K(5, 1)$

29. $P(-3, 7)$, $Q(4, 2)$

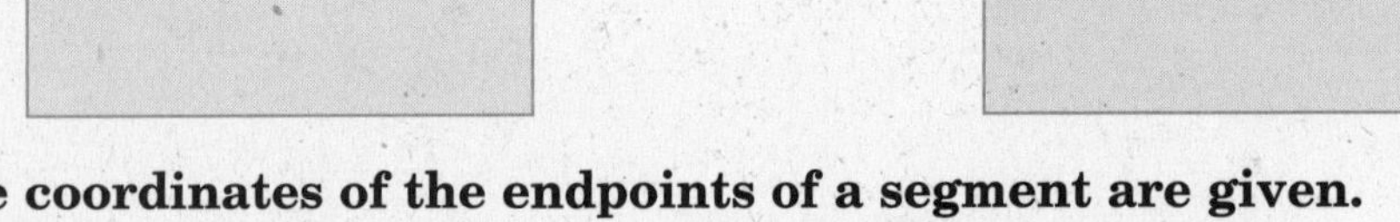

The coordinates of the endpoints of a segment are given. Find the coordinates of the midpoint of each segment.

30. $S(2, 4)$, $T(0, -2)$

31. $C(-5, -1)$, $D(3, -4)$

9-7 Similar Triangles and Indirect Measurement

For Questions 32 and 33, use the triangles at the right. $\triangle PQR \sim \triangle UVW$.

32. Name an angle with the same measure as $\angle W$.

33. Find the value of x.

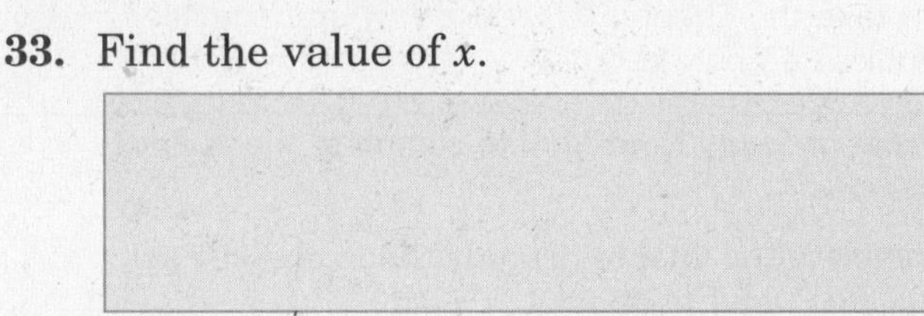

34. In the figure at the right, the triangles are similar. How far is the waterfall from the grove of redwood trees?

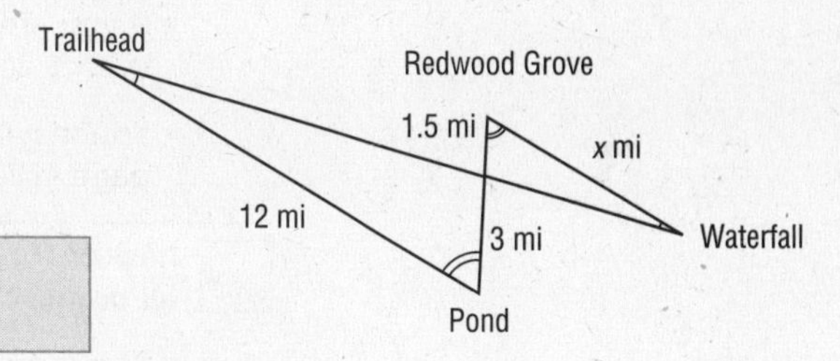

9-8 Sine, Cosine, and Tangent Ratios

Decide whether each statement is true or false.

35. Trigonometric ratios can be used with acute and obtuse triangles.

36. The value of the trigonometric ratio does not depend on the size of the triangle.

37. To determine tangent, you must know the measure of the hypotenuse.

For each triangle, find the missing measure to the nearest tenth.

38.

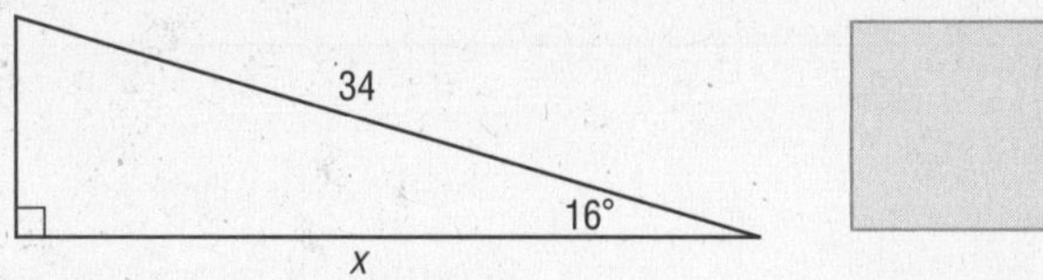

39.

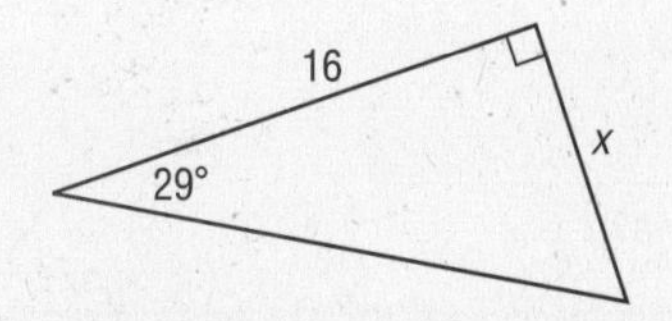

ARE YOU READY FOR THE CHAPTER TEST?

Visit **pre-alg.com** to access your textbook, more examples, self-check quizzes, and practice tests to help you study the concepts in Chapter 9.

Check the one that applies. Suggestions to help you study are given with each item.

☐ **I completed the review of all or most lessons without using my notes or asking for help.**

- You are probably ready for the Chapter Test.
- You may want take the Chapter 9 Practice Test on page 487 of your textbook as a final check.

☐ **I used my Foldable or Study Notebook to complete the review of all or most lessons.**

- You should complete the Chapter 9 Study Guide and Review on pages 483–486 of your textbook.
- If you are unsure of any concepts or skills, refer back to the specific lesson(s).
- You may also want to take the Chapter 9 Practice Test on page 487.

☐ **I asked for help from someone else to complete the review of all or most lessons.**

- You should review the examples and concepts in your Study Notebook and Chapter 9 Foldable.
- Then complete the Chapter 9 Study Guide and Review on pages 483–486 of your textbook.
- If you are unsure of any concepts or skills, refer back to the specific lesson(s).
- You may also want to take the Chapter 9 Practice Test on page 487.

Student Signature

Parent/Guardian Signature

Teacher Signature

Two-Dimensional Figures

Use the instructions below to make a Foldable to help you organize your notes as you study the chapter. You will see Foldable reminders in the margin of this Interactive Study Notebook to help you in taking notes.

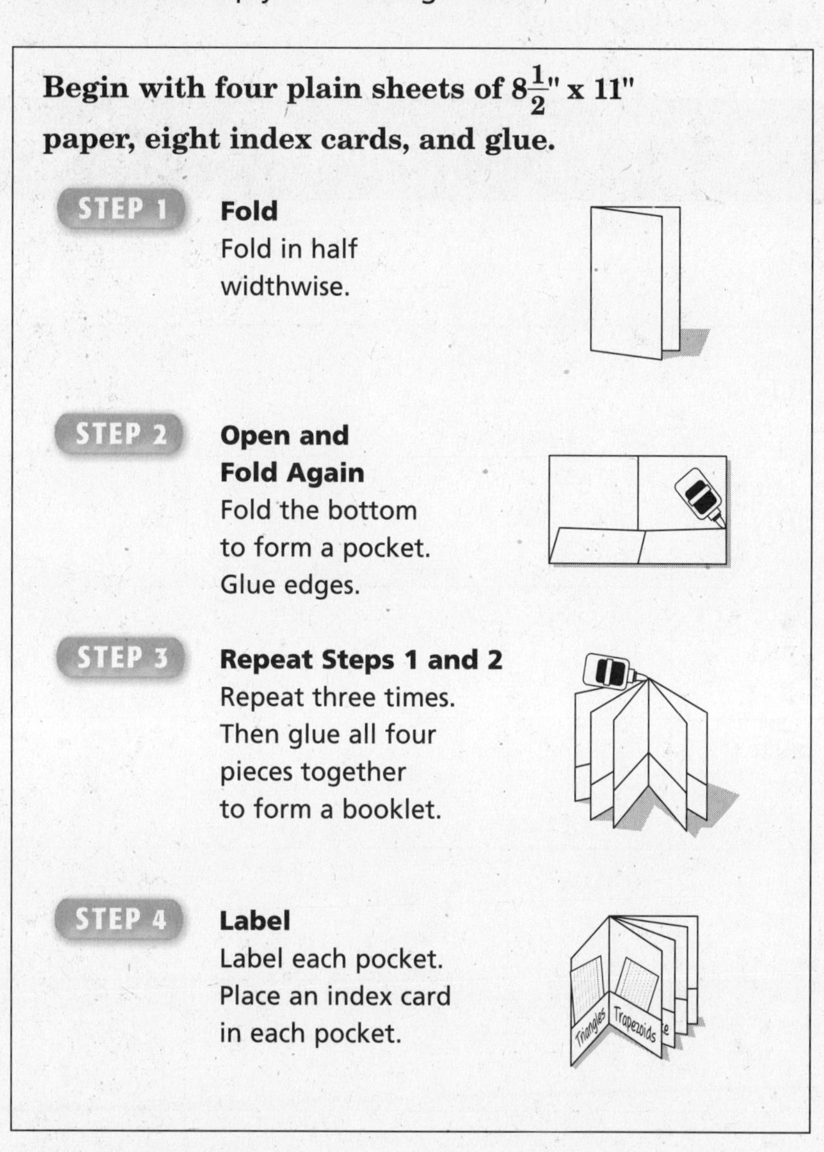

Begin with four plain sheets of $8\frac{1}{2}$" x 11" paper, eight index cards, and glue.

STEP 1 **Fold**
Fold in half widthwise.

STEP 2 **Open and Fold Again**
Fold the bottom to form a pocket. Glue edges.

STEP 3 **Repeat Steps 1 and 2**
Repeat three times. Then glue all four pieces together to form a booklet.

STEP 4 **Label**
Label each pocket. Place an index card in each pocket.

NOTE-TAKING TIP: To help you organize data, create study cards when taking notes, recording and defining vocabulary words, and explaining concepts.

BUILD YOUR VOCABULARY

This is an alphabetical list of new vocabulary terms you will learn in Chapter 10. As you complete the study notes for the chapter, you will see Build Your Vocabulary reminders to complete each term's definition or description on these pages. Remember to add the textbook page number in the second column for reference when you study.

Vocabulary Term	Found on Page	Definition	Description or Example
adjacent angles [uh-JAY-suhnt]			
circumference [suhr-KUHMP-fuhrnts]			
complementary angles [kahm-pluh-MEHN-tuh-ree]			
congruent [kuhn-GROO-uhnt]			
corresponding parts			
diagonal			
diameter			
parallel lines			
perpendicular lines			

Vocabulary Term	Found on Page	Definition	Description or Example
π (pi)			
polygon			
quadrilateral [KWAH-druh-LA-tuh-ruhl]			
radius			
reflection			
rotation			
supplementary angles [SUH-pluh-MEHN-tuh-ree]			
transformation			
translation			
transversal			
vertical angles			

10–1 Line and Angle Relationships

WHAT YOU'LL LEARN

- Identify the relationships of angles formed by two parallel lines and a transversal.
- Identify the relationships of vertical, adjacent, complementary, and supplementary angles.

BUILD YOUR VOCABULARY (pages 238–239)

Two lines in a plane that never intersect are **parallel lines**.

A line that ________ two parallel lines is called a **transversal**.

When two lines intersect, they form two pairs of ________ angles called **vertical angles**.

When two angles have the same ________, share a common side, and do not overlap, they are **adjacent angles**.

If the sum of the measures of two angles is ________, the angles are **complementary**.

If the sum of the measures of two angles is ________, the angles are **supplementary**.

Lines that ________ to form a ________ are **perpendicular lines**.

KEY CONCEPT

Parallel Lines Cut by a Transversal If two parallel lines are cut by a transversal, then the following pairs of angles are congruent.

- Corresponding angles are congruent.
- Alternate interior angles are congruent.
- Alternate exterior angles are congruent.

EXAMPLE Find Measures of Angles

1 **In the figure, $m \parallel n$ and t is a transversal. If $m\angle 7 = 123°$, find $m\angle 2$ and $m\angle 8$.**

Since $\angle 7$ and $\angle 2$ are alternate

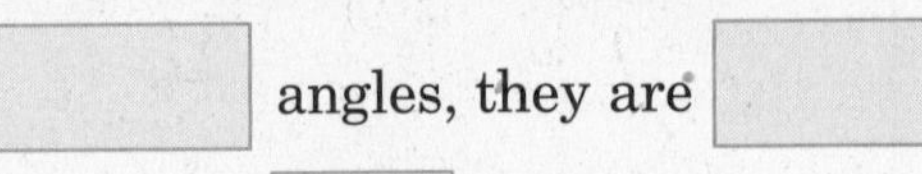

________ angles, they are ________.

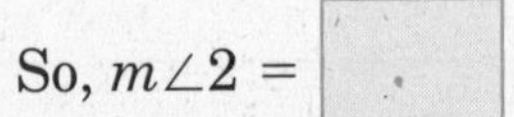

So, $m\angle 2 =$ ________.

Since $\angle 7$ and $\angle 8$ are ________ angles,

they are ________. So, $m\angle 8 =$ ________.

KEY CONCEPT

Names of Special Angles The eight angles formed by parallel lines and a transversal have special names.

- **Interior angles**
- **Exterior angles**
- **Alternate interior angles**
- **Alternate exterior angles**
- **Corresponding angles**

Your Turn In the figure in Example 1, $m \parallel n$ and t is a transversal. If $m\angle 4 = 57°$, find $m\angle 5$ and $m\angle 1$.

EXAMPLE Find a Missing Angle Measure

2 **If $m\angle D = 53°$ and $\angle D$ and $\angle E$ are complementary, what is $m\angle E$?**

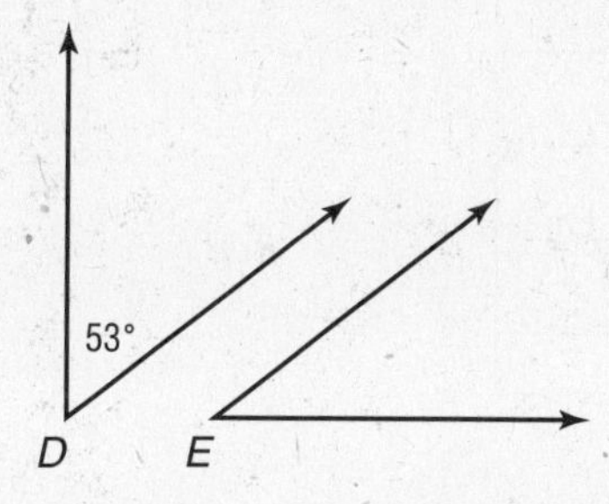

Since $\angle D$ and $\angle E$ are ______,

$m\angle D + m\angle E =$ ______.

$m\angle D + m\angle E =$ ______   ______ angles

$53° + m\angle E =$ ______   Replace $m\angle D$ with 53°.

$53° -$ ______ $+ m\angle E =$ ______ $-$ ______   Subtract ______ from each side.

$m\angle E =$ ______

Your Turn If $m\angle G = 104°$ and $\angle G$ and $\angle H$ are supplementary, what is $m\angle H$?

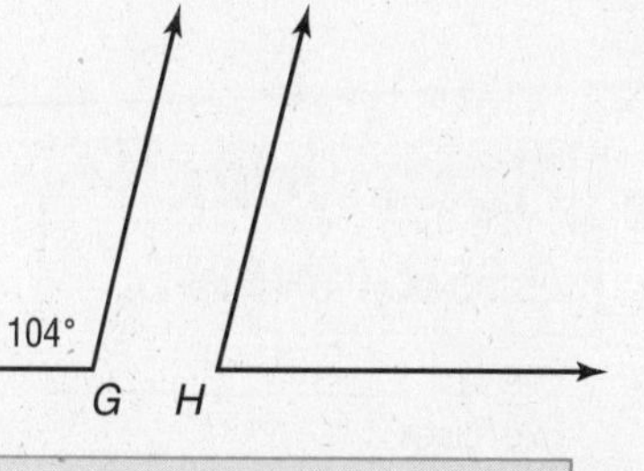

EXAMPLE Find Measures of Angles

WRITE IT

What is the difference between complementary angles and supplementary angles?

____

____

____

____

____

3 **Angles *PQR* and *STU* are supplementary. If $m\angle PQR = x - 15$ and $m\angle STU = x - 65$, find the measure of each angle.**

Step 1 Find the value of x.

$m\angle PQR + m\angle STU =$ ____ Supplementary angles

$(x - 15) + (x - 65) =$ ____ Substitution

____ − ____ = ____ Combine like terms.

____ = ____ Add ____ to each side.

$x =$ ____ Divide each side by ____.

Step 2 Replace x with ____ to find the measure of each angle.

$m\angle PQR = x - 15$

= ____ − 15 or ____

$m\angle STU = x - 65$

= ____ − 65 or ____

Your Turn Angles *ABC* and *DEF* are complementary. If $m\angle ABC = x + 12$ and $m\angle DEF = 2x - 9$, find the measure of each angle.

HOMEWORK ASSIGNMENT

Page(s):

Exercises:

10–2 Congruent Triangles

WHAT YOU'LL LEARN

- Identify congruent triangles and corresponding parts of congruent triangles.

KEY CONCEPT

Corresponding Parts of Congruent Triangles
If two triangles are congruent, their corresponding sides are congruent and their corresponding angles are congruent.

BUILD YOUR VOCABULARY (page 238)

Figures that have the same ______ and ______ are **congruent**.

The parts of congruent triangles that  are **corresponding parts**.

EXAMPLE Name Corresponding Parts

1 Name the corresponding parts in the congruent triangles shown. Then complete the congruence statement.

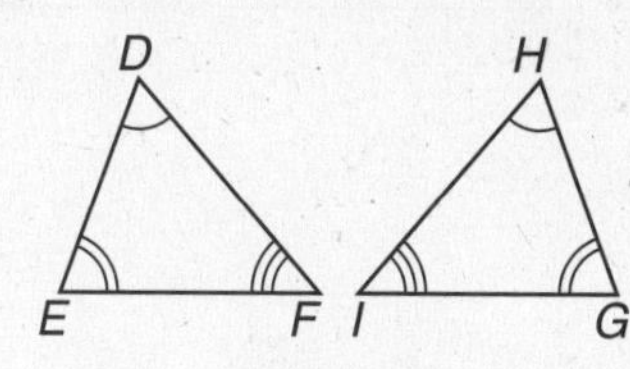

$\triangle DEF \cong ?$

Corresponding Angles

$\angle D \cong$ ______, $\angle E \cong$ ______, $\angle F \cong$ ______

Corresponding Sides

$\overline{DE} \cong$ ______, $\overline{DF} \cong$ ______, $\overline{EF} \cong$ ______

One congruence statement is 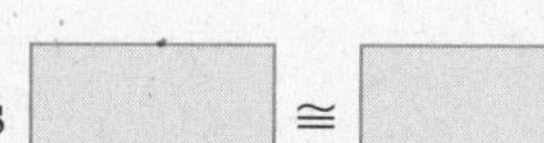.

Your Turn Name the corresponding parts in the congruent triangles shown. Then complete the congruence statement.

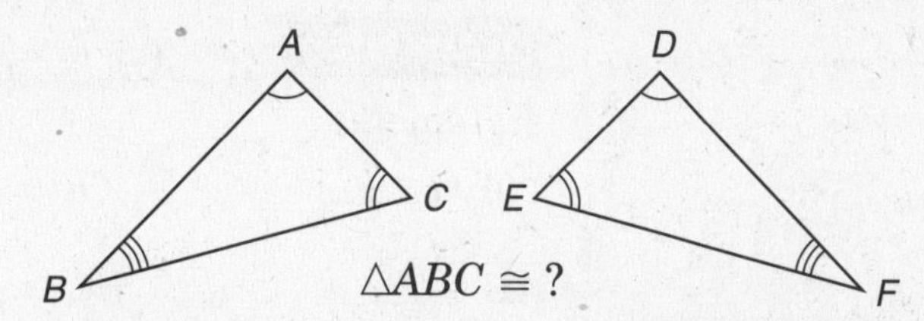

$\triangle ABC \cong ?$

EXAMPLE Use Congruence Statements

2 **If $\triangle MNO \cong \triangle QPR$, complete each congruence statement.**

$\angle M \cong$ ___?___ $\angle P \cong$ ___?___ $\angle O \cong$ ___?___

$\overline{QP} \cong$ ___?___ $\overline{NO} \cong$ ___?___ $\overline{MO} \cong$ ___?___

REVIEW IT

What do we call a triangle with at least two congruent sides? Three congruent sides? *(Lesson 9-4)*

____

____

____

____

____

Use the order of the vertices in $\triangle MNO \cong \triangle QPR$ to identify the corresponding parts.

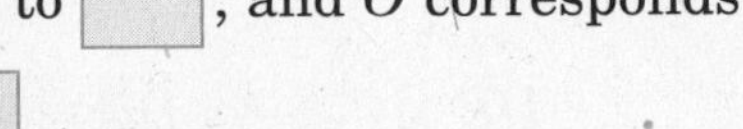

$\angle M$ corresponds to ☐, so $\angle M \cong$ ☐.

$\angle N$ corresponds to ☐, so $\angle N \cong$ ☐.

$\angle O$ corresponds to ☐, so $\angle O \cong$ ☐.

☐ corresponds to Q, and ☐ corresponds to P,

so ☐ $\cong \overline{QP}$.

N corresponds to ☐, and O corresponds to ☐,

so $\overline{NO} \cong$ ☐.

M corresponds to ☐, and O corresponds to ☐,

so $\overline{MO} \cong$ ☐

A drawing can be used to verify the congruent angles and sides.

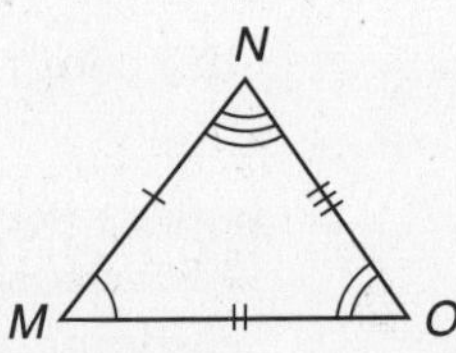

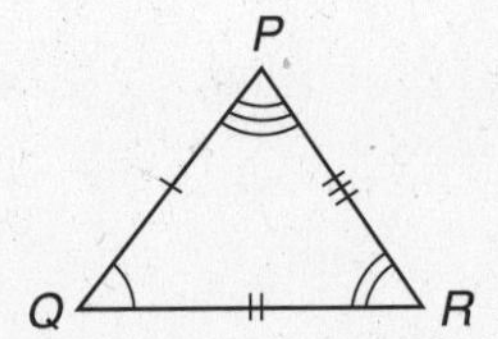

Your Turn If $\triangle ABC \cong \triangle DEF$, complete each congruence statement.

$\angle A \cong$ ___?___ $\angle E \cong$ ___?___ $\angle C \cong$ ___?___

$\overline{AB} \cong$ ___?___ $\overline{BC} \cong$ ___?___ $\overline{AC} \cong$ ___?___

EXAMPLE Find Missing Measures

3 CONSTRUCTION A brace is used to support a tabletop. In the figure, $\triangle ABC \cong \triangle DEF$.

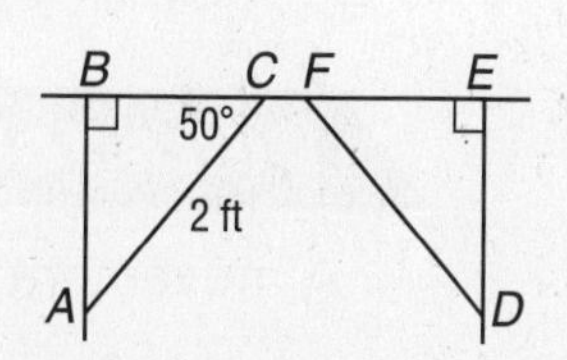

a. What is the measure of $\angle F$?

$\angle F$ and $\angle C$ are ______ angles. So, they are ______. Since $m\angle C =$ ______, $m\angle F =$ ______.

b. What is the length of $\overline{DF}$?

$\overline{DF}$ corresponds to ______. So, $\overline{DF}$ and ______ are ______. Since $AC =$ ______, $DF =$ ______.

Your Turn **In the figure, $\triangle ACB \cong \triangle DEF$.**

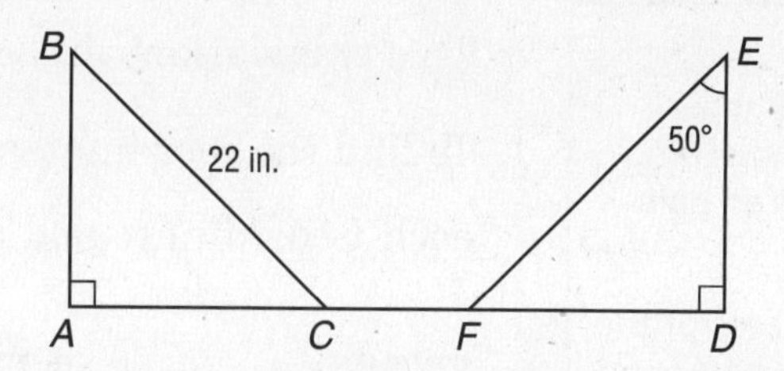

a. What is the measure of $\angle B$?

______

b. What the length of EF?

______

HOMEWORK ASSIGNMENT

Page(s):

Exercises:

10–3 Transformations on the Coordinate Plane

BUILD YOUR VOCABULARY (page 239)

A movement of a geometric figure is a **transformation**.

In a **translation**, you ______ a figure from one position to another without turning it.

In a **reflection**, you ______ a figure over a line.

In a **rotation**, you ______ the figure around a fixed point.

WHAT YOU'LL LEARN

- Draw translations, rotations, and reflections on a coordinate plane.

EXAMPLE Translation in a Coordinate Plane

1 **The vertices of $\triangle ABC$ are $A(-3, 7)$, $B(-1, 0)$, and $C(5, 5)$. Graph the triangle and the image of $\triangle ABC$ after a translation 4 units right and 5 units down.**

This translation can be written as the ordered pair ______.

To find the coordinates of the translated image, add ______ to each x-coordinate and add ______ to each y-coordinate.

vertex		4 right, 5 down		translation
$A(-3, 7)$	+		$\rightarrow$	A'
$B(-1, 0)$	+		$\rightarrow$	B'
$C(5, 5)$	+		$\rightarrow$	C'

The coordinates of the vertices of $\triangle A'B'C'$ are A' ______, B' ______, and C' ______.

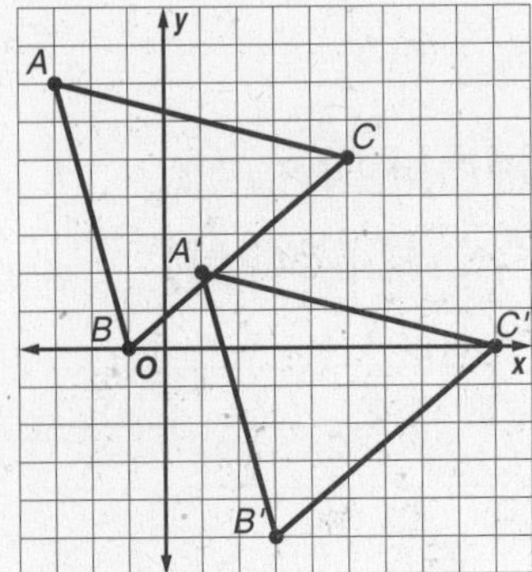

KEY CONCEPT

Translation

Step 1 Describe the translation using an ordered pair.

Step 2 Add the coordinates of the ordered pair to the coordinates of the original point.

Your Turn The vertices of $\triangle DEF$ are $D(-1, 5)$, $E(-3, 1)$, and $F(4, -4)$. Graph the triangle and the image of $\triangle DEF$ after a translation 3 units left and 2 units up.

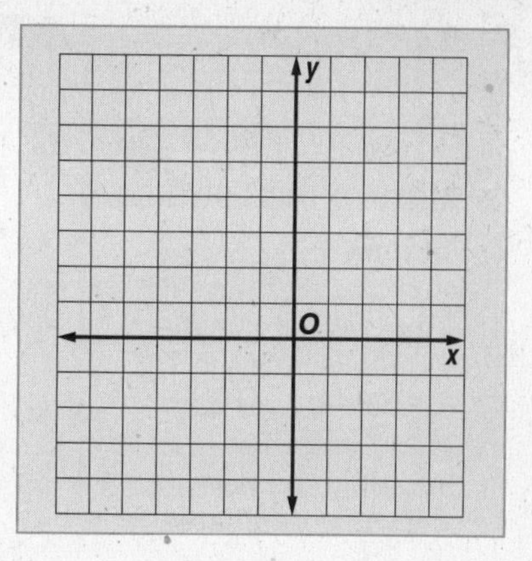

EXAMPLE Reflection in a Coordinate Plane

KEY CONCEPT

Reflection

- To reflect a point over the x-axis, use the same x-coordinate and multiply the y-coordinate by -1.
- To reflect a point over the y-axis, use the same y-coordinate and multiply the x-coordinate by -1.

2 The vertices of a figure are $M(-8, 6)$, $N(5, 9)$, $O(2, 1)$, and $P(-10, 3)$. Graph the figure and the image of the figure after a reflection over the y-axis.

To find the coordinates of the vertices of the image after a reflection over the y-axis, multiply the x-coordinate by ______ and use the same y-coordinate.

vertex				reflection
$M(-8, 6)$	→		→	M'
$N(5, 9)$	→		→	N'
$O(2, 1)$	→		→	O'
$P(-10, 3)$	→		→	P'

The coordinates of the vertices of the reflected figure are M' ______,

N' ______, O' ______, and

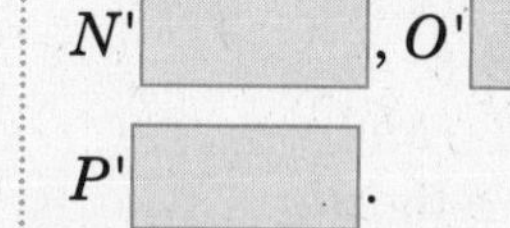

P' ______.

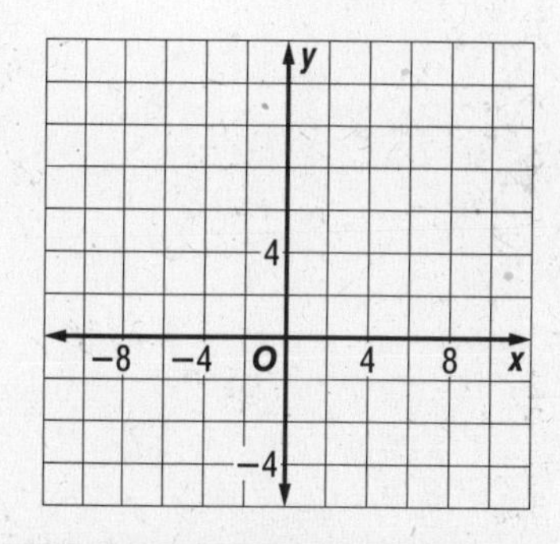

Your Turn The vertices of a figure are $Q(-2, 4)$, $R(-3, 1)$, $S(3, -2)$, and $T(4, 3)$. Graph the figure and the image of the figure after a reflection over the y-axis.

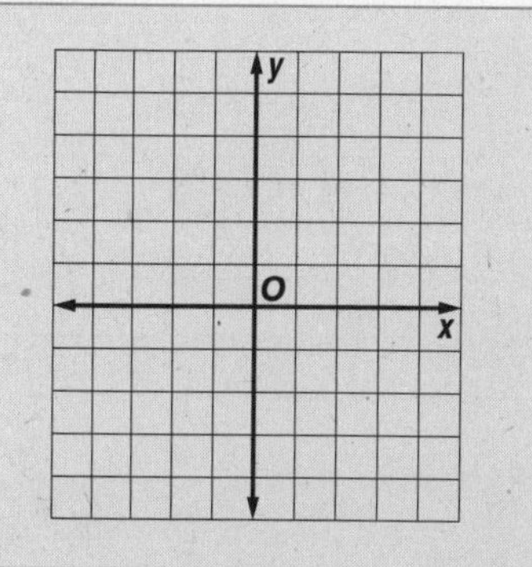

EXAMPLE Rotations in a Coordinate Plane

KEY CONCEPT

Rotation

- To rotate a figure 90° clockwise about the origin, switch the coordinates of each point and then multiply the new second coordinate by −1.
- To rotate a figure 90° counterclockwise about the origin, switch the coordinates of each point and then multiply the new first coordinate by −1.
- To rotate a figure 180° about the origin, multiply both coordinates of each point by −1.

3 **A figure has vertices $A(-4, 5)$, $B(-2, 4)$, $C(-1, 2)$, $D(-3, 1)$, and $E(-5, 3)$. Graph the figure and the image of the figure after a rotation of 180°.**

To rotate the figure, multiply both coordinates of each point by ______.

$A(-4, 5) \rightarrow A'$ ______

$B(-2, 4) \rightarrow B'$ ______

$C(-1, 2) \rightarrow C'$ ______

$D(-3, 1) \rightarrow D'$ ______

$E(-5, 3) \rightarrow E'$ ______

The coordinates of the vertices of the rotated figure are A' ______, B' ______, C' ______, D' ______, and E' ______.

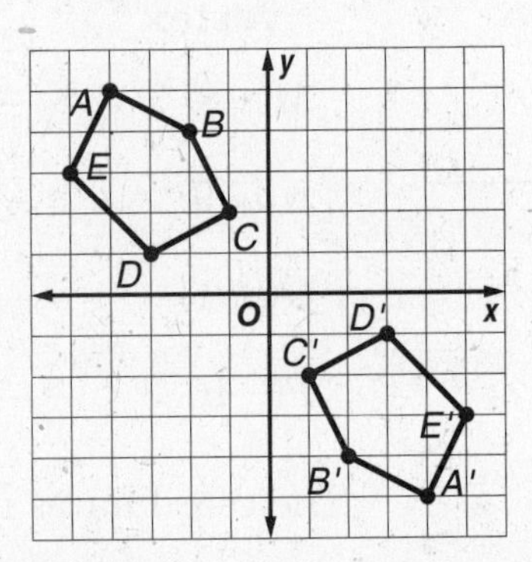

Your Turn A figure has vertices $A(2, -1)$, $B(3, 4)$, $C(-3, 4)$, $D(-5, -1)$, and $E(1, -4)$. Graph the figure and the image of the figure after a rotation of 180°.

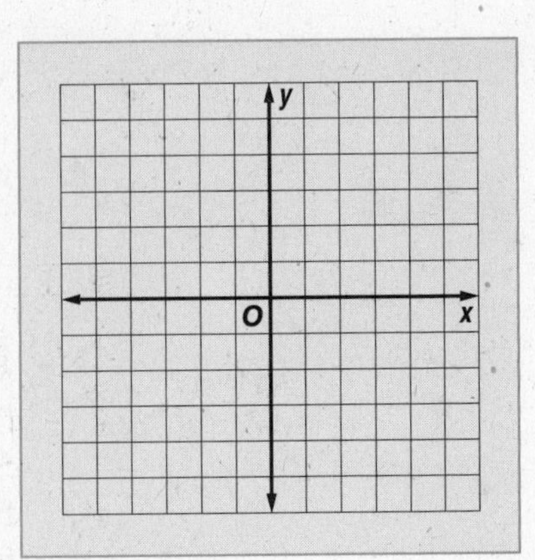

HOMEWORK ASSIGNMENT

Page(s):

Exercises:

10–4 Quadrilaterals

WHAT YOU'LL LEARN

- Find the missing angle measures of a quadrilateral.
- Classify quadrilaterals.

BUILD YOUR VOCABULARY (page 239)

A **quadrilateral** is a closed figure with ______ sides and ______ vertices.

KEY CONCEPT

Angles of a Quadrilateral The sum of the measures of the angles of a quadrilateral is 360°.

EXAMPLE Find Angle Measures

1 **Find the value of x. Then find each missing angle measure.**

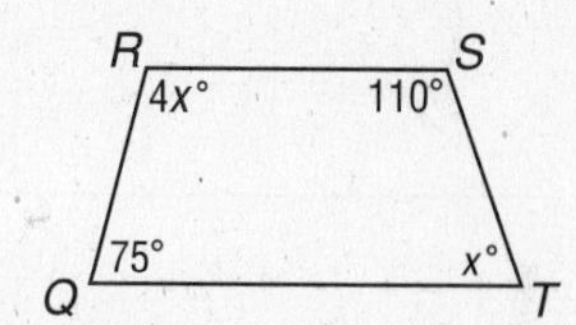

Words

The sum of the measures of the angles is 360°.

Variable
Let $m\angle Q$, $m\angle R$, $m\angle S$, and $m\angle T$ represent the measures of the angles.

Equation

$m\angle Q + m\angle R + m\angle S + m\angle T =$ ______

$m\angle Q + m\angle R + m\angle S + m\angle T =$ ______	Angles of a quadrilateral
$75 + 4x + 110 + x =$ ______	Substitution
______ + ______ = ______	Combine like terms.
______ + ______ − ______ = ______ − ______	Subtract.
______ $= 175$	Simplify.
$x =$ ______	Divide.

The value of x is ______.

So, $m\angle T =$ ______ and $m\angle R =$ ______ or ______.

Your Turn Find the value of x. Then find each missing angle measure.

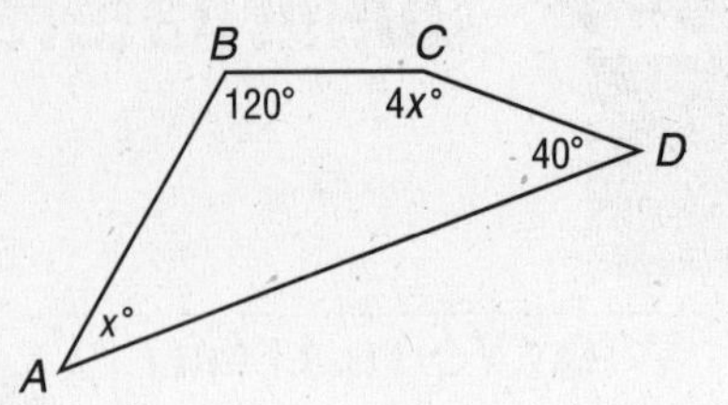

FOLDABLES

ORGANIZE IT

On your *Quadrilaterals* index card, draw three examples of quadrilaterals, and describe how to find the sum of the measures of the angles in a quadrilateral.

EXAMPLE Classify Quadrilaterals

2 **Classify each quadrilateral using the name that best describes it.**

a.

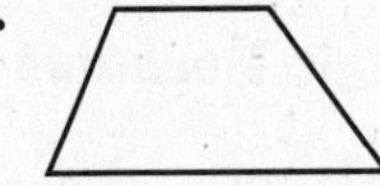

The quadrilateral has ______ of ______ ______. It is a trapezoid.

b.

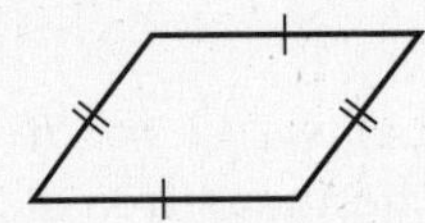

The quadrilateral has ______ of ______ ______ and ______. It is a ______.

Your Turn **Classify each quadrilateral using the name that best describes it.**

a.

b.

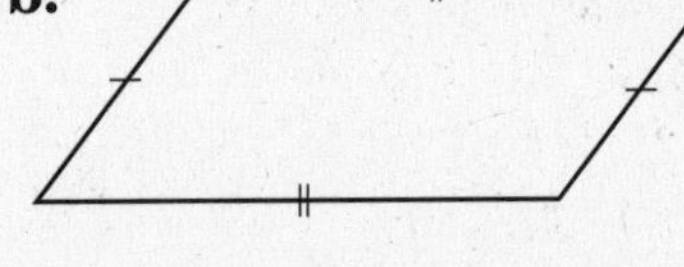

HOMEWORK ASSIGNMENT

Page(s):

Exercises:

10–5 Area: Parallelograms, Triangles, and Trapezoids

WHAT YOU'LL LEARN

- Find area of parallelograms.
- Find the areas of triangles and trapezoids.

KEY CONCEPT

Area of a Parallelogram
If a parallelogram has a base of b units and a height of h units, then the area A is bh square units.

EXAMPLE Find Areas of Parallelograms

1 **Find the area of each parallelogram.**

a.

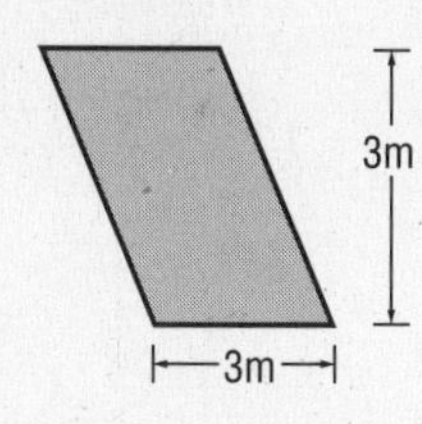

The base is ________.

The height is ________.

$A = bh$ Area of a parallelogram

$A =$ ________ $b =$ ____, $h =$ ____

$A =$ ____ Multiply.

The area is ________.

b.

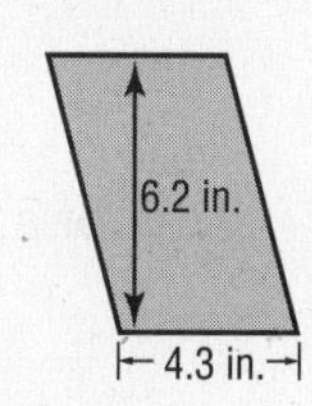

The base is ________.

The height is ________.

$A = bh$ Area of a parallelogram

$A =$ ________ $b =$ ____, $h =$ ____

$A =$ ____ Multiply.

The area is ________.

Your Turn **Find the area of each parallelogram.**

a.

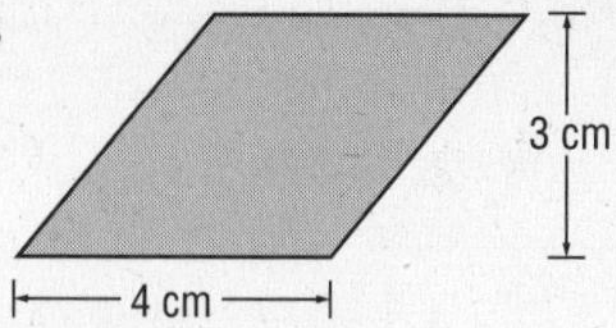

b.

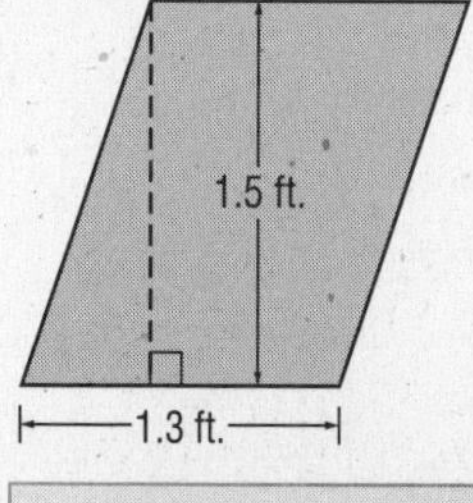

KEY CONCEPT

Area of a Triangle If a triangle has a base of b units and a height of h units, then the area A is $\frac{1}{2}bh$ square units.

EXAMPLE Find Areas of Triangles

2 **Find the area of each triangle.**

a.

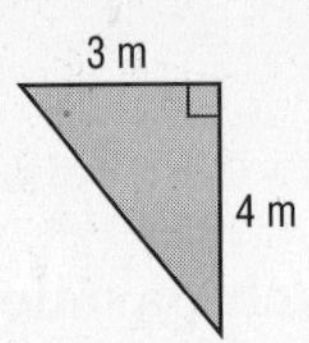

The base is ______.

The height is ______.

$A = \frac{1}{2}bh$ Area of a triangle

$A =$ ______ $b =$ ___, $h =$ ___

$A =$ ___ Multiply.

The area of the triangle is ______.

b.

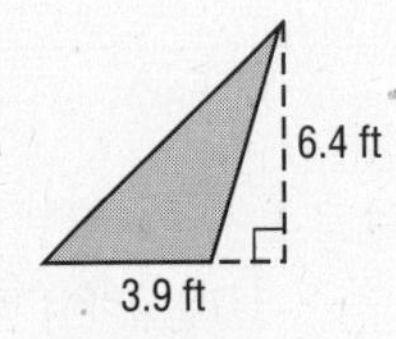

The base is ______.

The height is ______.

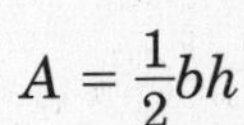

$A = \frac{1}{2}bh$ Area of a triangle

$A =$ ______ $b =$ ___, $h =$ ___

$A =$ ___ Multiply.

The area of the triangle is ______.

FOLDABLES™ ORGANIZE IT

Add diagrams, labels, and area formulas to the index cards for parallelograms, triangles, and trapezoids in your Foldable.

Your Turn Find the area of each triangle.

a.

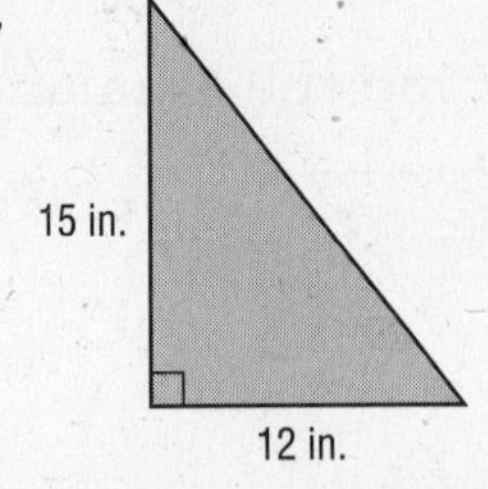

b.

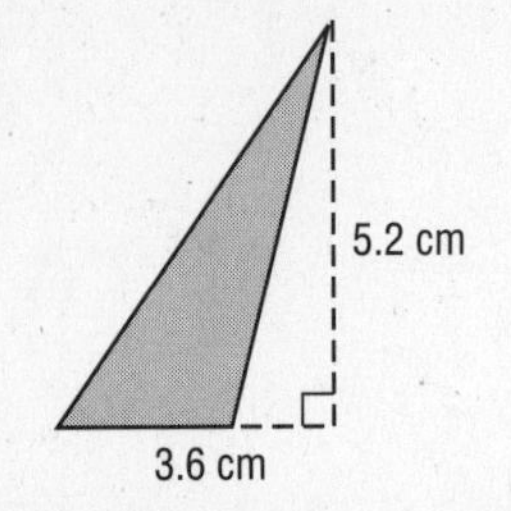

KEY CONCEPT

Area of a Trapezoid
If a trapezoid has bases of b units and a height of h units, then the area A of the trapezoid is $\frac{1}{2}h(a + b)$ square units.

EXAMPLE Find Area of a Trapezoid

3 **Find the area of the trapezoid.**

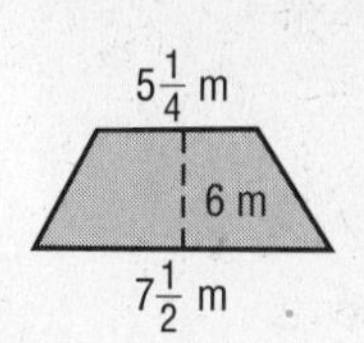

The height is ______.

The bases are ______ and ______.

$A =$ ______ Area of a ______

$A =$ ______ $h =$ ___, $a =$ ___, and $b =$ ___

$A =$ ______ Add.

$A =$ ______ Divide out the common factors.

$A =$ ___ or ___ Simplify.

The area of the trapezoid is ______.

Your Turn Find the area of the trapezoid.

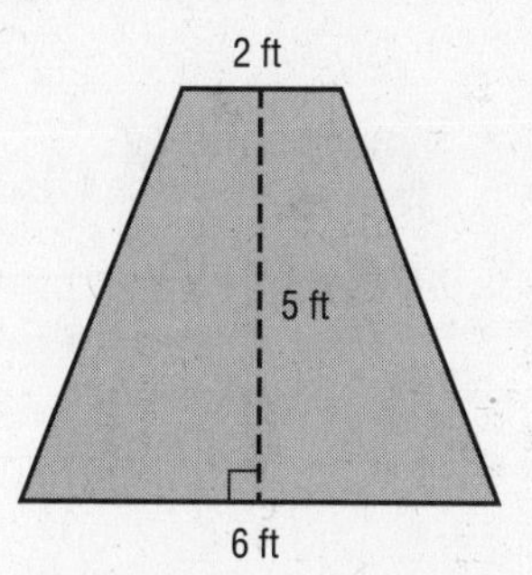

HOMEWORK ASSIGNMENT

Page(s):

Exercises:

10–6 Polygons

WHAT YOU'LL LEARN

- Classify polygons.
- Determine the sum of the measures of the interior and exterior angles of a polygon.

BUILD YOUR VOCABULARY (pages 238–239)

A **polygon** is a simple, closed figure formed by ______ or more ______.

A **diagonal** is a line segment in a polygon that ______ two nonconsecutive ______.

EXAMPLE Classify Polygons

1 Classify each polygon.

a.

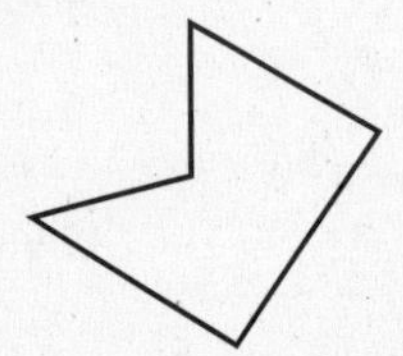

This polygon has ______ sides. It is a ______.

b.

This polygon has ______ sides. It is a ______.

Your Turn **Classify each polygon.**

a.

b.

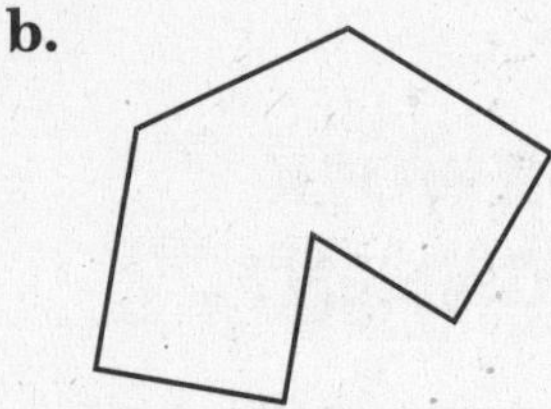

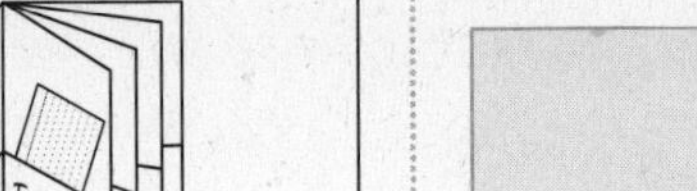

FOLDABLES ORGANIZE IT

On your index card for polygons, draw several polygons and label them with their name and number of sides.

KEY CONCEPT

Interior Angles of a Polygon If a polygon has n sides, then $n - 2$ triangles are formed. The sum of the degree measures of the interior angles of the polygon is $(n - 2)180$.

EXAMPLE Measures of Interior Angles

2 Find the sum of the measures of the interior angles of a quadrilateral.

A quadrilateral has ____ sides. Therefore, $n =$ ____.

$(n - 2)180 =$ ____ Replace n with ____.

$=$ ____ or ____ Simplify.

The sum of the measures of the interior angles of a quadrilateral is ____.

Your Turn Find the sum of the measures of the interior angles of a pentagon.

EXAMPLE Find Angle Measures of a Regular Polygon

3 TRAFFIC SIGNS **A stop sign is a regular octagon. What is the measure of one interior angle in a stop sign?**

Step 1 Find the sum of the measures of the angles.

An octagon has 8 sides. Therefore, $n =$ ____.

$(n - 2)180 =$ ____ Replace n with ____.

$=$ ____ or ____ Simplify.

The sum of the measures of the interior angles in ____.

Step 2 Divide the sum by 8 to find the measure of one angle.

____ $\div 8 =$ ____

So, the measure of one interior angle in a stop sign is ____.

Your Turn A picnic table in the park is a regular hexagon. What is the measure of one interior angle in the picnic table?

HOMEWORK ASSIGNMENT

Page(s):

Exercises:

10–7 Circumference and Area: Circles

BUILD YOUR VOCABULARY (pages 238–239)

The distance across the circle through its  is its **diameter**.

The distance from the  to any point on the circle is its **radius**.

The ______ of the **circumference** of a circle to the  of the circle is always equal to 3.1415926 . . . , represented by the Greek letter **π (pi).**

WHAT YOU'LL LEARN

- Find circumference of circles.
- Find area of circles.

KEY CONCEPT

Circumference of a Circle The circumference of a circle is equal to its diameter times π, or 2 times its radius times π.

EXAMPLE Find the Circumference of a Circle

1 Find the circumference of each circle to the nearest tenth.

a.

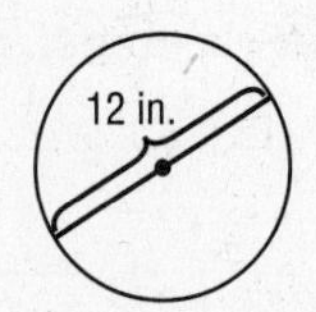

$C = \pi d$ Circumference of a circle

$C =$ 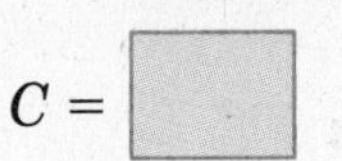 Replace d with ______.

$C =$ ______ Simplify. This is the *exact* circumference.

Using a calculator, you find that the circumference is about .

b.

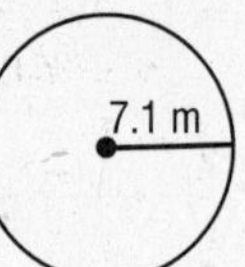

$C = 2\pi r$ Circumference of a circle

$C =$ 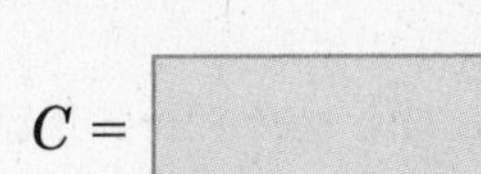 Replace r with 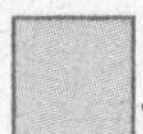.

$C \approx$ ______ Simplify. Use a calculator.

The circumference is about 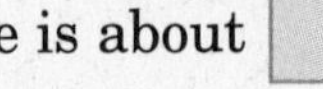.

Your Turn **Find the circumference of each circle to the nearest tenth.**

a. 4 ft

b. 1.6 cm

EXAMPLE Use Circumference to Solve a Problem

2 LANDSCAPING **A landscaper has a tree whose roots form a ball-shaped bulb with a circumference of about 110 inches. How wide will the landscaper have to dig the hole in order to plant the tree?**

Use the formula for the circumference of a circle to find the diameter.

$C = \pi d$	Circumference of a circle
____ $= \pi \cdot d$	Replace C with ____.
____ $= d$	Divide each side by ____.
____ $\approx$ ____	Simplify. Use a calculator.

The diameter of the hole should be at least ____.

Your Turn A circular swimming pool has a circumference of 24 feet. Matt must swim across the diameter of the pool. How far will Matt swim?

EXAMPLE Find Areas of Circles

KEY CONCEPT

Area of a Circle The area of a circle is equal to π times the square of its radius.

FOLDABLES

Add a diagram of a circle to your *Circles* index card. Label the center, diameter, radius and circumference. Then write the formulas for the circumference and area of a circle.

3 Find the area of each circle. Round to the nearest tenth.

a.

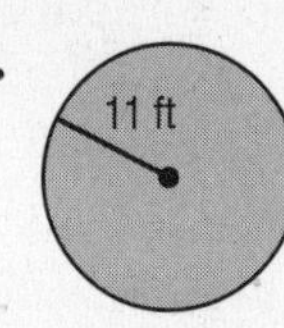

$A = \pi r^2$ Area of a circle

$A =$ ______ Replace r with ______.

$A =$ ______ Evaluate ______.

$A \approx$ ______ Use a calculator.

The area is about ______.

b.

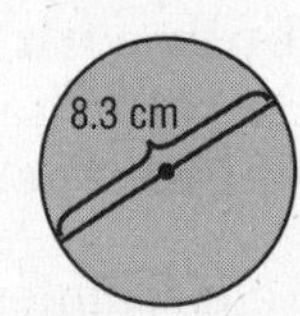

$A = \pi r^2$ Area of a circle

$A =$ ______ Replace r with ______.

$A =$  Evaluate ______.

$A \approx$  Use a calculator.

The area is about ______.

Your Turn **Find the area of each circle. Round to the nearest tenth.**

a.

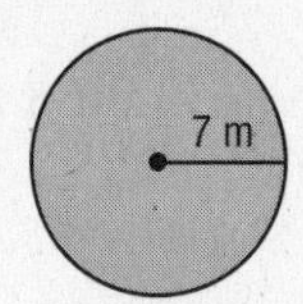

b.

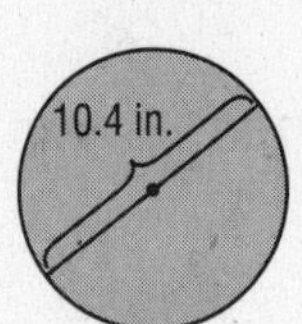

HOMEWORK ASSIGNMENT

Page(s):

Exercises:

10–8 Area: Irregular Figures

WHAT YOU'LL LEARN

- Find the area of irregular figures.

EXAMPLE Find Area of Irregular Figures

1 Find the area of the figure to the nearest tenth.

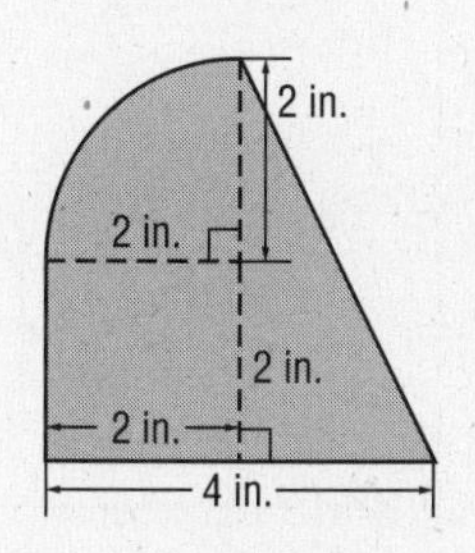

Separate the figure into a triangle, square, and a quarter-circle. Then find the sum of the areas of the figure.

Area of Square

$A = bh$ — Area of a square

$A =$ ____ or ____ — $b = h =$ ____

Area of Triangle

$A = \frac{1}{2}bh$ — Area of a triangle

$A =$ ____ or ____ — $b =$ ____, $h =$ ____

Area of Quarter-circle

$A = \frac{1}{4}\pi r^2$ — Area of a quarter-circle

$A =$ ____ — $r =$ ____

The area of the figure is ____ + ____ + ____ or about

square inches.

REVIEW IT

What is the difference between πr^2 and $(\pi r)^2$? *(Lesson 4-2)*

______

______

______

______

______

Your Turn Find the area of the figure to the nearest tenth.

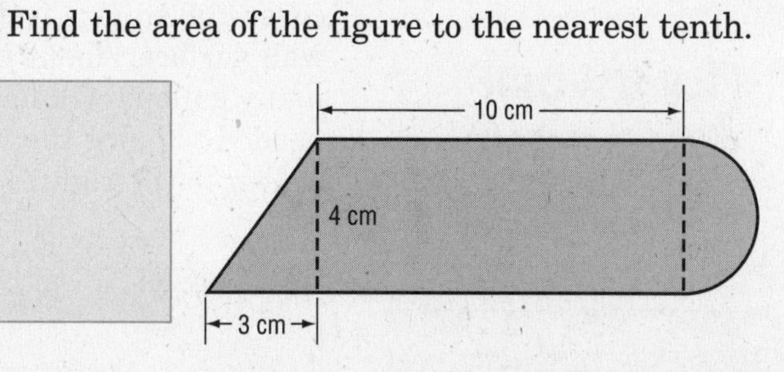

FOLDABLES™

ORGANIZE IT

On your *Irregular Figures* index card, describe how to find the area of an irregular figure.

EXAMPLE Use Area of Irregular Figures

2 CARPETING **Carpeting costs $2 per square foot. How much will it cost to carpet the area shown?**

14 ft
10 ft
3 ft
3 ft
12 ft

Step 1 Find the area to be carpeted.

Area of Rectangle

$A = bh$ — Area of a rectangle

$A =$ ☐ or ☐ — $b =$ ☐, $h =$ ☐

Area of Square

$A = bh$ — Area of a square

$A =$ ☐ or ☐ — $b = h =$ ☐

Area of Triangle

$A = \frac{1}{2}bh$ — Area of a triangle

$A =$ ☐ or ☐ — $b =$ ☐, $h =$ ☐

The area to be carpeted is ☐ + ☐ + ☐ or

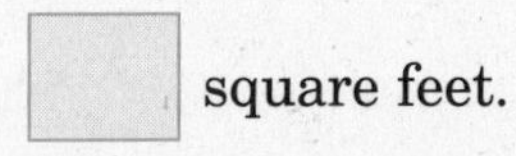

☐ square feet.

Step 2 Find the cost of the carpeting.

☐ × ☐ = ☐

So, it will cost ☐ to carpet the area.

Your Turn One gallon of paint is advertised to cover 100 square feet of wall surface. About how many gallons will be needed to paint the wall shown to the right?

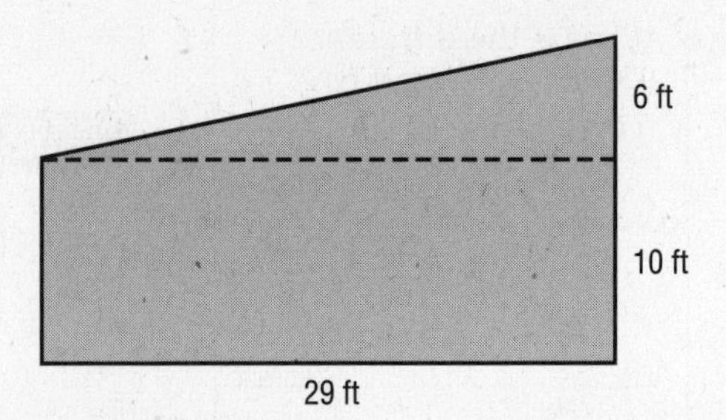

HOMEWORK ASSIGNMENT

Page(s):

Exercises:

CHAPTER 10

BRINGING IT ALL TOGETHER

STUDY GUIDE

FOLDABLES	VOCABULARY PUZZLEMAKER	BUILD YOUR VOCABULARY
Use your **Chapter 10 Foldable** to help you study for your chapter test.	To make a crossword puzzle, word search, or jumble puzzle of the vocabulary words in Chapter 10, go to: www.glencoe.com/sec/math/t_resources/free/index.php	You can use your completed **Vocabulary Builder** (pages 238–239) to help you solve the puzzle.

10-1 Line and Angle Relationships

Complete.

1. Two angles are ______ if the sum of their measures is 90°.

2. When two lines intersect, they form two pairs of opposite angles called ______.

In the figure at the right, $\ell \parallel m$ and p is a transversal. If $m\angle 5 = 96°$, find the measure of each angle.

3. $\angle 2$ ______ 4. $\angle 3$ ______ 5. $\angle 8$ 

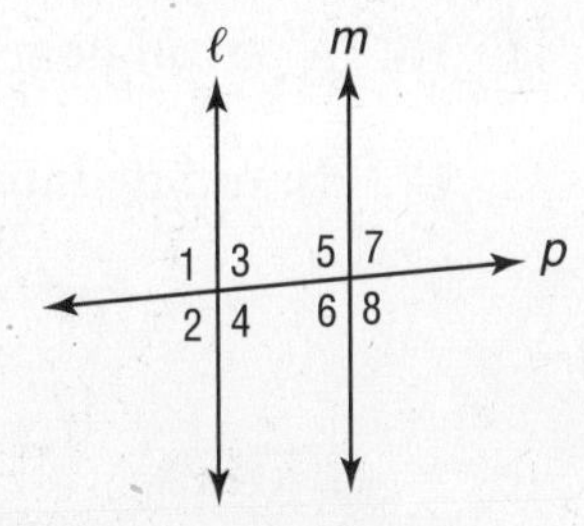

10-2 Congruent Triangles

In the figure shown, the triangles are congruent. Complete each congruence statement.

6. $\angle J \cong$ ______ 7. $\overline{JH} \cong$ ______

8. $\overline{HK} \cong$ ______ 9. $\angle K \cong$ ______

10. $\angle H \cong$ ______ 11. $\overline{KJ} \cong$ ______

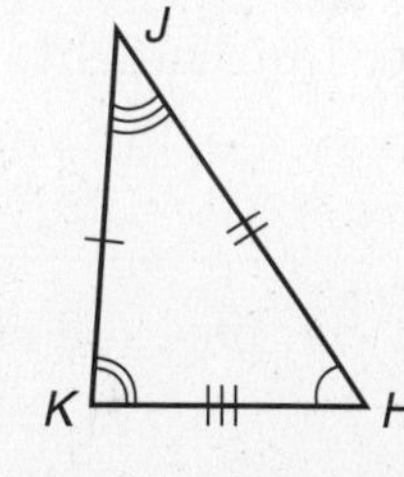

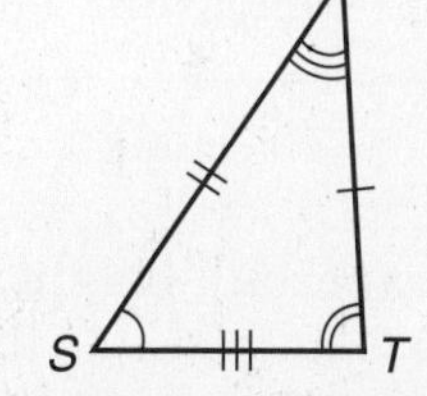

10-3 Transformations on the Coordinate Plane

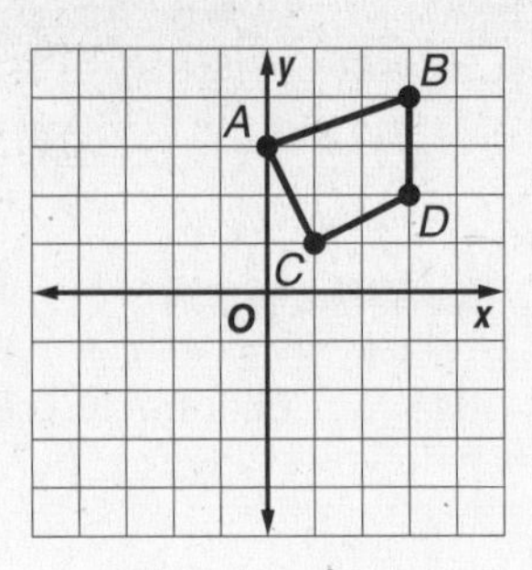

12. Suppose the figure graphed is reflected over the y-axis. Find the coordinates of the vertices after the reflection.

13. A figure has the vertices $P(4, -2)$, $Q(3, -4)$, $R(1, -4)$, $S(2, -1)$. Find the coordinates of the vertices of the figure after a rotation of 180°.

10-4 Quadrilaterals

For Exercises 14-16, match each description with a quadrilateral.

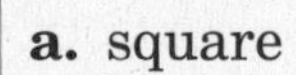

a. square
b. trapezoid
c. rectangle
d. rhombus

14. a parallelogram with four congruent sides and four right angles

15. one pair of opposite sides is parallel

16. a parallelogram with four congruent sides

17. In quadrilateral $EFGH$, $m\angle E = 90°$, $m\angle F = 120°$, and $m\angle G = 70°$. Find $m\angle H$.

10-5 Area: Parallelograms, Triangles, and Trapezoids

Find the area of each figure described.

18. triangle: base, 6 ft; height, 4 ft

19. parallelogram: base, 13m; height, 7m

20. trapezoid: height, 4 cm; bases, 3 cm and 9 cm

10-6 Polygons

Find the sum of the measures of the interior angles of each polygon.

21. decagon **22.** heptagon **23.** 15-gon 

Find the measure of an interior angle of each polygon.

24. regular octagon  **25.** regular nonagon 

10-7 Circumference and Area: Circles

Complete.

26. The distance around a circle is called the .

27. The  is the distance across a circle through its center.

Find the circumference and area of each circle. Round to the nearest tenth.

28.

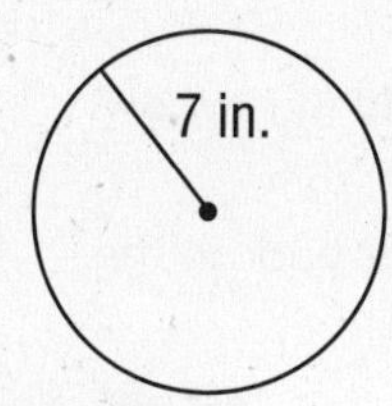

29.

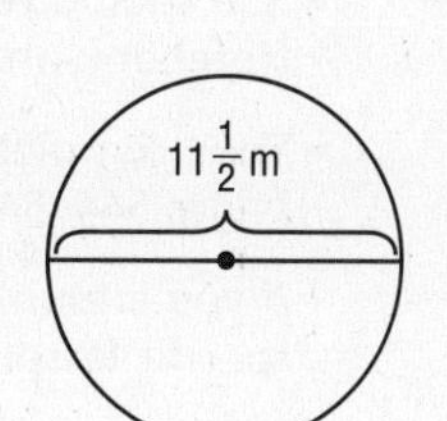

10-8 Area: Irregular Figures

Find the area of each figure. Round to the nearest tenth, if necessary.

30.

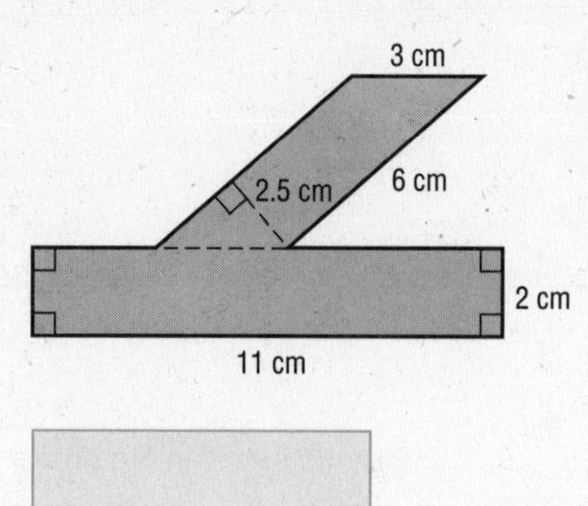

31.

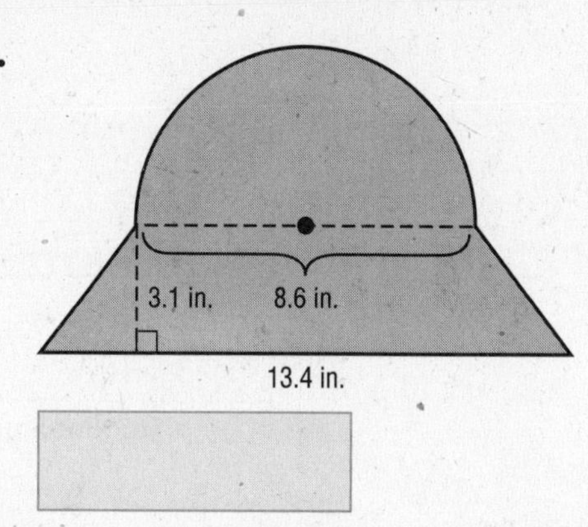

CHAPTER 10

Checklist

ARE YOU READY FOR THE CHAPTER TEST?

Visit **pre-alg.com** to access your textbook, more examples, self-check quizzes, and practice tests to help you study the concepts in Chapter 10.

Check the one that applies. Suggestions to help you study are given with each item.

☐ **I completed the review of all or most lessons without using my notes or asking for help.**

- You are probably ready for the Chapter Test.
- You may want take the Chapter 10 Practice Test on page 549 of your textbook as a final check.

☐ **I used my Foldable or Study Notebook to complete the review of all or most lessons.**

- You should complete the Chapter 10 Study Guide and Review on pages 544–548 of your textbook.
- If you are unsure of any concepts or skills, refer back to the specific lesson(s).
- You may also want to take the Chapter 10 Practice Test on page 549.

☐ **I asked for help from someone else to complete the review of all or most lessons.**

- You should review the examples and concepts in your Study Notebook and Chapter 10 Foldable.
- Then complete the Chapter 10 Study Guide and Review on pages 544–548 of your textbook.
- If you are unsure of any concepts or skills, refer back to the specific lesson(s).
- You may also want to take the Chapter 10 Practice Test on page 549.

Student Signature

Parent/Guardian Signature

Teacher Signature

Three-Dimensional Figures

Use the instructions below to make a Foldable to help you organize your notes as you study the chapter. You will see Foldable reminders in the margin of this Interactive Study Notebook to help you in taking notes.

Begin with a plain piece of 11" x 17" paper.

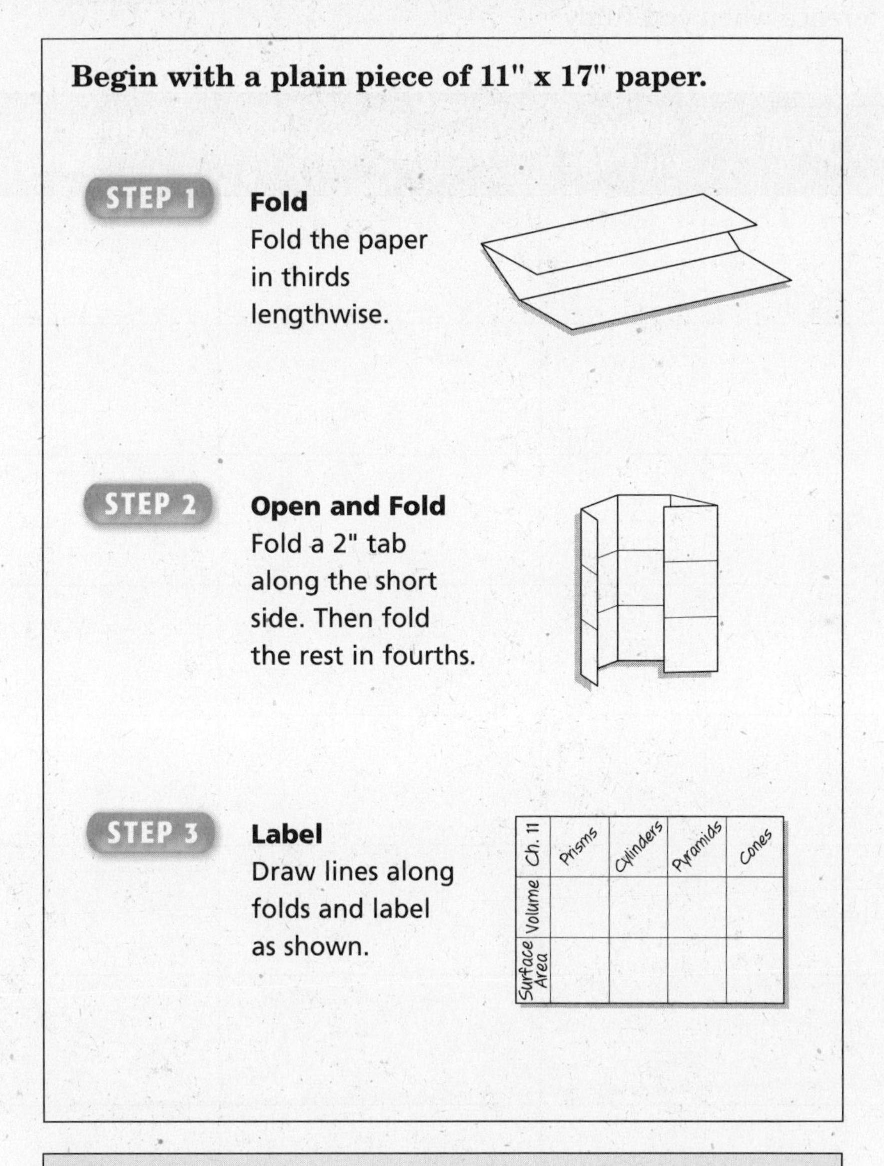

NOTE-TAKING TIP: When taking notes, use a table to make comparisons about the new material. Determine what will be compared, decide what standards will be used, and then use what is known to find similarities and differences.

BUILD YOUR VOCABULARY

This is an alphabetical list of new vocabulary terms you will learn in Chapter 11. As you complete the study notes for the chapter, you will see Build Your Vocabulary reminders to complete each term's definition or description on these pages. Remember to add the textbook page number in the second column for reference when you study.

Vocabulary Term	Found on Page	Definition	Description or Example
base			
cone			
cylinder [SIH-luhn-duhr]			
edge			
face			
lateral [LA-tuh-ruhl] area			
lateral face			
plane			
polyhedron [PAH-lee-HEE-druhn]			

Vocabulary Term	Found on Page	Definition	Description or Example
precision [prih-SIH-ZHUHN]			
prism			
pyramid			
significant digits [sihg-NIH-fih-kuhnt]			
similar solids			
skew lines [SKYOO]			
slant height			
solid			
surface area			
vertex			
volume			

11-1 Three-Dimensional Figures

WHAT YOU'LL LEARN

- Identify three-dimensional figures.
- Identify diagonals and skew lines.

KEY CONCEPT

Polyhedrons

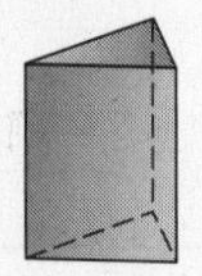

triangular prism

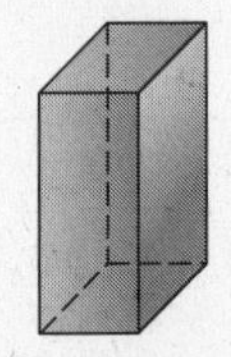

rectangular prism

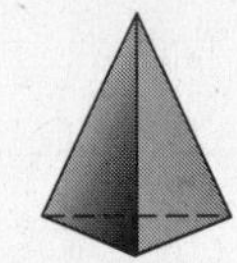

triangular pyramid

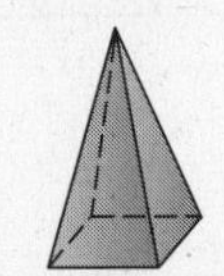

rectangular pyramid

BUILD YOUR VOCABULARY (pages 266–267)

A **plane** is a two-dimensional ______ surface that extends in all directions.

Intersecting planes can form ______ figures or **solids**. A **polyhedron** is a solid with flat surfaces that are ______.

In a polyhedron, an **edge** is where two planes intersect in a ______. A **face** is a ______ surface. A **vertex** is where ______ or more planes ______ in a point.

A **prism** is a polyhedron with two ______, congruent faces called **bases**.

A **pyramid** is a polyhedron with one base that is any polygon. Its other faces are ______.

EXAMPLE Identify Prisms and Pyramids

1 **Identify each solid. Name the bases, faces, edges, and vertices.**

a.

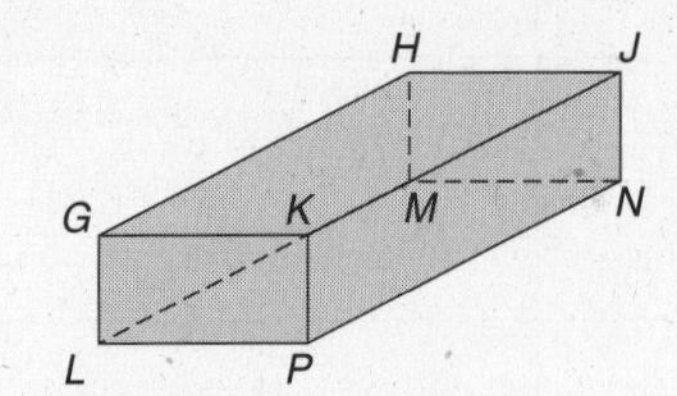

This figure has two parallel congruent bases that are ______, *GHJK* and *LMNP*, so it is a ______.

FOLDABLES™

ORGANIZE IT

In your table, write the type and number of bases of the different prisms and pyramids introduced in this lesson.

Ch. 11	Prisms	Cylinders	Pyramids	Cones
Volume				
Surface Area				

faces:

edges:

vertices:

b.

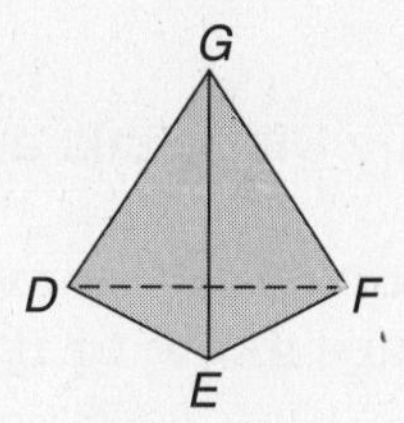

This figure has one ______ base, *DEF*, so it is a ______.

faces:

edges:

vertices:

Your Turn **Identify the solid. Name the bases, faces, edges, and vertices.**

a.

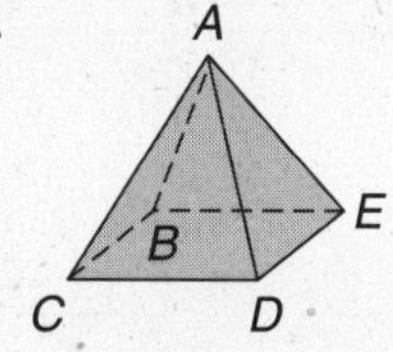

b.

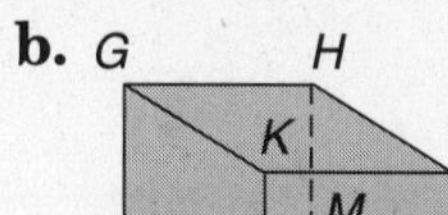
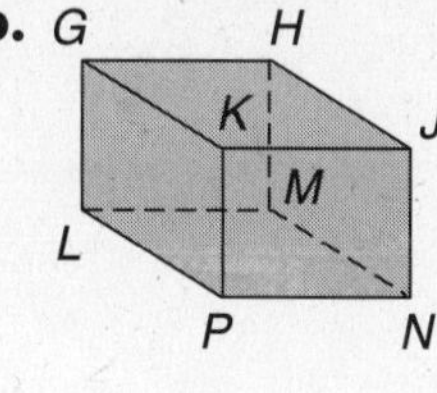

REMEMBER IT

In a rectangular prism, any two parallel rectangles are bases, and any face is a base in a triangular pyramid. Bases do not have to be on the bottom of a figure.

BUILD YOUR VOCABULARY (page 267)

Skew lines are lines that are ________ intersecting nor ________. They lie in ________ planes.

EXAMPLE Identify Diagonals and Skew Lines

2 Identify a diagonal and name all segments that are skew to it.

$\overline{QW}$ is a diagonal because vertex ____ and vertex ____ do not intersect any of the same faces.

The segments skew to $\overline{QW}$ are ________________.

Your Turn Identify a diagonal and name all segments that are skew to it.

____

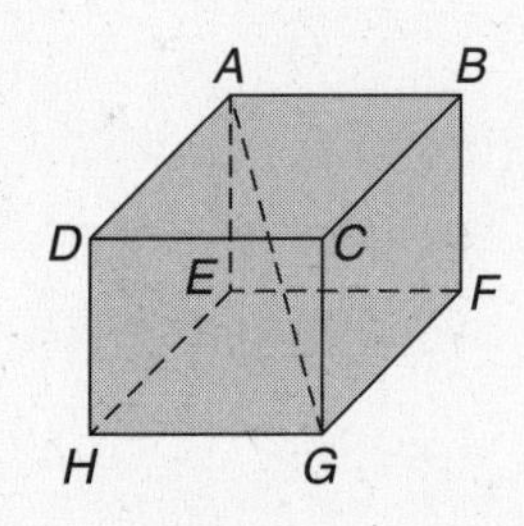

HOMEWORK ASSIGNMENT

Page(s):

Exercises:

11–2 Volume: Prisms and Cylinders

What You'll Learn

- Find volumes of prisms.
- Find volumes of circular cylinders.

BUILD YOUR VOCABULARY (page 267)

Volume is the ________ of ________ occupied by a solid region.

Key Concept

Volume of a Prism
The volume *V* of a prism is the area of the base *B* times the height *h*.

EXAMPLE Volume of a Rectangular Prism

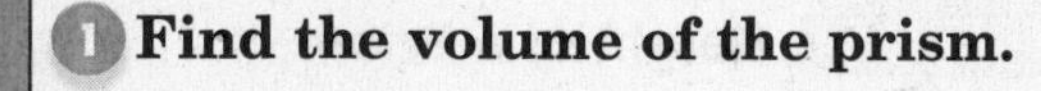

1 Find the volume of the prism.

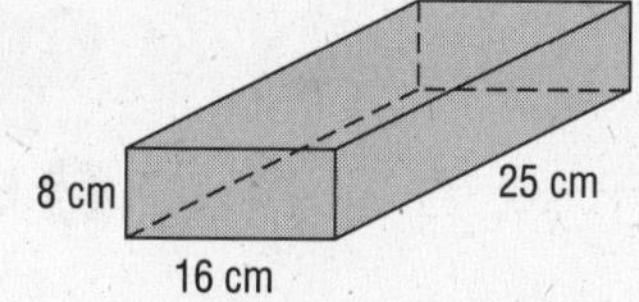

$V = Bh$ Formula for volume of a prism

$V =$ ________ The base is a ________, so $B =$ ________.

$V =$ ________ ____ $= 25$, ____ $= 16$, ____ $= 8$

$V =$ ________ Simplify.

The volume is ________ cubic centimeters.

EXAMPLE Volume of a Triangular Prism

2 Find the volume of the triangular prism.

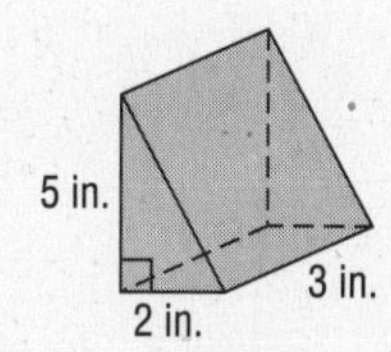

$V = Bh$ Formula for volume of a prism

$V =$ ________ $B =$ area of base or ________.

$V =$ ________ The ________ of the prism is ________.

$V =$ ________ Simplify.

The volume is ________ cubic inches.

Your Turn **Find the volume of each prism.**

a.

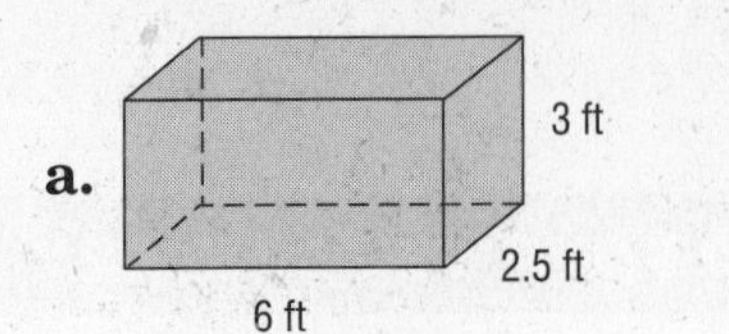

b.

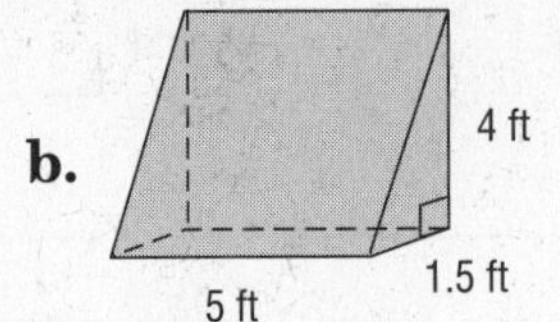

EXAMPLE **Height of a Prism**

3 **BAKING** **Cake batter is poured into a pan that is a rectangular prism whose base is an 8-inch square. If the cake batter occupies 192 cubic inches, what will be the height of the batter?**

$V = Bh$ — Formula for volume of a prism

$V = \ell \cdot w \cdot h$ — Formula for volume of a rectangular prism

☐ = ☐ — Replace V with ☐,

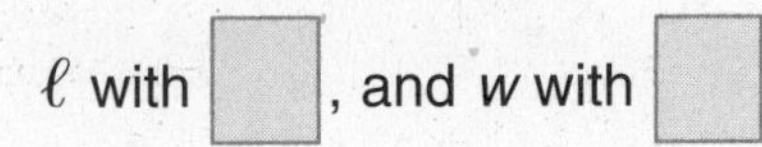

ℓ with ☐, and w with ☐.

☐ = ☐ — Simplify.

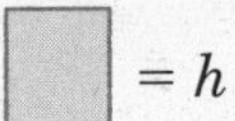

☐ $= h$ — Divide each side by ☐.

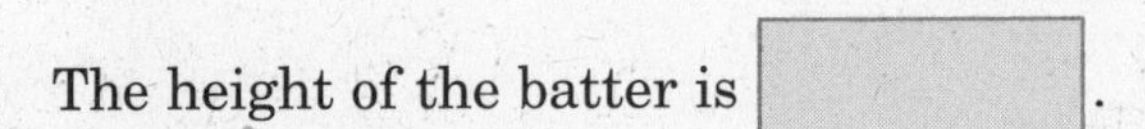

The height of the batter is ☐.

Your Turn A swimming pool is filled with 960 cubic feet of water. The pool is a rectangular prism 20 feet long and 12 feet wide and is the same depth throughout. Find the depth of the water.

BUILD YOUR VOCABULARY (page 266)

A **cylinder** is a ______ whose bases are congruent, parallel ______, connected with a ______ side.

KEY CONCEPT

Volume of a Cylinder The volume V of a cylinder with radius r is the area of the base B times the height h.

FOLDABLES Write the formulas for the volume of a prism and the volume of a cylinder in your table.

EXAMPLE Volume of a Cylinder

4 Find the volume of the cylinder. Round to the nearest tenth.

7 ft

14 ft

$V =$ ______ Formula for volume of a ______

$V =$ ______ Replace r with ______ and h with ______.

$V \approx$ ______ Simplify.

The volume is about ______ cubic feet.

Your Turn Find the volume of the cylinder. Round to the nearest tenth.

4 in.

7 in.

______

HOMEWORK ASSIGNMENT

Page(s):

Exercises:

11–3 Volume: Pyramids and Cones

What You'll Learn

- Find volumes of pyramids.
- Find volumes of cones.

Key Concepts

Volume of a Pyramid
The volume *V* of a pyramid is one-third the area of the base *B* times the height *h*.

Volume of a Cone
The volume *V* of a cone with radius *r* is one-third the area of the base *B* times the height *h*.

FOLDABLES Write these formulas in your table.

EXAMPLE Volumes of Pyramids

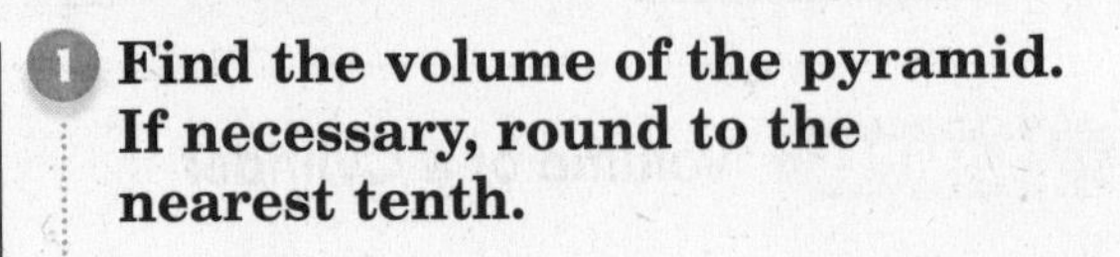

1 Find the volume of the pyramid. If necessary, round to the nearest tenth.

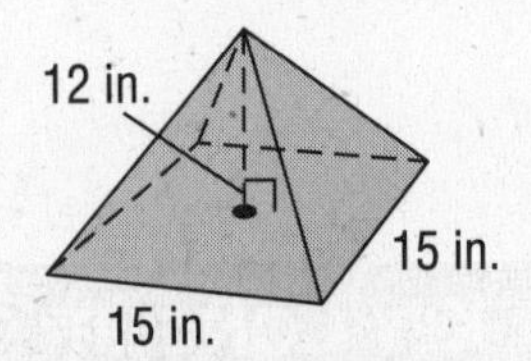

$V = \frac{1}{3}Bh$ Formula for volume of a pyramid

$V =$ ______ Replace *B* with ______ · ______.

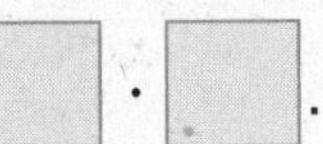

$V =$ ______ The height is ______ inches.

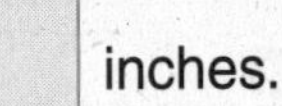

$V =$ ______ Simplify.

The volume is ______ cubic inches.

Build Your Vocabulary (page 266)

A **cone** is a three-dimensional figure with one ______ base. A curved surface connects the base and the vertex.

EXAMPLE Volume of a Cone

2 Find the volume of the cone. Round to the nearest tenth.

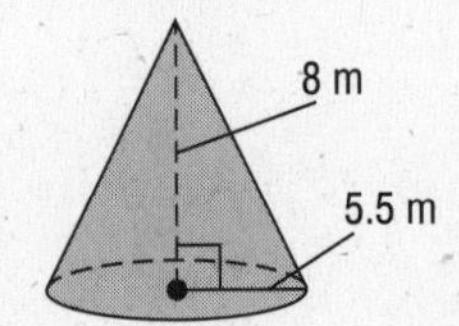

$V = \frac{1}{3}\pi r^2 h$ Formula for volume of a cone

$V =$ ______ $r =$ ______ and $h =$ ______

$V \approx$ ______ Simplify.

The volume is about ______ cubic meters.

Your Turn **Find the volume of each solid. Round to the nearest tenth.**

a.

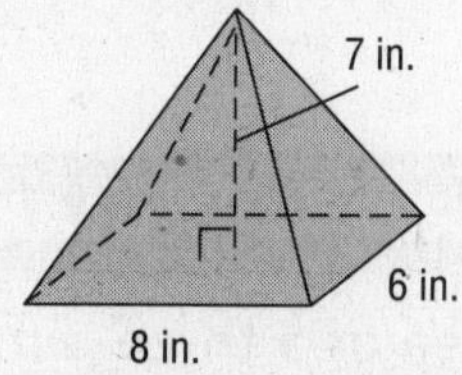

b. 

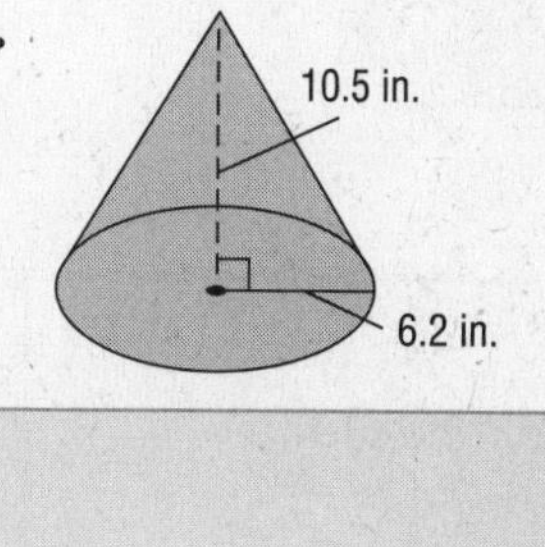

EXAMPLE Use Volume to Solve Problems

3 LANDSCAPING **When mulch was dumped from a truck, it formed a cone-shaped mound with a diameter of 15 feet and a height of 8 feet.**

a. What is the volume of the mulch?

$V = \frac{1}{3}\pi r^2 h$ Formula for volume of a cone

$V =$ ______ $r =$ ___ , $h =$ ___

$V \approx$ ___ cubic feet

b. How many square feet can be covered with this mulch if 1 cubic foot covers 4 square feet of ground?

ft² of ground = ______ mulch × $\frac{\text{\_\_\_ of ground}}{\text{\_\_\_ mulch}}$

= ______ of ground

Your Turn **A load of wood chips for a playground was dumped and formed a cone-shaped mound with a diameter of 10 feet and a height of 6 feet.**

a. What is the volume of the wood chips?

b. How many square feet of the playground can be covered with wood chips if 1 cubic foot of wood chips can cover 3 square feet of the playground?

HOMEWORK ASSIGNMENT

Page(s):

Exercises:

11-4 Surface Area: Prisms and Cylinders

WHAT YOU'LL LEARN

- Find the surface areas of prisms.
- Find surface areas of cylinders.

BUILD YOUR VOCABULARY (page 267)

The **surface area** of a three-dimensional figure is the ______ of the ______ of all of the ______ of the figure.

EXAMPLE Surface Area of a Rectangular Prism

1 **Find the surface area of the rectangular prism.**

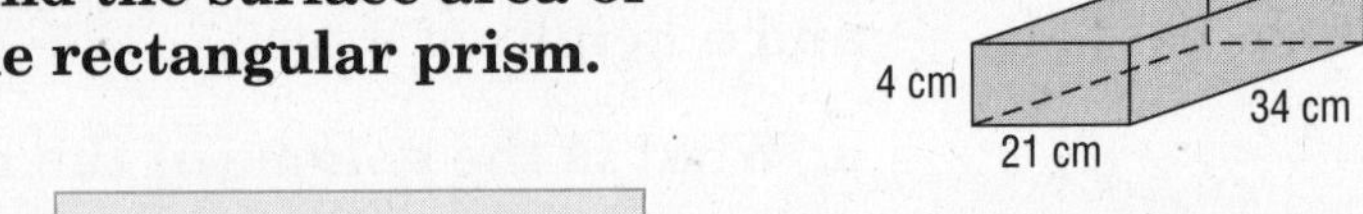

$S =$ ______ Write the formula.

$S =$ ______ + ______ + ______ Substitution

$S =$ ______ square centimeters Simplify.

The surface area is ______ square centimeters.

KEY CONCEPT

Surface Area of Rectangular Prisms
The surface area S of a rectangular prism with length l, width w, and height h is the sum of the areas of the faces.

EXAMPLE Surface Area of a Triangular Prism

2 **Find the surface area of the triangular prism.**

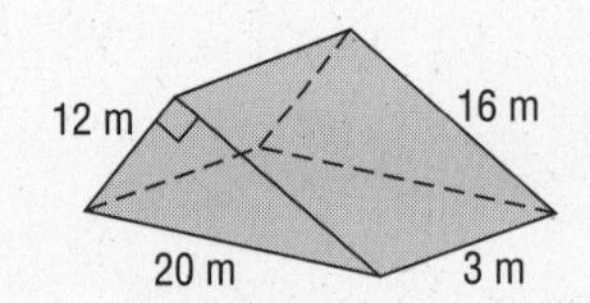

Find the area of each face.

Bottom: ______ = ______ Right Side: ______ = ______

Left Side: ______ = ______

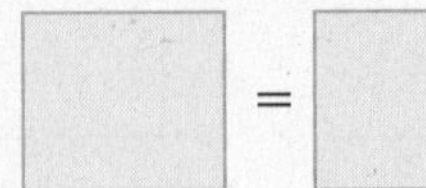

Two Bases: ______ = ______

Add to find the total surface area.

______ + ______ + ______ + ______ = ______

Your Turn **Find the surface area of each triangular prism.**

a.

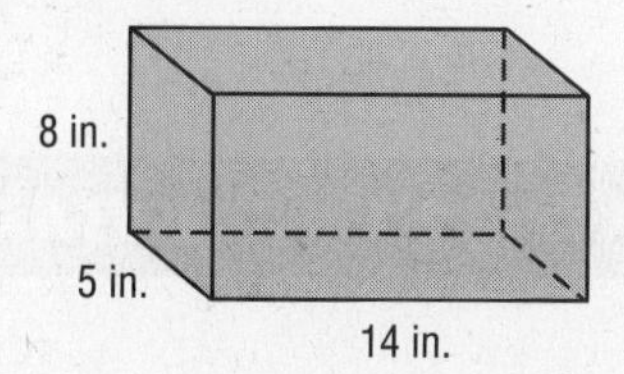

b.

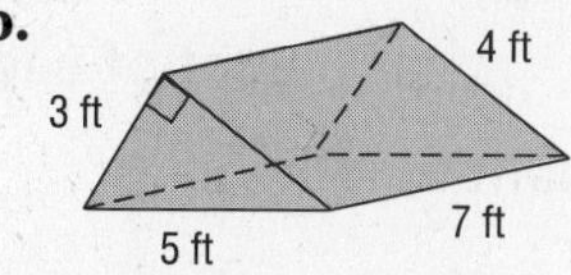

EXAMPLE Surface Area of a Cylinder

KEY CONCEPT

Surface Area of Cylinders The surface area S of a cylinder with height h and radius r is the area of the two bases plus the area of the curved surface.

FOLDABLES Write the formulas for the surface area of a prism and the surface area of a cylinder in your table.

3 Find the surface area of the cylinder. Round to the nearest tenth.

2.5 m

8 m

$S =$  Formula for surface area of a cylinder

$S =$  Replace r with 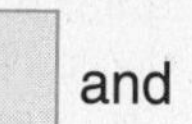 and h with ____.

$S \approx$ ____ Simplify.

The surface area is about ____ square meters.

Your Turn Find the surface area of the cylinder. Round to the nearest tenth.

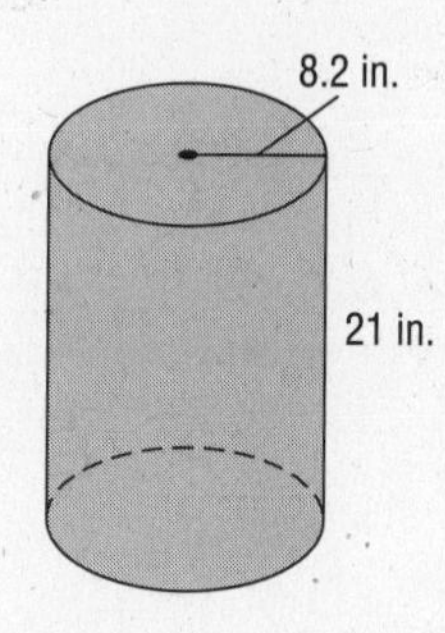

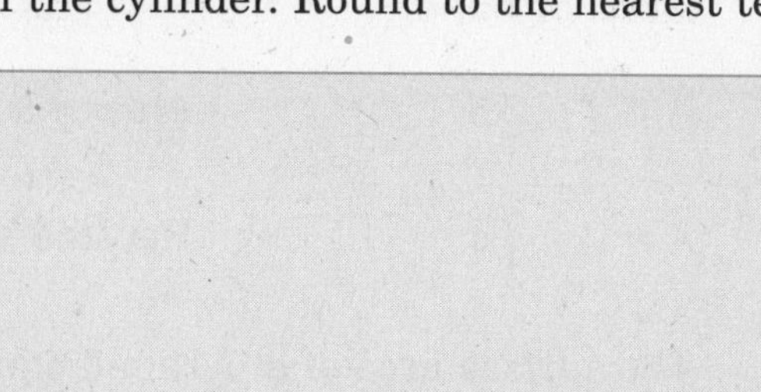

HOMEWORK ASSIGNMENT

Page(s):

Exercises:

11-5 Surface Area: Pyramids and Cones

What You'll Learn

- Find the surface areas of pyramids.
- Find surface areas of cones.

BUILD YOUR VOCABULARY (pages 266–267)

The ______ of a pyramid are called **lateral faces.**

The ______ or height of each ______ is called the **slant height.**

The ______ of the ______ of the lateral faces is the **lateral area** of a pyramid.

EXAMPLE Surface Area of a Pyramid

1 Find the surface area of the square pyramid.

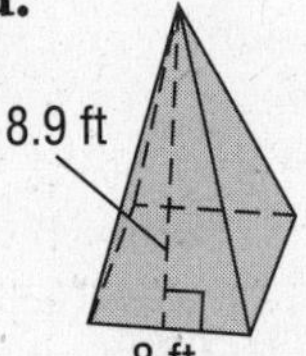

Find the lateral area and the base area.

Area of each lateral face

$A = \frac{1}{2}bh$ — Area of a triangle

$A =$ ______ — Replace b with ______ and h with ______.

$A =$ ______ — Simplify.

There are 4 faces, so the lateral area is 4 ______ or

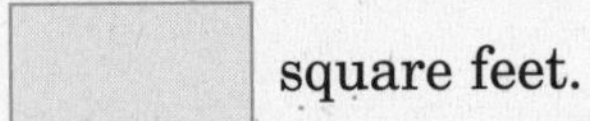

______ square feet.

Area of base

$A = s^2$ — Area of a square

$A =$ ______ or ______ — Replace s with ______ and simplify.

The surface area of a pyramid equals the lateral area plus the area of the base. So, the surface area of the square pyramid is ______ + ______ or ______ square feet.

WRITE IT

If the base of a pyramid is a regular polygon, what do you know about its lateral faces?

______

______

______

______

Your Turn Find the surface area of the square pyramid.

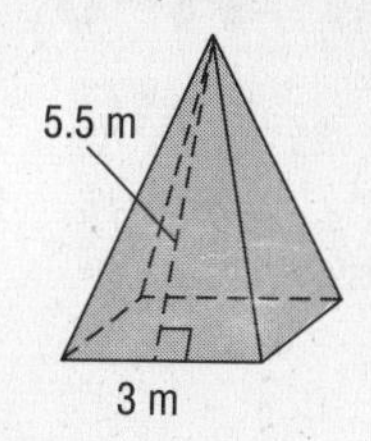

EXAMPLE Surface Area of a Cone

2 Find the surface area of the cone. Round to the nearest tenth.

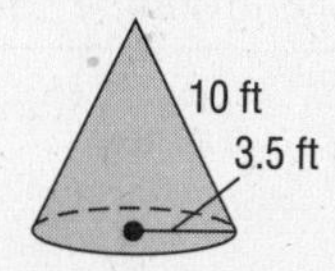

KEY CONCEPT

Surface Area of a Cone
The surface area S of a cone with slant height l and radius r is the lateral area plus the area of the base.

FOLDABLES Write the formulas for the surface area of a pyramid and the surface area of a cone in your table.

$S = \pi r \ell + \pi r^2$ Formula for surface area of a cone

$S =$ ______ + ______ Replace ______ with ______ and ℓ with ______.

$S \approx$ ______ square feet Simplify.

The surface area is about ______ square feet.

Your Turn Find the surface area of the cone. Round to the nearest tenth.

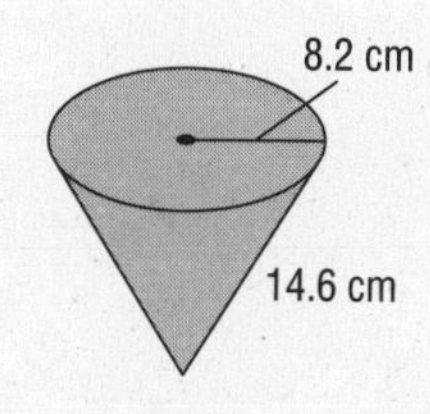

HOMEWORK ASSIGNMENT

Page(s):

Exercises:

11–6 Similar Solids

BUILD YOUR VOCABULARY (page 267)

Two solids are **similar solids** if they have the same ______ and their corresponding ______ measures are ______.

WHAT YOU'LL LEARN

- Identify similar solids.
- Solve problems involving similar solids.

EXAMPLE Identify Similar Solids

1 Determine whether the pair of solids is similar.

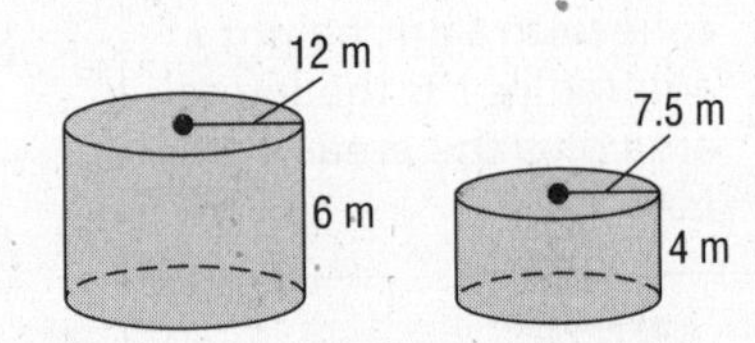

______ $\stackrel{?}{=}$ ______ Write a proportion comparing radii and heights.

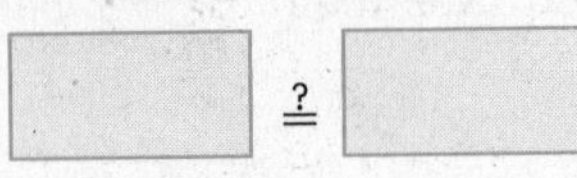

______ $\stackrel{?}{=}$ ______ Find the cross products.

Simplify.

The radii and heights are  proportional, so the cylinders are ______ similar.

Your Turn Determine whether the pair of solids is similar.

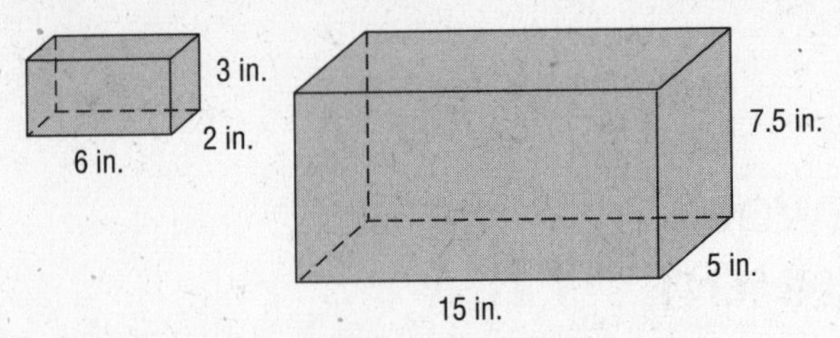

EXAMPLE Find Missing Measures

2 The cylinders shown are similar. Find the radius of cylinder A.

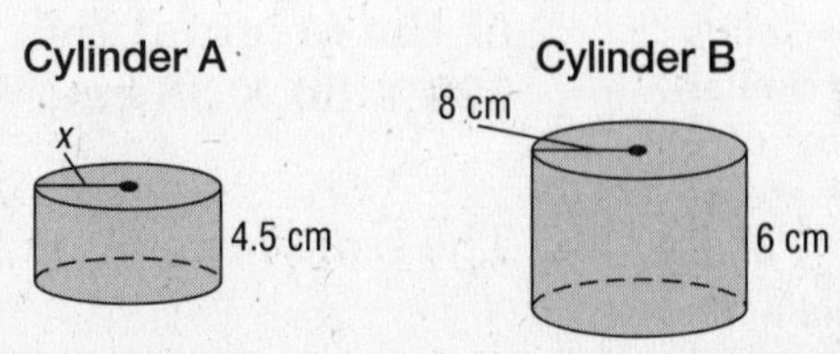

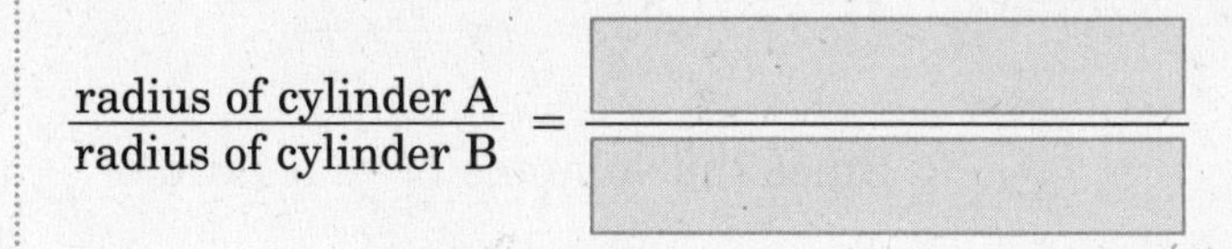
$\frac{\text{radius of cylinder A}}{\text{radius of cylinder B}} = \frac{\square}{\square}$

$\square = \square$ Substitute the known values.

$\square x = 8(\square)$ Find the cross products.

$\square = \square$ Simplify.

$x = \square$ Divide each side by $\square$.

The radius of cylinder A is $\square$ centimeters.

Your Turn The rectangular prisms below are similar. Find the height of prism B.

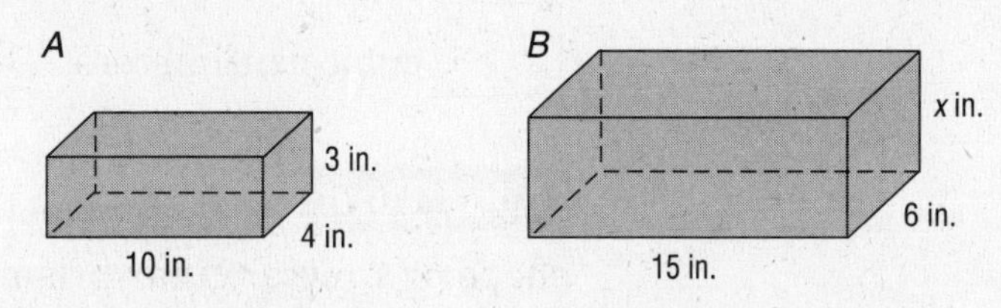

EXAMPLE Use Similar Solids to Solve a Problem

KEY CONCEPT

Ratios of Similar Solids
If two solids are similar with a scale factor of $\frac{a}{b}$, then the surface areas have a ratio of $\frac{a^2}{b^2}$ and the volumes have a ratio of $\frac{a^3}{b^3}$.

3 DOLL HOUSES **Lita made a model of her fish tank for her doll house. The model is exactly $\frac{1}{25}$ the size of the original fish tank, whose dimensions are 120 × 30 × 38 cm. What is the volume of the model?**

The scale factor $\frac{a}{b}$ is ______ and the volume of the fish tank is

$120 \cdot 30 \cdot 38$ or ______ cm^3.

Since the volumes have a ratio of $\frac{a^3}{b^3}$ and $\frac{a}{b}$ = ______,

replace a with ______ and b with ______ in $\frac{a^3}{b^3}$.

$\frac{\text{volume of model}}{\text{volume of tank}} = \frac{a^3}{b^3}$ Write the ratio of volumes.

= ______ / ______ Replace a with ______ and b with ______.

= ______ / ______ Simplify.

So, the volume of the tank is ______ times the volume of the model.

The volume of the model is 136,800 ÷ ______ or about ______ cubic centimeters.

Your Turn A scale model of a railroad boxcar is in the shape of a rectangular prism and is $\frac{1}{50}$ the size of the actual boxcar. The scale model has a volume of 72 cubic inches. What is the volume of the actual boxcar?

HOMEWORK ASSIGNMENT

Page(s):
Exercises:

11-7 Precision and Significant Digits

WHAT YOU'LL LEARN

- Describe measurements using precision and significant digits.
- Apply precision and significant digits in problem-solving situations.

BUILD YOUR VOCABULARY (page 267)

The **precision** of a ______ is the ______ to which a measurement is made.

Significant digits are the digits recorded from ______, indicating the ______ of the measurement.

EXAMPLE Identify Precision Units

1 **Identify the precision unit of the thermometer.**

The precision unit is ______.

Your Turn Identify the precision unit of the ruler.

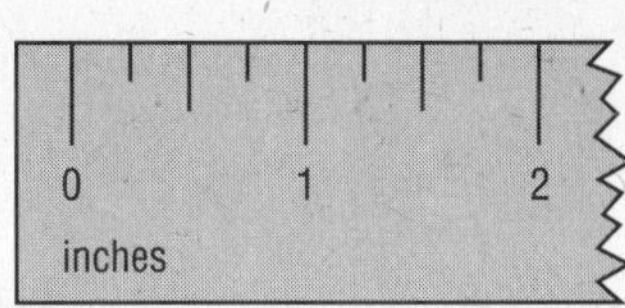

______

EXAMPLE Identify Significant Digits

2 **Determine the number of significant digits in each measure.**

a. 1040 miles

______ significant digits

b. 0.003 centimeter

______ significant digit

Your Turn **Determine the number of significant digits in each measure.**

a. 34.70 inches ______

b. 0.000003 mile ______

WRITE IT

Explain why the numbers 20, 20.0, and 20.00 have different numbers of significant digits.

______

______

______

______

EXAMPLE Add Measurements

3 The sides of a quadrilateral measure 0.6 meter, 0.044 meter, 0.024 meter, and 0.103 meter. Use the correct number of significant digits to find the perimeter.

0.6 ⟵ ____ decimal place
0.044 ⟵ ____ decimal places
0.024 ⟵ ____ decimal places
+ 0.103 ⟵ ____ decimal places
____

The least precise measurement, 0.6, has ____ decimal place.

So, round ____ to ____ decimal place. The perimeter of the quadrilateral is about ____ meter.

EXAMPLE Multiply Measurements

4 What is the area of the bedroom shown here?

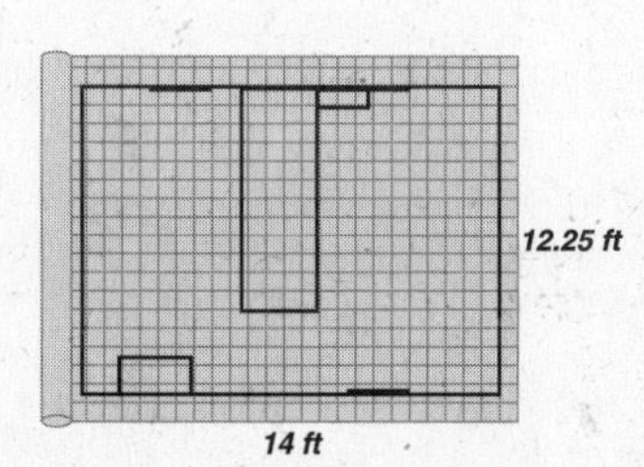

To find the area, multiply the length and the width.

12.25 ⟵ ____ significant digits
× 14 ⟵ ____ significant digits
____ ⟵ ____ significant digits

The answer cannot have more significant digits than the measurements of the length and width. So, round ____ square feet to ____ significant digits. The area of the bedroom is about ____ square feet.

Your Turn

a. The sides of a triangle measure 2.04 centimeters, 3.2 centimeters, and 2.625 centimeters. Use the correct number of significant digits to find the perimeter.

____

b. Suppose a bedroom was 13.75 feet wide and 12.5 feet long. What would be the area of the bedroom?

____

Page(s):
Exercises:

CHAPTER 11

BRINGING IT ALL TOGETHER

STUDY GUIDE

FOLDABLES	VOCABULARY PUZZLEMAKER	BUILD YOUR VOCABULARY
Use your **Chapter 11 Foldable** to help you study for your chapter test.	To make a crossword puzzle, word search, or jumble puzzle of the vocabulary words in Chapter 11, go to: www.glencoe.com/sec/math/t_resources/free/index.php	You can use your completed **Vocabulary Builder** (pages 266–267) to help you solve the puzzle.

11-1 Three-Dimensional Figures

1. Identify the solid. Name the faces, edges, and vertices.

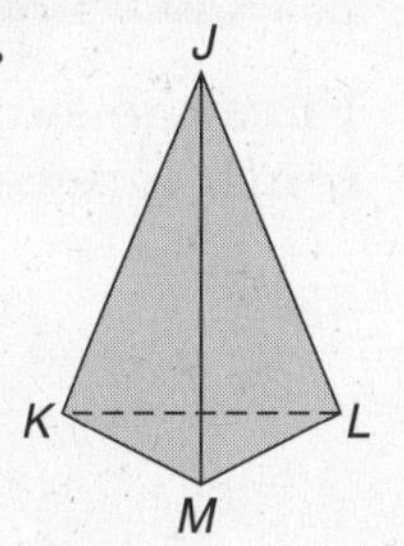

State whether each sentence is true or false. If false, replace the underlined word to make a true sentence.

2. A pyramid is a solid with two bases.

3. Skew lines are neither intersecting nor vertical.

11-2 Volume: Prisms and Cylinders

Find the volume of each prism or cylinder. If necessary, round to the nearest tenth.

4.

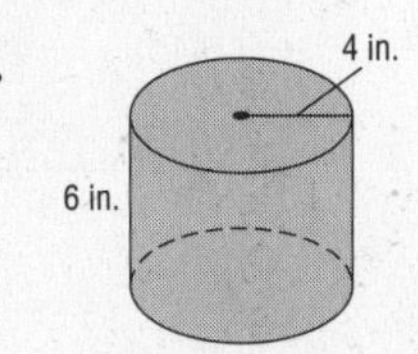

5.

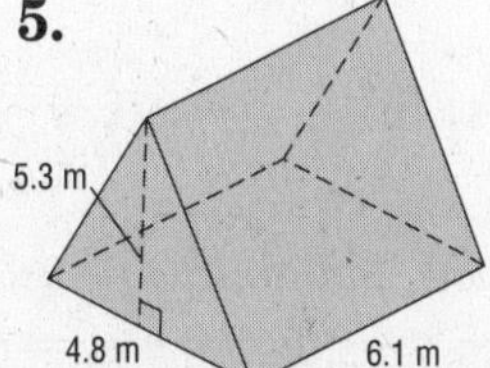

11-3 Volume: Pyramids and Cones

Find the volume of each solid. If necessary, round to the nearest tenth.

6. cone: diameter 14 ft, height 11 ft

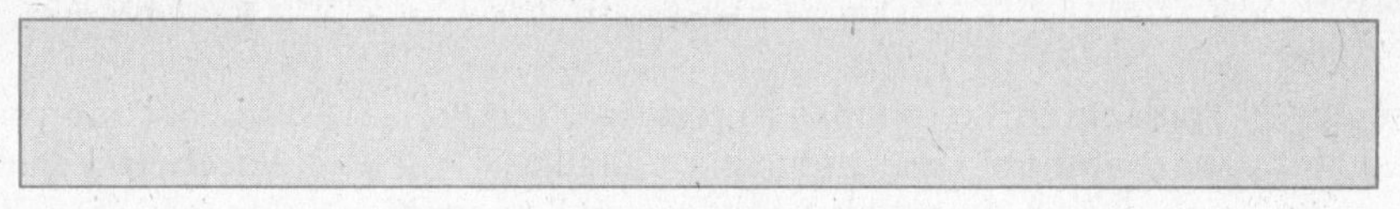

7. square pyramid: length 4.5 m, height 6.8 m

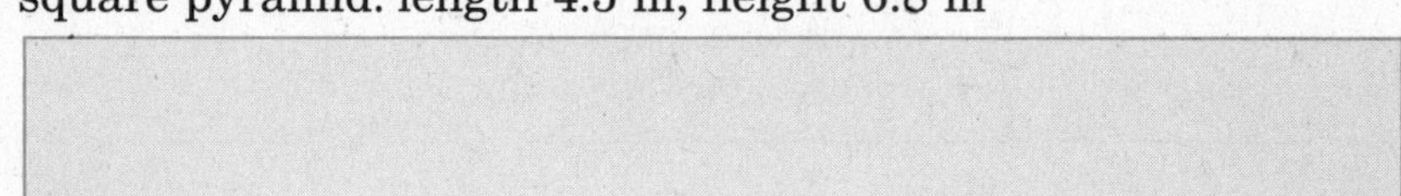

11-4 Surface Area: Prisms and Cylinders

Find the surface area of each solid. Round to the nearest tenth if necessary.

8.

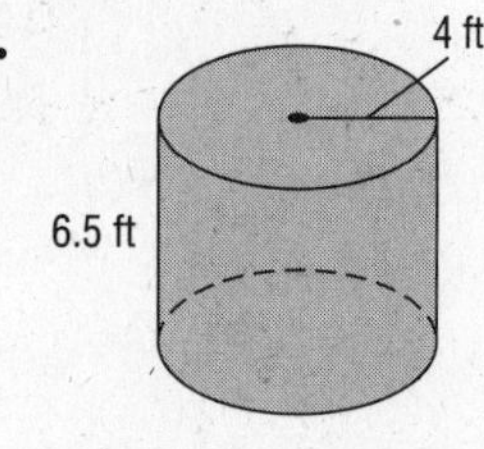

9.

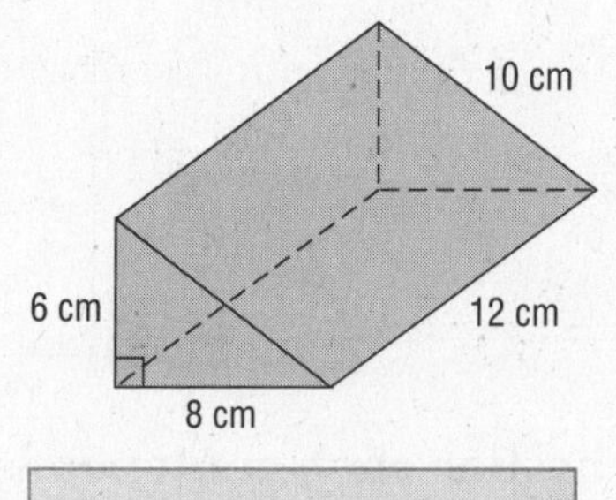

11-5 Surface Area: Pyramids and Cones

Find the surface area of each solid. Round to the nearest tenth if necessary.

10.

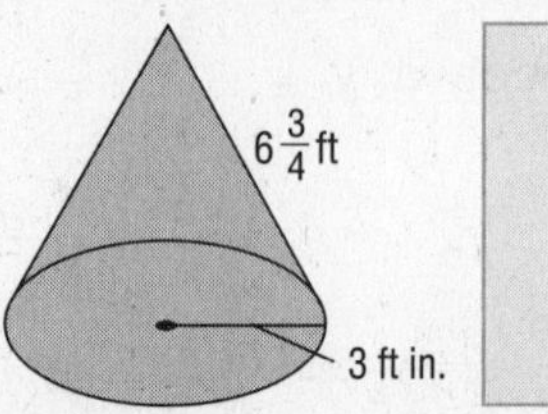

11.

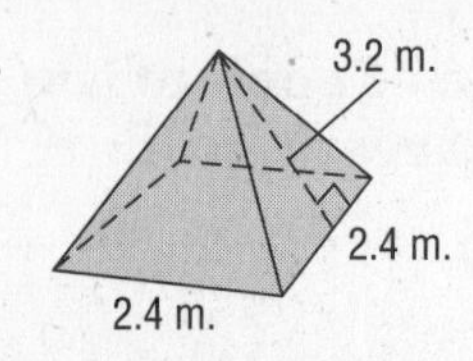

12. square pyramid: base side lengths 5 in., slant height 8 in.

11-6

Similar Solids

Determine whether each pair of solids is similar.

13.

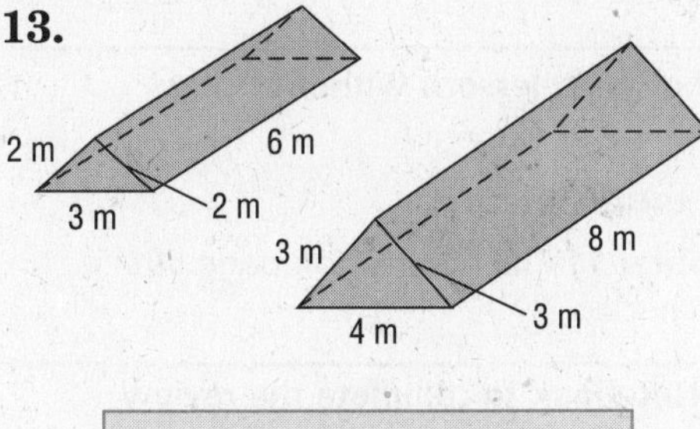

14.

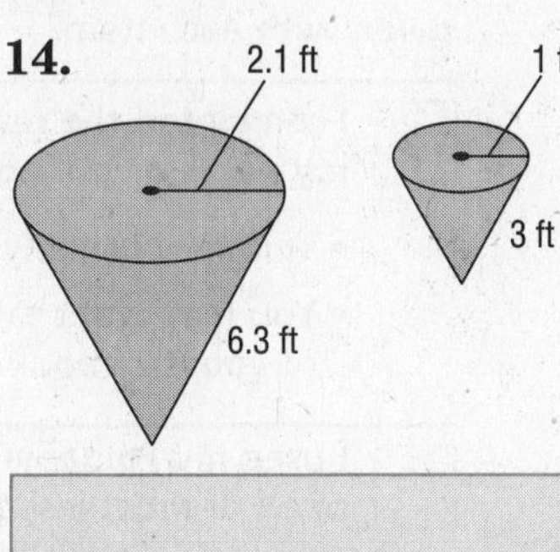

Find the missing measure for the pair of similar solids.

15.

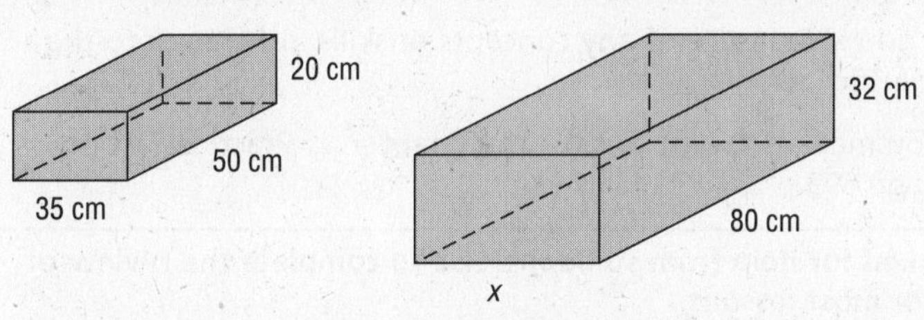

11-7

Precision and Significant Digits

State whether each sentence is true or false. If false, replace the underlined word to make a true sentence.

16. When adding or subtracting measurements, the sum or differences should have the same precision as the <u>least</u> precise measurement.

17. When multiplying or dividing measurements, the product or quotient should have the same number of significant digits as the measurement with the <u>greatest</u> number of significant digits.

Determine the number of significant digits in each measure.

18. 0.061 m　　**19.** 150.1 in.　　**20.** 9000 m

Calculate. Round to the correct number of significant digits.

21. 5.30 km + 0.425 km　　**22.** 31.2 m × 0.015 m

ARE YOU READY FOR THE CHAPTER TEST?

Visit **pre-alg.com** to access your textbook, more examples, self-check quizzes, and practice tests to help you study the concepts in Chapter 11.

Check the one that applies. Suggestions to help you study are given with each item.

☐ **I completed the review of all or most lessons without using my notes or asking for help.**

- You are probably ready for the Chapter Test.
- You may want take the Chapter 11 Practice Test on page 599 of your textbook as a final check.

☐ **I used my Foldable or Study Notebook to complete the review of all or most lessons.**

- You should complete the Chapter 11 Study Guide and Review on pages 595–598 of your textbook.
- If you are unsure of any concepts or skills, refer back to the specific lesson(s).
- You may also want to take the Chapter 11 Practice Test on page 599.

☐ **I asked for help from someone else to complete the review of all or most lessons.**

- You should review the examples and concepts in your Study Notebook and Chapter 11 Foldable.
- Then complete the Chapter 11 Study Guide and Review on pages 595–598 of your textbook.
- If you are unsure of any concepts or skills, refer back to the specific lesson(s).
- You may also want to take the Chapter 11 Practice Test on page 599.

Student Signature

Parent/Guardian Signature

Teacher Signature

More Statistics and Probability

Use the instructions below to make a Foldable to help you organize your notes as you study the chapter. You will see Foldable reminders in the margin of this Interactive Study notebook to help you in taking notes.

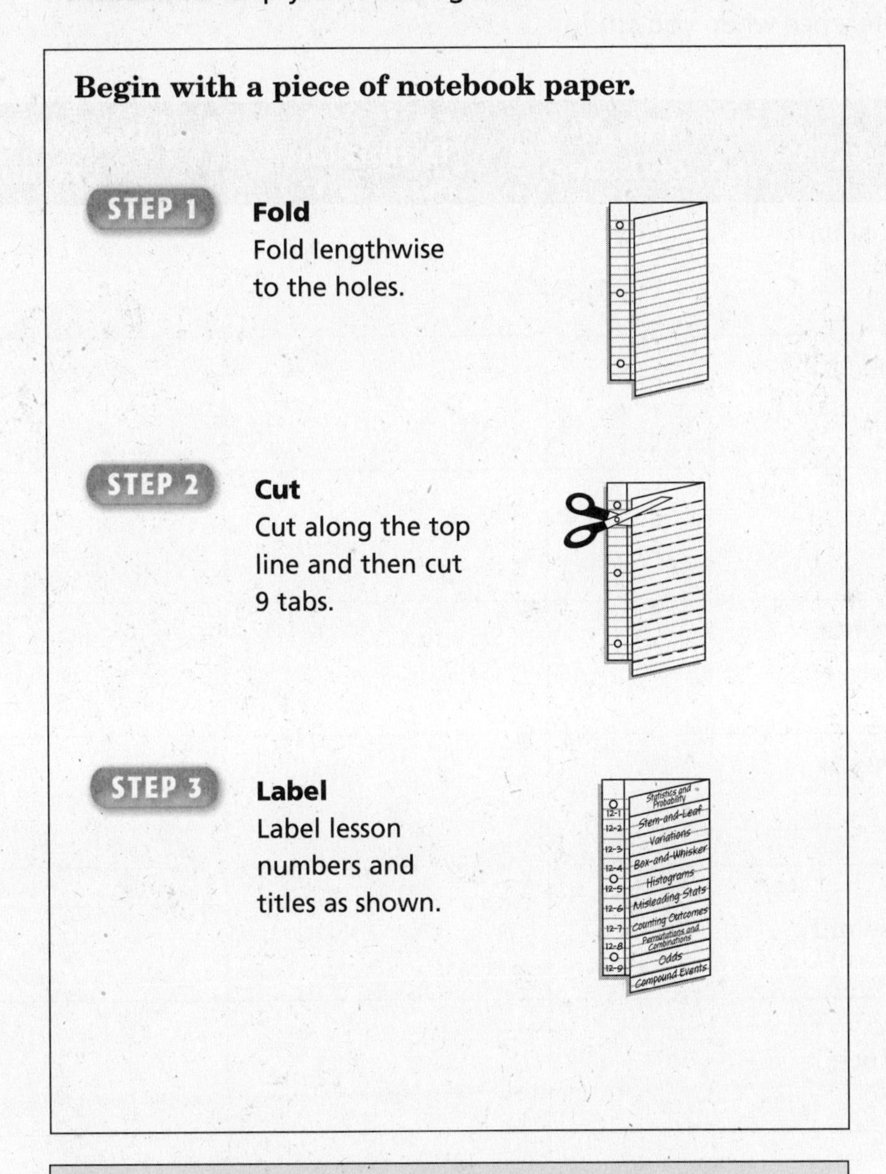

Begin with a piece of notebook paper.

STEP 1 **Fold**
Fold lengthwise to the holes.

STEP 2 **Cut**
Cut along the top line and then cut 9 tabs.

STEP 3 **Label**
Label lesson numbers and titles as shown.

NOTE-TAKING TIP: When taking notes on statistics, include your own statistical examples as you write down concepts and definitions. This will help you to better understand statistics.

BUILD YOUR VOCABULARY

This is an alphabetical list of new vocabulary terms you will learn in Chapter 12. As you complete the study notes for the chapter, you will see Build Your Vocabulary reminders to complete each term's definition or description on these pages. Remember to add the textbook page number in the second column for reference when you study.

Vocabulary Term	Found on Page	Definition	Description or Example
back-to-back stem-and-leaf plot			
box-and-whisker plot			
combination			
compound events			
dependent events			
factorial [fak-TOHR-ee-uhl]			
Fundamental Counting Principle			
histogram			
independent events			

Vocabulary Term	Found on Page	Definition	Description or Example
interquartile range [IN-tuhr-KWAWR-tyl]			
measures of variation			
mutually exclusive events			
odds			
outliers			
permutation [PUHR-myoo-TAY-shuhn]			
quartiles			
range			
stem-and-leaf plot			
tree diagram			
upper and lower quartiles			

12–1 Stem-and-Leaf Plots

WHAT YOU'LL LEARN

- Display data in stem-and-leaf plots.
- Interpret data in stem-and-leaf plots.

BUILD YOUR VOCABULARY (page 291)

In a **stem-and-leaf plot**, numerical data are listed in ascending or descending ______.

Stem	Leaf
5	1 2 3 6
6	0 5
7	1

6 | 0 = 60

The ______ place value of the data is used for the **stems**. The ______ place value forms the **leaves**.

FOLDABLES™

ORGANIZE IT

Write a description of a stem-and-leaf plot under the tab for this lesson.

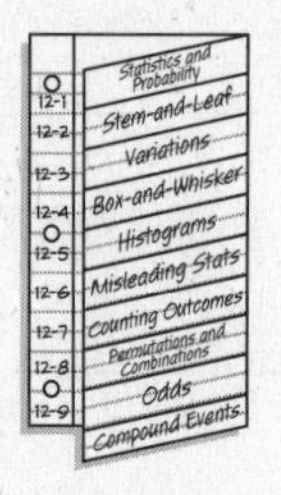

EXAMPLE Draw a Stem-and-Leaf Plot

1 FOOD **Display the data in a stem-and-leaf plot.**

Peanuts Harvest, 2001	
State	**Amount (lb)**
Alabama	2400
Florida	2800
Georgia	2800
New Mexico	2400
North Carolina	2900
Oklahoma	2200
South Carolina	2900
Texas	2600
Virginia	3000

Step 1 Find the least and the greatest number. Then identify the greatest place value digit in each number.

2200

The least number has ____ in the ______ place.

3000

The greatest number has ____ in the ______ place.

REVIEW IT

What are the mean, the median, and the mode of a set data? *(Lesson 5–8)*

____

____

____

____

____

Step 2 Draw a vertical line and write the stems ____ and ____ to the left of the line.

Step 3 Write the leaves to the right of the line, with the corresponding stem.

Step 4 Rearrange the leaves so they are ordered from least to greatest. Then include a key.

Stem	Leaf
____	____
____	____

2|4 = 2400 lb

Your Turn Display the speeds 65, 72, 59, 68, 75, 70, 68, 64, 67, 69, 72, and 55 given in miles per hour in a stem-and-leaf plot.

EXAMPLE Interpret Data

2 VOTING **The stem-and-leaf plot lists the percent of voters in each state that voted for U.S. representatives in 1998.** **Source:** U.S. Census Bureau

Stem	Leaf
1	0 1
2	2 5 5 8 8
3	0 0 0 2 2 2 3 3 3 4 4 5 5 6 6 7 7 7 8 8 9 9
4	0 0 0 1 1 3 3 3 4 4 4 4 5 6 7 9 9
5	0 1 4 9

3|4 = 34%

a. Which interval contains the most percentages?

Most of the data occurs in the ____ interval.

b. What is the greatest percent of voters that voted for U.S. representatives?

The greatest percent is ____.

c. What is the median percent of voters that voted for U.S. representatives?

The median in this case is the mean of the middle two numbers or ____.

Your Turn

The stem-and-leaf plot lists the amount of allowance students are given each month.

Stem	Leaf
0	0 5
1	0 2 2 5 8 8 8
2	0 0 0 4 4 5 5 5 5
3	0 0 2 2 2 4 4 5 5 6 6
4	0 2 4 4 5 5 5 5 8 8 9 9
5	0 0

2|5 = $25

a. In which interval do most of the monthly allowances occur?

b. What is the greatest monthly allowance given?

c. What is the median monthly allowance given?

BUILD YOUR VOCABULARY (page 290)

(page 290)

Two sets of data can be ______ using a **back-to-back stem-and-leaf plot**. The leaves for one set of data are on one side of the ______ and the leaves for the other set of data are on the other side.

EXAMPLE Compare Two Sets of Data

3 AGRICULTURE **The yearly production of honey in California and Florida is shown for the years 1993 to 1997, in millions of pounds.**

Source: USDA

California		Stem	Florida		
		1	6	9	
7	4	2	0	3	5
9	2	3			
	5	4			

9|3 = 39 million lb   *2|5 = 25* million lb

a. Which state produces more honey?

______; it produces between ______ and ______ million pounds per year.

b. Which state has the most varied production? Explain.

______; the data are more spread out.

Your Turn **The exam score earned on the first test in a particular class is shown for male and female students.**

Male		Female
8 2	6	
9 6 4	7	4 8 8 9
7 4 2 2 0	8	1 3 4 8 9
6 5 3	9	2 5 9

2|8 = 82 *7|4 = 74*

a. Which group of students had the higher test scores?

b. Which group of students had more varied test scores?

HOMEWORK ASSIGNMENT

Page(s):

Exercises:

12–2 Measures of Variation

BUILD YOUR VOCABULARY (page 291)

Measures of variation are used to describe the ________ of the data.

The **range** of a set of data is the ________ between the greatest and the least values of the set.

The **quartiles** are the values that divide a set of data into ________ equal parts.

The ________ of the lower half of a set of data is the **lower quartile**.

The median of the ________ of a set of data is the **upper quartile**.

WHAT YOU'LL LEARN

- Find measures of variation.
- Use measures of variation to interpret and compare data.

WRITE IT

What does the range describe about a set of data?

____

____

____

____

EXAMPLE Range

1 Find the range of each set of data.

a. {$79, $42, $38, $51, $63, $91}

The greatest value is ________, and the least value is ________.

So, the range is ________ – ________ or ________.

b.

Stem	Leaf
3	3 3 5 7 7 8
4	0 3 3 4 9
5	4 9

3|5 = 35

The greatest value is ________ and the least value is ________.

So, the range is ________ – ________ or ________.

Your Turn **Find the range of each set of data.**

a. {14, 37, 82, 45, 24, 10, 75}

b.

Stem	Leaf
5	2 3 5 5 9
6	4 8 9
7	0 1 8 9

6|8 = 68

KEY CONCEPT

Interquartile Range The interquartile range is the range of the middle half of a set of data. It is the difference between the upper quartile and the lower quartile.

EXAMPLE Interquartile Range

2 Find the interquartile range for the each of data.

a. {38, 40, 32, 34, 36, 45, 33}

Step 1 List the data from least to greatest. Then find the median.

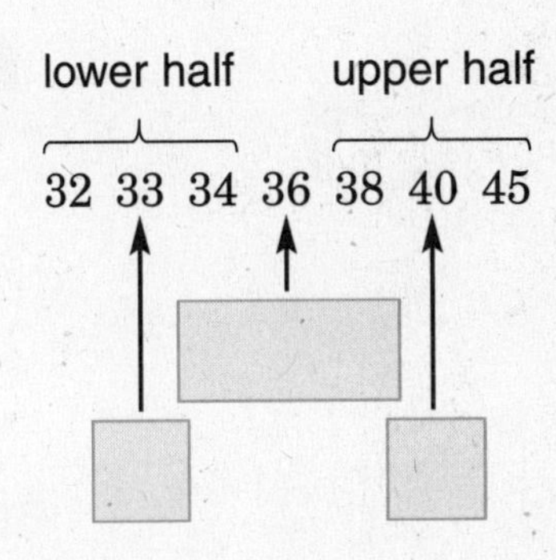

Step 2 Find the upper and lower quartiles.

The interquartile range

b. {2, 27, 17, 14, 14, 22, 15, 32, 24, 25}

Step 1 List the data from least to greatest. Then find the median.

Step 2 Find the upper and lower quartiles.

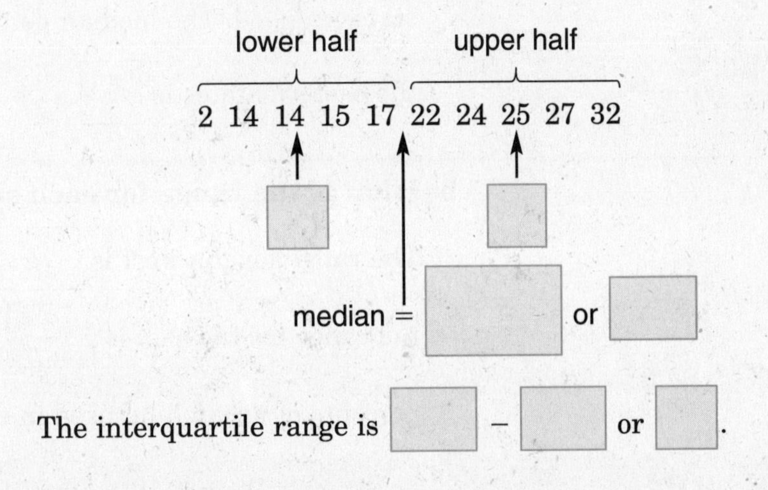

ORGANIZE IT

Explain the difference between the range and the interquartile range of a set of data under the tab for Lesson 12–2.

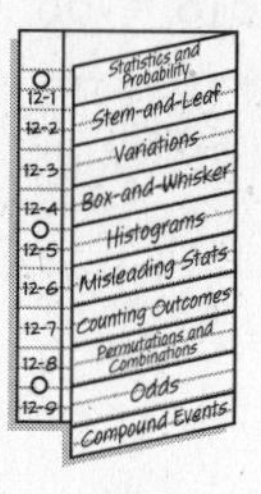

Your Turn **Find the interquartile range for each set of data.**

a. {52, 74, 98, 80, 63, 84, 77}

b. {12, 18, 25, 31, 23, 19, 16, 22, 28, 32}

EXAMPLE Interpret and Compare Data

3 LAND USE **The urban land in certain western and eastern states is listed below as the percent of each state's total land, rounded to the nearest percent.**

Western States		Eastern States
1 1 1 1 1 0 0	0	
3 2 2 2 1 1 1	0	3 3 4 5 6 6 8
5 4 4	0	8 9 9 9 9 9 9
	1	1 3 3 4 4 5
	2	3 6 7
2|0 = 2%	3	5 *2|7 = 27%*

Source: U.S. Census Bureau

a. What is the median percent of urban land use for each region?

The median percent of urban land use for the western states is ______. The median percent of urban land use for the eastern states is ______.

b. What is the range for each set of data.

The range for the west is ______ – ______ or ______, and the range for the east is ______ – ______ or ______. The percents of urban land used in the ______ vary more.

Your Turn **The hours per week spent exercising for teenagers and people in their twenties are listed in the stem-and-leaf plot.**

Teens		Twenties
5 4 2 0	0	0 4 6 7 9
7 3	1	0 2 2 5
1	2	0 3 4 5 8
3|1 = 13 hr		*1|5 = 15 hr*

a. What is the median time spent exercising for each group?

b. Compare the range for each set of data.

HOMEWORK ASSIGNMENT

Page(s):

Exercises:

12–3 Box-and-Whisker Plots

WHAT YOU'LL LEARN

- Display data in a box-and-whisker plot.
- Interpret data in a box-and-whisker plot.

BUILD YOUR VOCABULARY (page 290)

A **box-and-whisker plot** ______ a set of data into ______ using the medial and quartiles. A *box* is drawn around the ______, and *whiskers* extend from each quartile to the ______ data points.

REMEMBER IT

The median does not necessarily divide the box in half. Data clustered toward one quartile will shift the median its direction.

EXAMPLE Draw a Box-and-Whisker Plot

1 JOBS **The projected number of employees in 2008 in the fastest-growing occupations is shown. Display the data in a box-and-whisker plot.**

Fastest-Growing Jobs			
Occupation	Jobs (1000s)	Occupation	Jobs (1000s)
Computer Engineer	622	Desktop Publishing	44
Computer Support	869	Paralegal/Legal Assistant	220
Systems Analyst	1194	Home Health Aide	1179
Database Administrator	155	Medical Assistant	398

Source: U.S. Census Bureau

Step 1 Find the ______ and ______ number. Then draw a number line that covers the ______ of the data.

Step 2 Find the ______, the extremes, and the upper and lower ______. Mark these points above the number line.

Step 3 Draw a box and the whiskers.

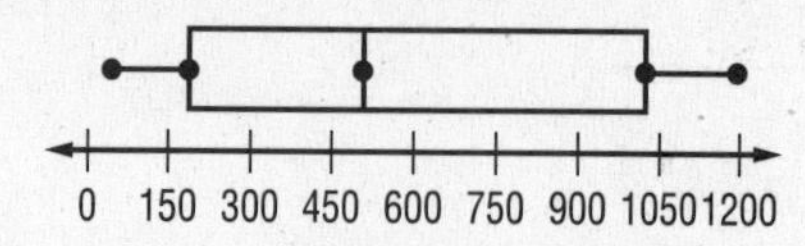

ORGANIZE IT

Write a description of a box-and-whisker plot under the tab for this lesson.

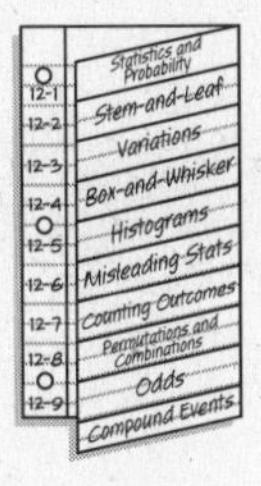

Your Turn The data listed below represents the time, in minutes, required for students to travel from home to school each day. Display the data in a box-and-whisker plot.

14 32 7 45 18 22 26 9 4 18 15

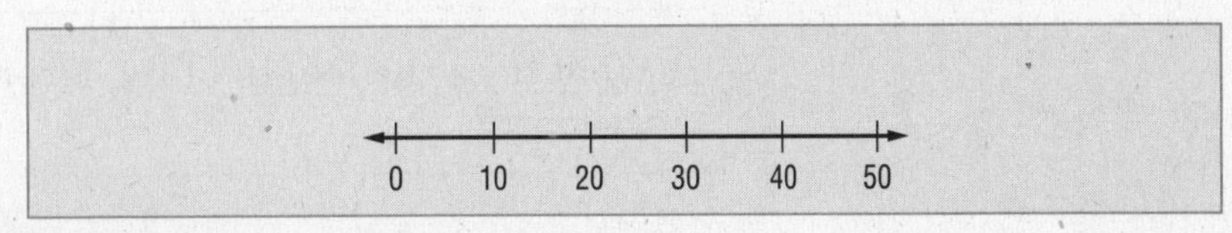

EXAMPLE Interpret Data

2 WEATHER **The box-and-whisker plot below shows the average percent of sunny days per year for selected cities in each state.**

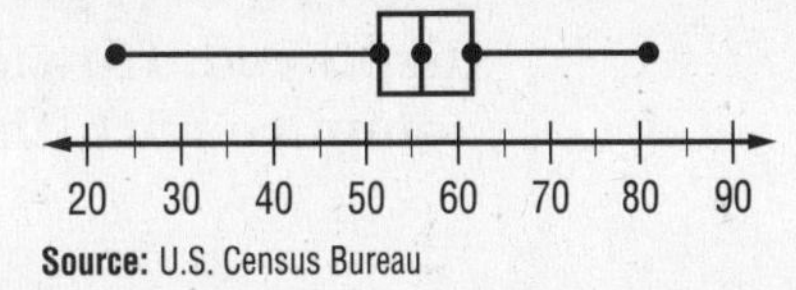

a. What is the smallest percent of sunny days in any state?

The smallest percent of sunny days in any state is .

b. Half of the selected cities have an average percent of sunny days under what percent?

Half of the selected cities have an average percent of sunny days under ____.

c. What does the length of the box in the box-and-whisker plot tell about the data?

The length of the box is ____. This tells us that the data values are ____.

Your Turn **The box-and-whisker plot below shows the average amount spent per month on clothing.**

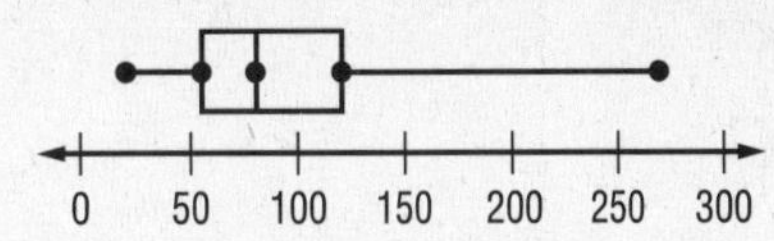

a. What is the smallest amount spent per month on clothing?

b. Half of the monthly expenditures on clothing are under what amount?

c. What does the length of the box-and-whisker plot tell about the data?

EXAMPLE Compare Two Sets of Data

3 TREES **The average maximum height, in feet, for selected evergreen trees and deciduous trees is displayed. How do the heights of evergreen trees compare with the heights of deciduous trees?**

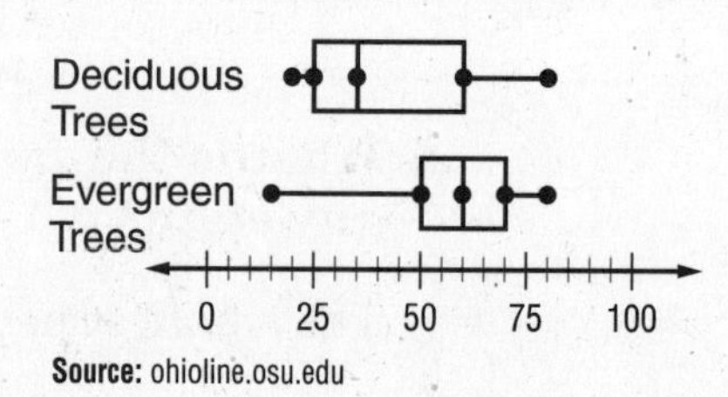

Most deciduous trees range in height between ☐ and ☐ feet. However, some are as tall as ☐ feet. Most evergreen trees range in height between ☐ and ☐ feet. However, some are as tall as ☐ feet. Most evergreen trees are ☐ than most deciduous trees.

Your Turn The average gas mileage, in miles per gallon, for selected compact cars and sedans is displayed. How do the gas mileages of compact cars compare with the gas mileages for sedans?

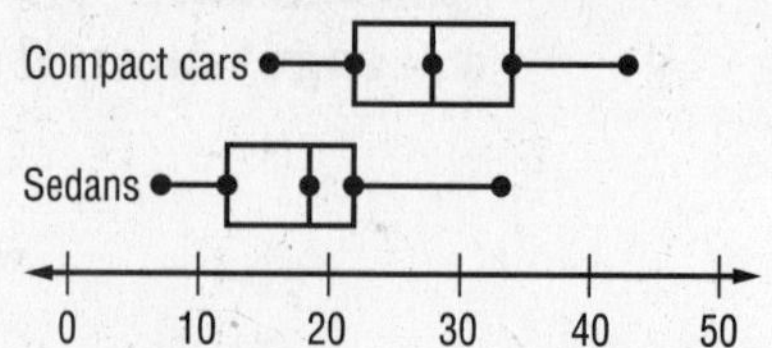

HOMEWORK ASSIGNMENT

Page(s):

Exercises:

12–4 Histograms

What You'll Learn

- Display data in a histogram.
- Interpret data in a histogram.

Build Your Vocabulary (page 290)

A **histogram** uses ______ to display numerical data that have been organized into ______ intervals.

Write It

What type of data does a histogram display?

______

______

______

______

______

Example Draw a Histogram

1 TOURISM **The frequency table shows the number of overseas visitors to certain U.S. cities in 1999. Display the data in a histogram.**

Overseas Travelers		
Number of Visitors (1000s)	Tally	Frequency
0–1000	𝍸	5
1001–2000	III	3
2001–3000	𝍸	5
3001–4000	I	1
4001–5000		0
5001–6000	I	1

Source: U.S. Department of Commerce

Step 1 Draw and label a horizontal and vertical axis. Include a ______.

Step 2 Show the intervals from the frequency table on the ______ axis and an interval of 1 on the ______ axis.

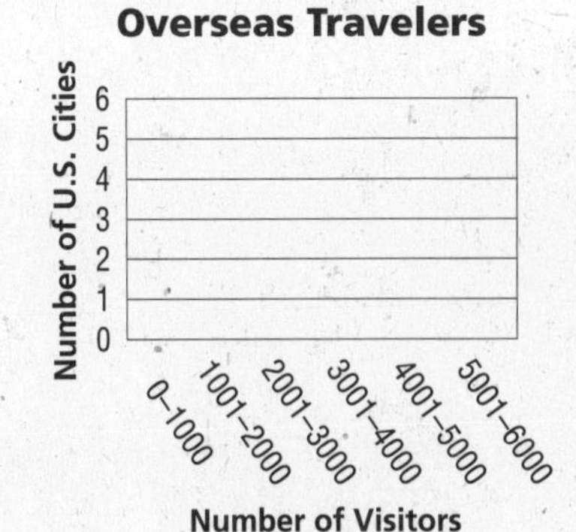

Step 3 For each interval, draw a bar whose height is given by the ______.

ORGANIZE IT

Describe how to display data in a histogram under the tab for this lesson.

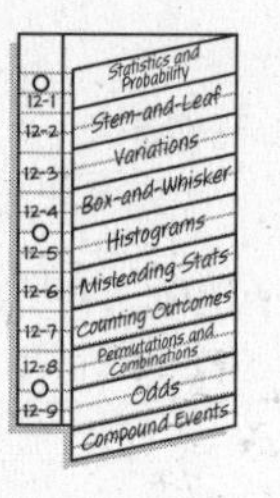

Your Turn The frequency table show the number of daily customers a new grocery store has during its first 30 days in business. Display the data in a histogram.

Daily Customers		
Number of Customers	Tally	Frequency
0–49	𝍸 I	6
50–99	𝍸 𝍸 II	12
100–149	𝍸 IIII	9
150–199	III	3

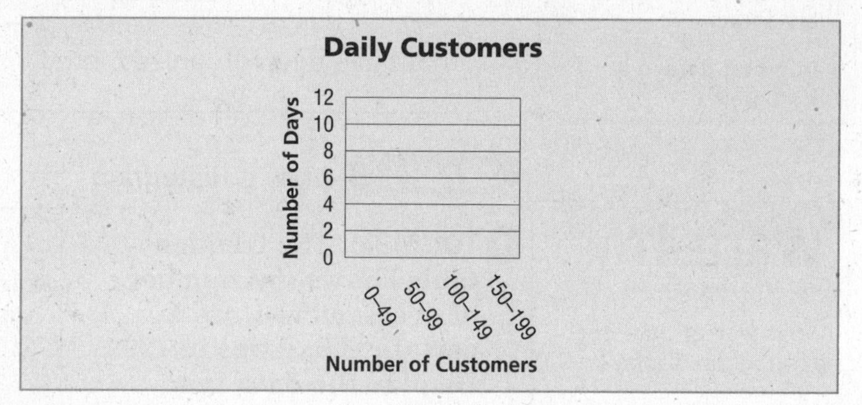

EXAMPLE Interpret Data

2 ELEVATIONS Use the histogram.

a. How many states have highest points with elevations at least 3751 meters?

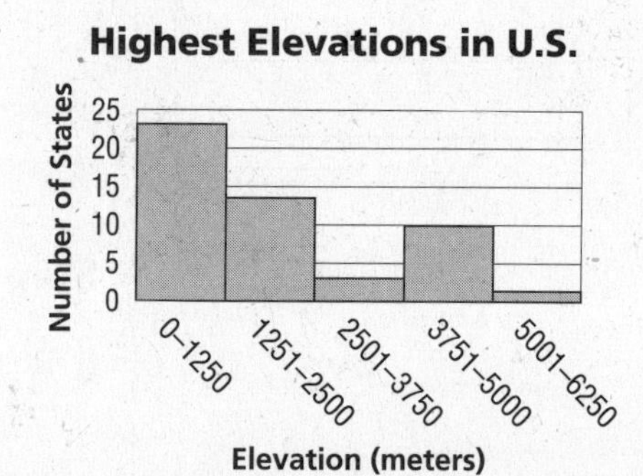

Since ☐ states have elevations in the 3751–5000 range and 2 states have elevations in the ☐ range, ☐ + ☐ or ☐ states have highest points with elevations at least 3751 meters.

b. Is it possible to tell the height of the tallest point?

No, you can only tell that the highest point is between ☐ and ☐ meters.

Your Turn **Use this histogram.**

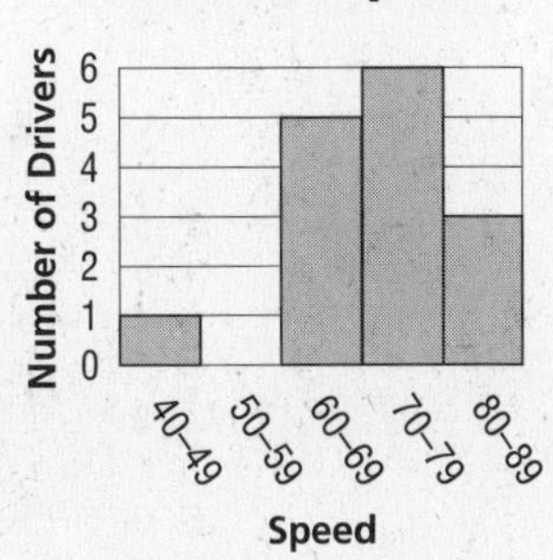

a. How many drivers had a speed of at least 70 miles per hour?

b. Is it possible to tell the lowest speed driven?

EXAMPLE Compare Two Sets of Data

3 EMPLOYMENT **Use the histograms.**

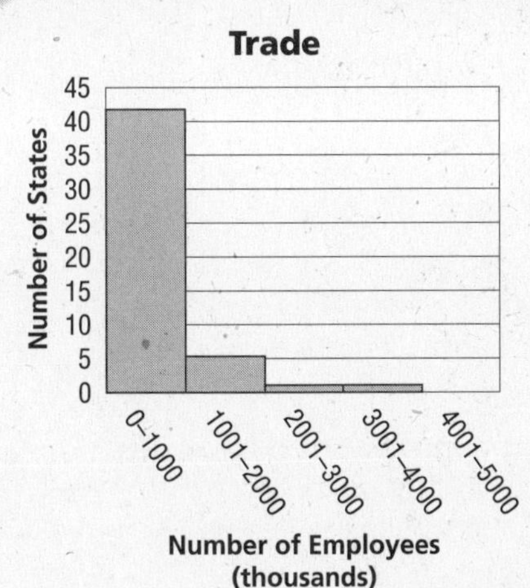

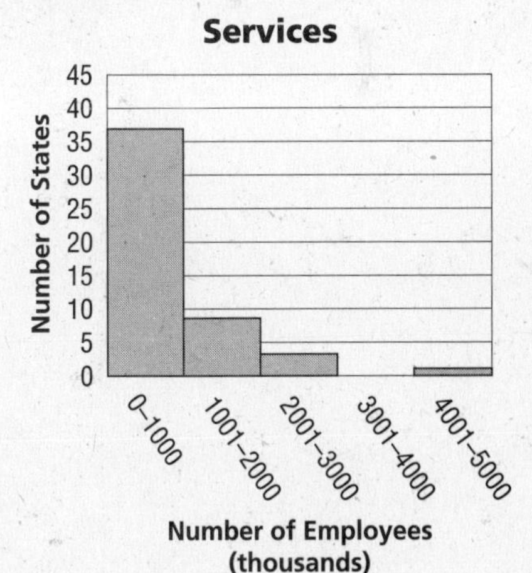

Which business sector has more states with between 1,001,000 and 3,000,000 employees?

By comparing the graphs, you find that ________ has more states with between 1,001,000 and 3,000,000 employees.

Your Turn Use the histograms that show weekly dining expenses.

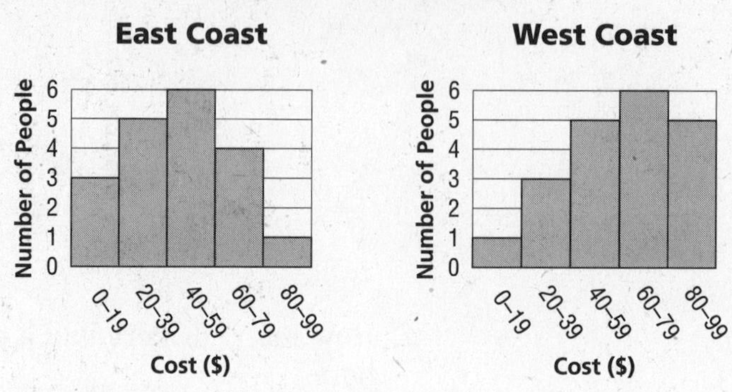

Which coast has more people spending at least $60 weekly?

HOMEWORK ASSIGNMENT

Page(s):

Exercises:

12–5 Misleading Statistics

WHAT YOU'LL LEARN

- Recognize when statistics are misleading.

EXAMPLE Misleading Graphs

1 FOOD **The graphs show the increase in the price of lemons.**

Price of Lemons Graph A

Dollars per Pound: 0, 0.50, 0.70, 0.90, 1.10, 1.30, 1.50
Year: '93 '94 '95 '96 '97 '98 '99

Price of Lemons Graph B

Dollars per Pound: 0, 0.50, 0.70, 0.90, 1.10, 1.30, 1.50
Year: '93 '95 '97 '99

a. Why do the graphs look different?

The ______ scales differ.

b. Which graph appears to show a more rapid increase in the price of lemons after 1997? Explain.

Graph B; the slope of the line from ______ to ______ is steeper in Graph B.

Your Turn **The graphs show the increase in attendance at a public elementary school.**

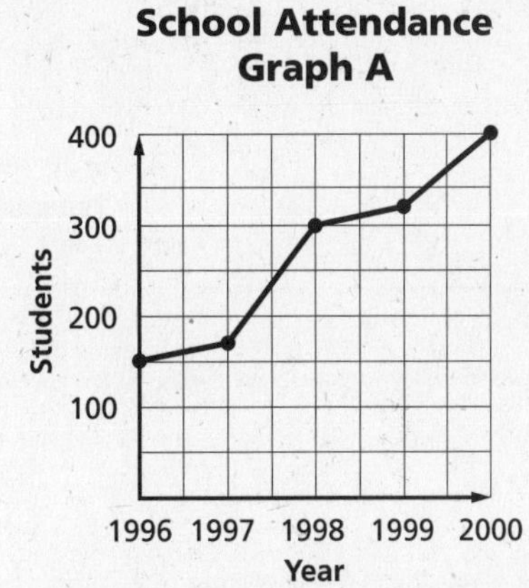

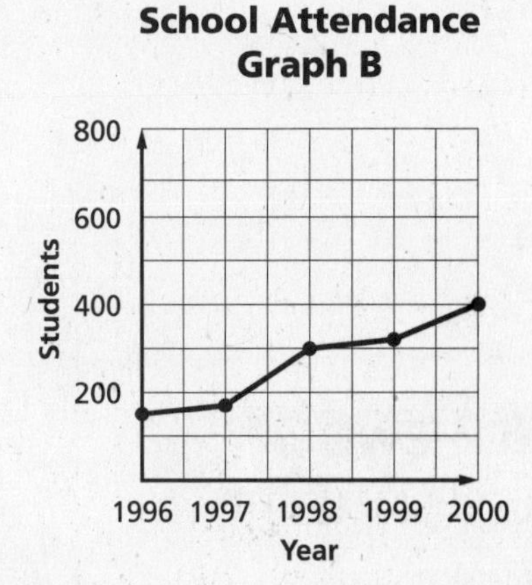

REMEMBER IT

Carefully read the labels and the scales when interpreting a graph.

a. Why do the graphs look different?

b. Which graph appears to show a more rapid increase in attendance between 1997 and 1998? Explain.

______

FOLDABLES™

ORGANIZE IT

Under the Lesson 12-5 tab, draw an example of a misleading graph, and explain why it is misleading

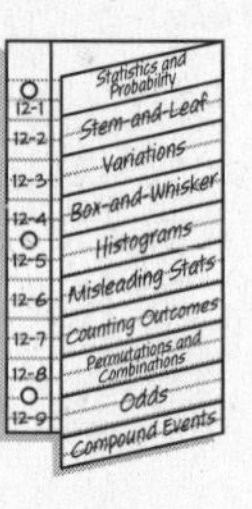

EXAMPLE Misleading Bar Graphs

2 INTERNET **Explain why the graph on the right is misleading.**

Internet Users (%)

Percent: 0, 20, 40, 50, 60

Age: 18–24 (58.7), 25–34 (53.3), 35–44 (53.8), 45–54 (54.8), 55–64 (35.1), 65–up (10.7)

The inconsistent ______ and ______ scales cause the data to be misleading. The graph gives the impression that the percentages of people aged 18–24 who use the Internet is ______ more than those aged 65 and up. By using the scale, you can see that it is only about ______ more.

Your Turn Explain why the graph is misleading.

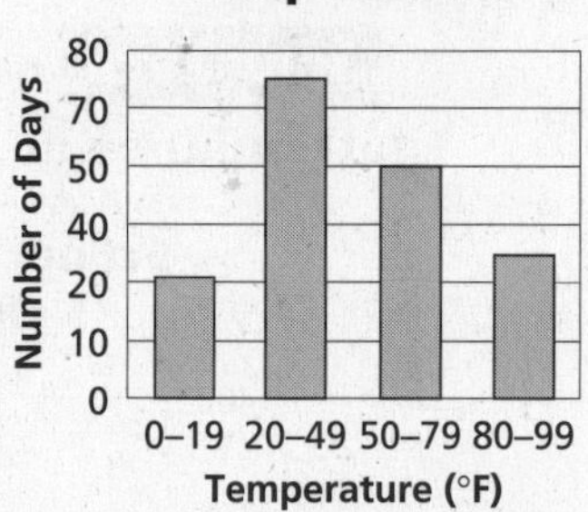

______

HOMEWORK ASSIGNMENT

Page(s):

Exercises:

12–6 Counting Outcomes

BUILD YOUR VOCABULARY (page 291)

A **tree diagram** is a diagram used to show the total number of possible outcomes.

WHAT YOU'LL LEARN

- Use tree diagrams or the Fundamental Counting Principle to count outcomes.
- Use the Fundamental Counting Principle to find the probability of an event.

REVIEW IT

What is the name for the set of all possible outcomes of a probability event? *(Lesson 6-9)*

____

____

____

____

____

____

EXAMPLE Use a Tree Diagram to Count Outcomes

1 GREETING CARDS **A greeting-card maker offers four birthday greeting in five possible colors as shown in the table below. How many different cards can be made from four greeting choices and five color choices?**

You can draw a diagram to find the number of possible cards.

Greeting	Color
Humorous	Blue
Traditional	Green
Romantic	Orange
"From the Group"	Purple
	Red

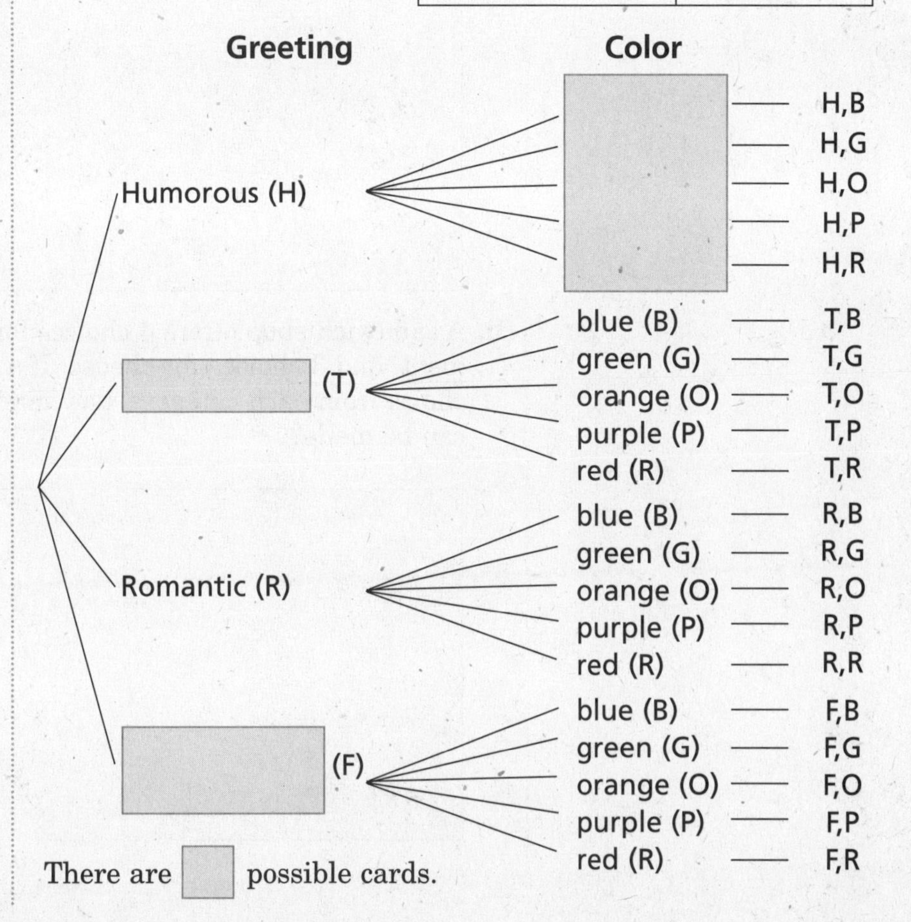

There are ______ possible cards.

KEY CONCEPT

Fundamental Counting Principle If even M can occur in m ways and is followed by event N that can occur in n ways, then the event M followed by N can occur in $m \cdot n$ ways.

FOLDABLES Explain how to determine the number of possible outcomes using a tree diagram and the Fundamental Counting Principle.

EXAMPLE Use the Fundamental Counting Principle

2 CELL PHONES **A cell phone company offers 3 payment plans, 4 styles of phones, and 6 decorative phone wraps. How many phone options are available?**

Use the Fundamental Counting Principle.

The number of types of payment plans	times	the number of styles of phones	times	the number of decorative wraps	equals	the number of possible outcomes.
☐	×	☐	×	☐	=	☐

There are ☐ possible phone options.

Your Turn

a. An ice cream parlor offers a special on one-scoop sundaes with one topping. The ice cream parlor has 5 different flavors of ice cream and three different choices for toppings. How many different sundaes can be made?

b. A sandwich shop offers 4 choices for bread, 5 choices for meat, and 3 choices for cheese. If a customer can make one choice from each category, how many different sandwiches can be made?

EXAMPLE Find Probabilities

3 Henry rolls a number cube and tosses a coin.

a. What is the probability that he will roll a 3 and toss heads?

First find the number of outcomes.

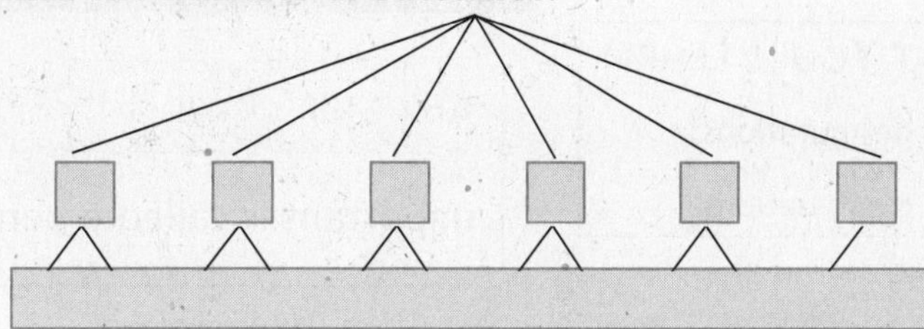

There are ___ possible outcomes. There is one outcome that has a 3 and a head.

$P(3 \text{ and head}) = \frac{\text{number of favorable outcomes}}{\text{number of possible outcomes}} = $ ___

The probability that Henry will roll a 3 and toss heads is ___.

b. What is the probability of winning a multi-state lottery game where the winning number is made up of 6 numbers from 1 to 50 chosen at random? Assume all numbers are eligible each draw.

First, find the number of possible outcomes. Use the Fundamental Counting Principle.

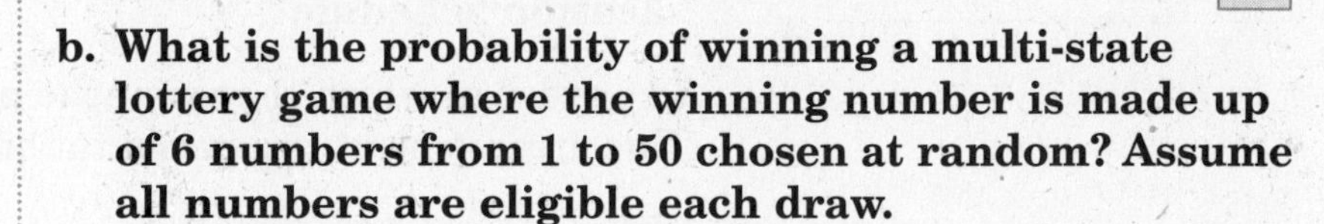

The total number of outcomes is ___ × ___ × ___ × ___ × ___ × ___ or 15,625,000,000.

There is 1 winning number. So, the probability of winning with 1 ticket is ___.

Your Turn

a. Bob rolls a number cube and tosses a coin. What is the probability that he will roll an odd number and toss tails?

b. What is the probability of winning a lottery where the winning number is made up of 5 numbers from 1 to 20 chosen at random? Assume all numbers are eligible each draw.

HOMEWORK ASSIGNMENT

Page(s):

Exercises:

12–7 Permutations and Combinations

BUILD YOUR VOCABULARY (page 291)

An ______ or listing in which ______ is important is called a **permutation**.

WHAT YOU'LL LEARN

- Use permutations.
- Use combinations.

EXAMPLE Use a Permutation

1 TRAVEL **The Reyes family will visit a complex of theme parks during their summer vacation. They have a four-day pass good at one park per day; they can choose from seven parks.**

REMEMBER IT

The first factor in a permutation is the number of things you are choosing from.

a. How many different ways can they arrange their vacation schedule?

The order in which they visit the parks is important. This arrangement is a permutation.

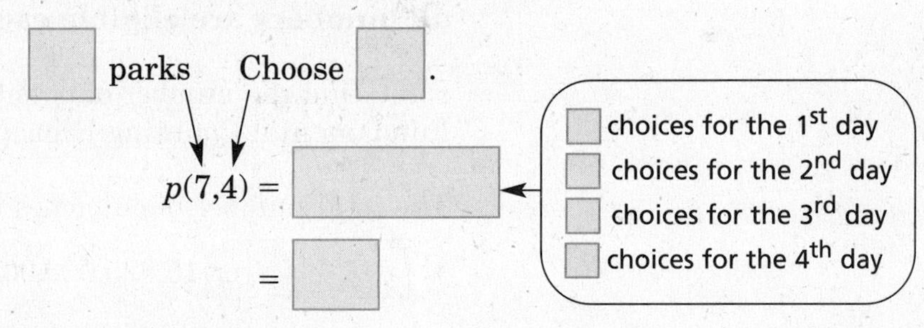

There are ______ possible arrangements.

b. How many five-digit numbers can be made from the digits 2, 4, 5, 8, and 9 if each digit is used only once?

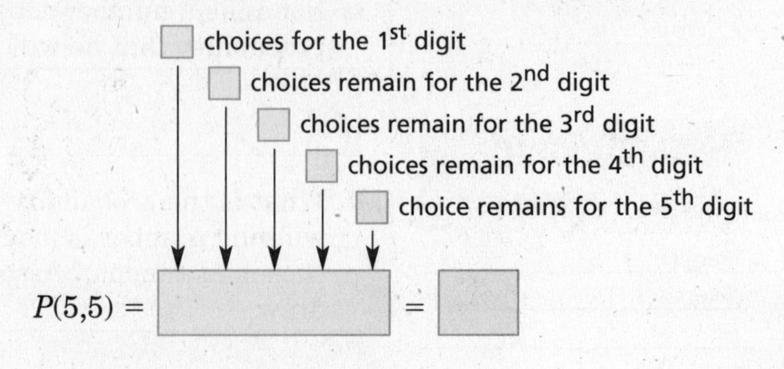

There are ______ numbers that can be made.

Your Turn

a. How may ways can five runners be arranged on a three-person relay team?

b. How many six-digit numbers can be made from the digits 1, 2, 3, 4, 5, and 6 if each digit is used only once?

BUILD YOUR VOCABULARY (page 290)

The notation *n* **factorial** means the product of all counting numbers beginning with *n* and counting backward to 1.

An arrangement or ______ in which ______ is *not* important is called a **combination**.

EXAMPLE Factorial Notation

2 Find the value of 12!.

12! = ______

= ______

EXAMPLE Use a Combination

3 HATS **How many ways can a window dresser choose two hats out of a fedora, a bowler, and a sombrero?**

Since order is not important, this arrangement is a combination.

First, list all of the permutations of the types of hats taken ______ at a time. Then cross off arrangements that are the same as another one.

There are ______ ways to choose two hats from three possible hats.

FOLDABLES™

ORGANIZE IT

Explain the difference between a permutation and a combination under the Lesson 12-7 tab.

Your Turn

a. Find the value of 7!.

b. How many ways can two shirts be selected from a display having a red shirt, a blue shirt, a green shirt, and a white shirt?

EXAMPLE Use a Combination to Solve a Problem

4 GEOMETRY Find the number of line segments that can be drawn between any two vertices of a hexagon.

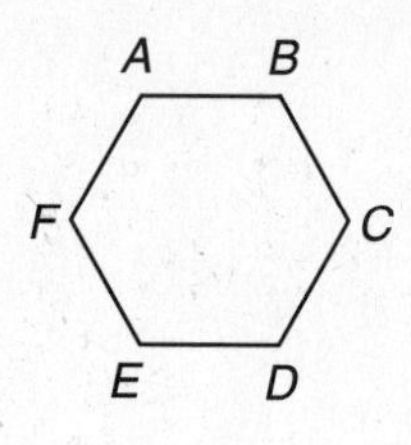

A hexagon has 6 vertices. The segment connecting vertex A to vertex C is the same as the segment connecting C to A, so

this is a ______. Find the ______ of ______ vertices taken ______ at a time.

$$C\_\_\_ = \frac{P(6, 2)}{2!} = \frac{\_ \cdot \_}{\_ \cdot \_} \text{ or } \_\_\_$$

Your Turn Find the number of line segments that can be drawn between any two vertices of a pentagon.

HOMEWORK ASSIGNMENT

Page(s):

Exercises:

12–8 Odds

WHAT YOU'LL LEARN

- Find the odds of a simple event.

BUILD YOUR VOCABULARY (page 291)

The **odds** in favor of an event is the ______ that compares the number of ______ the event can ______ to the ways that the event *cannot* occur.

KEY CONCEPT

Definition of Odds The odds in favor of an outcome is the ratio of the number of ways the outcome can occur to the number of ways the outcome cannot occur.

The odds against an outcome is the ratio of the number of ways the outcome cannot occur to the number of ways the outcome can occur.

EXAMPLE Find Odds

1 a. Find the odds of a sum greater than 5 if a pair of number cubes are rolled.

There are ______ · ______ or ______ sums possible for rolling a pair of number cubes.

There are ______ sums greater than 5.

There are ______ − ______ or ______ sums that are not greater than 5.

Odds of rolling a sum greater than 5.

= number of ways to roll a sum greater than 5 to number of ways to roll any other sum

= ______ ______ ______ or ______

The odds of rolling a sum greater than 5 are ______.

Your Turn Find the odds of a sum less than 10 if a pair of number cubes are rolled.

b. A bag contains 5 yellow marbles, 3 white marbles, and 1 black marble. What are the odds against drawing a white marble from the bag?

There are ☐ – ☐ or ☐ marbles that are not white.

Odds against drawing a white marble

=

= ☐ ☐ ☐ or ☐

The odds against drawing a white marble are ☐.

Your Turn A bag contains 4 blue marbles, 3 green marbles, and 6 yellow marbles. What are the odds against drawing a green marble from the bag?

FOLDABLES

ORGANIZE IT

Under the Lesson 12-8 tab, write two examples of when you would use odds to describe an event.

EXAMPLE Use Odds

2 After 2 weeks, 8 out of 20 sunflower seeds that Tamara planted had sprouted. Based on these results, what are the odds that a sunflower seed will sprout under the same conditions?

To find the odds, compare the number of successes to the number of failures.

Tamara planted ☐ seeds. ☐ seeds sprouted.

☐ – ☐ or ☐ seeds had not sprouted.

successes : failures = ☐ to ☐ or ☐ to ☐

The odds that a sunflower seed will sprout under the same conditions are ☐.

Your Turn Kyle took 15 free throw shots with a basketball. He made 9 of the shots. Based on these results, what are the odds Kyle will make a free throw shot?

HOMEWORK ASSIGNMENT

Page(s):

Exercises:

12–9 Probability of Compound Events

What You'll Learn

- Find the probability of independent and dependent events.
- Find the probability of mutually exclusive events.

Build Your Vocabulary (page 290)

A **compound event** consists of ______ simple events.

In **independent events**, the outcome of one event does *not* ______ the outcome of a second event.

Key Concept

Probability of Two Independent Events
The probability of two independent events is found by multiplying the probability of the first event by the probability of the second event.

EXAMPLE Probability of Independent Events

1 GAMES **In a popular dice game, the highest possible score in a single turn is a roll of five of a kind. After rolling one five of a kind, every other five of a kind you roll earns 100 points. What is the probability of rolling two five of a kind in a row?**

The events are ______ since each roll of the dice does not affect the outcome of the next roll.

There are ______ ways to roll five of a kind.

There are 6^5 or ______ ways to roll five dice. So, the probability of rolling five of a kind on a toss of the dice is ______ or ______.

P(two five of a kind)

$= P$(five of a kind on first roll) · P(five of a kind on second roll)

$=$ 

$=$ 

Your Turn Find the probability of rolling doubles four times in a row when rolling a pair of number cubes.

BUILD YOUR VOCABULARY (pages 290–291)

If the ______ of one event ______ the outcome of a second event, the events are called **dependent events**.

If two events ______ happen at the ______, they are said to be **mutually exclusive**.

KEY CONCEPT

Probability of Two Dependent Events
If two events, *A* and *B* are dependent, then the probability of both events occurring is the product of the probability of *A* and the probability of *B* after *A* occurs.

EXAMPLE Probability of Dependent Events

2 SHIRTS **Charlie's clothes closet contains 3 blue shirts, 10 white shirts, and 7 striped shirts. What is the probability that Charlie will reach in and randomly select a white shirt followed by a striped shirt?**

$P(\text{white shirt and striped shirt}) = \square \cdot \square$

$= \square$ or $\square$

10 of 20 shirts are white.

7 of 19 remaining shirts are striped.

The probability Charlie will select a white shirt followed by a striped shirt is ______.

Your Turn A plate has 6 chocolate chip cookies, 4 peanut butter cookies, and 5 sugar cookies. What is the probability of randomly selecting a chocolate chip cookie followed by a sugar cookie?

KEY CONCEPT

Probability of Two Mutually Exclusive Events
The probability of one or the other of two mutually exclusive events can be found by adding the probability of the first event to the probability of the second event.

FOLDABLES Under the 12–9 tab, explain when you would use each of the three probability formulas defined in this lesson.

EXAMPLE Probability of Mutually Exclusive Events

3 CARDS **You draw a card from a standard deck of playing cards. What is the probability that the card will be a black nine or any heart?**

The events are ________ because the card can not be both a black nine and a heart at the same time.

P(black nine or heart) = P(________) + P(________)

= ____ + ____

= ____

The probability that the card will be a black nine or any heart is ____.

Your Turn You draw a card from a standard deck of playing cards. What is the probability that the card will be a club or a red face card?

HOMEWORK ASSIGNMENT

Page(s):

Exercises:

CHAPTER 12

BRINGING IT ALL TOGETHER

STUDY GUIDE

FOLDABLES	VOCABULARY PUZZLEMAKER	BUILD YOUR VOCABULARY
Use your **Chapter 12 Foldable** to help you study for your chapter test.	To make a crossword puzzle, word search, or jumble puzzle of the vocabulary words in Chapter 12, go to: www.glencoe.com/sec/math/t_resources/free/index.php	You can use your completed **Vocabulary Builder** (pages 290–291) to help you solve the puzzle.

12-1 Stem-and-Leaf Plots

Use the table at the right.

1. Display the data set in a stem-and-leaf plot.

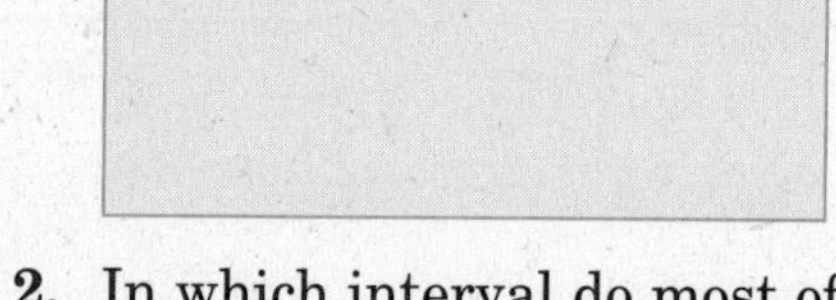

2. In which interval do most of the players fall?

Home runs scored by American League Leaders in 2002	
Chavez	34
Delgado	33
Giambi	41
Ordóñez	38
Palmeiro	43
Rodríguez	57
Soriano	39
Tejeda	34
Thome	52

Source: www.mlb.com

12-2 Measures of Variation

Find the range and interquartile range for each set of data.

3. {42, 22, 59, 82, 15, 37, 71, 24}

4.

Stem	Leaf
1	3 7
2	2 3 8
3	1 4 6 7

1|3 = 13

5.

Stem	Leaf
7	0 2 4 7 8
8	0 2 7
9	3 6

7|0 = 70

12-3 Box-and-Whisker Plots

Draw a box-and-whisker plot for each set of data.

6. 24, 40, 22, 15, 52, 46, 31, 22, 36

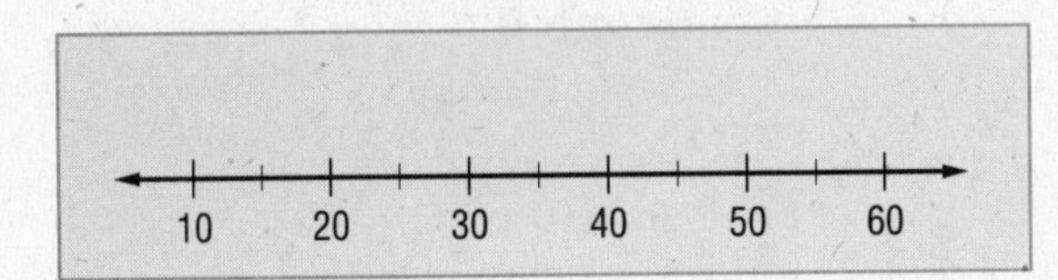

7. 342, 264, 289, 272, 245, 316, 331, 249, 270, 261

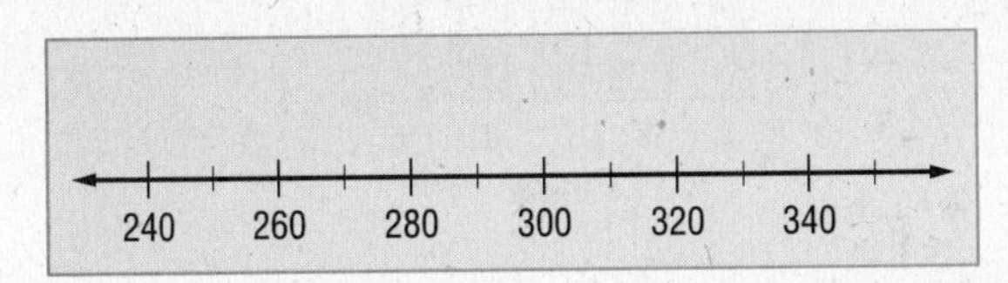

For exercises 8 and 9, use the box-and-whisker plot shown.

8. What is the warmest lowest recorded temperature?

Lowest Recorded Temperature (°C) in the US

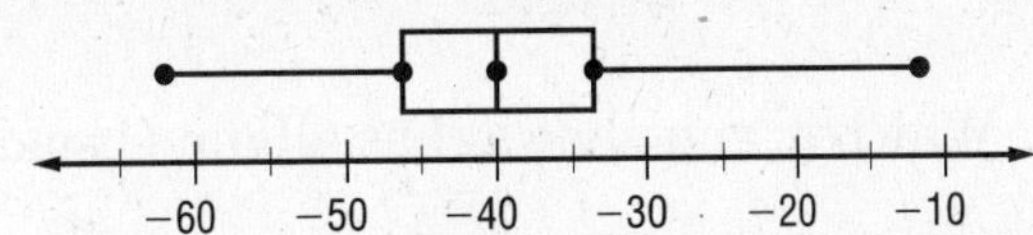

9. What percent of the temperatures range from 0°C to −40°C?

12-4 Histograms

For Exercises 10–12, use the frequency table shown.

10. Display the data in a histogram.

Movies Seen in the Last 12 Months

Number of People: 0, 5, 10, 15, 20, 25

Number of Movies: 1–4, 5–8, 9–12, 13–16

Movies Seen in the Last 12 Months

Movies	Tally	Frequency
1–4	卌 ||||	9
5–8	卌 卌 卌	15
9–12	卌 卌 卌 卌 ||	22
13–16	卌	5

11. How many people were surveyed?

12. How many people surveyed saw no more than 8 movies?

12-5 Misleading Statistics

For Exercise 13–15, use the graphs shown.

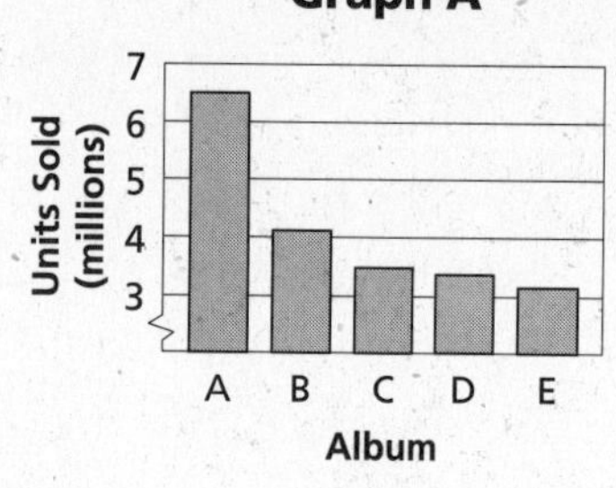

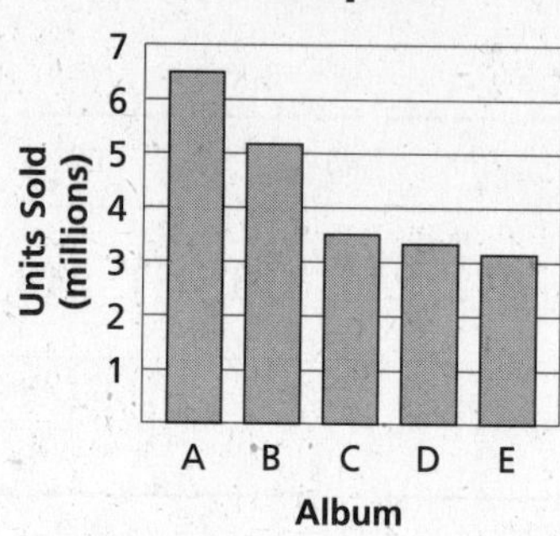

13. Which graph gives the impression that the top-selling 2003 album sold far more units than any other in the top five?

14. Which graph shows that album C sold more than half as many units as album A?

15. What causes the graphs to differ in appearance?

12-6 Counting Outcomes

Find the number of possible outcomes for each situation.

16. One part of a test has 7 true-false questions.

17. A bicycle is made with a choice of two seats, three frames, and five colors.

18. What is the probability of rolling exactly one 6 when two number cubes are rolled?

12-7 Permutations and Combinations

Tell whether each situation is a *permutation* or *combination*. Then solve.

19. How many ways can you choose 4 books from 15 on a shelf?

20. How many 4-digit numbers can you write using the digits 1, 2, 3, and 4 exactly once in each number?

12-8 Odds

Underline the correct term or phrase to complete each sentence.

21. The odds (in favor of, against) an outcome is the ratio of the number of successes to the number of failures.

22. The number of successes added to the number of failures equals the number of (favorable, total possible) outcomes.

23. What are the odds of getting a sum less than 6 when rolling a pair of number cubes?

12-9 Probability of Compound Events

Pens are drawn from a bag containing 6 red, 8 black, and 4 blue pens. Label each situation as *independent*, *dependent*, or *mutually exclusive* events. Then find each probability.

24. drawing a red pen, which is replaced, followed by a blue pen

25. drawing a black pen or a blue pen

26. What is $P(A \text{ and } B)$ if $P(A) = \frac{1}{6}$, $P(B) = \frac{2}{3}$, $P(B \text{ following } A) = \frac{4}{5}$, and A and B are dependent?

CHAPTER 12

Checklist

ARE YOU READY FOR THE CHAPTER TEST?

Visit **pre-alg.com** to access your textbook, more examples, self-check quizzes, and practice tests to help you study the concepts in Chapter 12.

Check the one that applies. Suggestions to help you study are given with each item.

☐ **I completed the review of all or most lessons without using my notes or asking for help.**

- You are probably ready for the Chapter Test.
- You may want take the Chapter 12 Practice Test on page 663 of your textbook as a final check.

☐ **I used my Foldable or Study Notebook to complete the review of all or most lessons.**

- You should complete the Chapter 12 Study Guide and Review on pages 658–662 of your textbook.
- If you are unsure of any concepts or skills, refer back to the specific lesson(s).
- You may also want to take the Chapter 12 Practice Test on page 663.

☐ **I asked for help from someone else to complete the review of all or most lessons.**

- You should review the examples and concepts in your Study Notebook and Chapter 12 Foldable.
- Then complete the Chapter 12 Study Guide and Review on pages 658–662 of your textbook.
- If you are unsure of any concepts or skills, refer back to the specific lesson(s).
- You may also want to take the Chapter 12 Practice Test on page 663.

Student Signature

Parent/Guardian Signature

Teacher Signature

Polynomials and Nonlinear Functions

Use the instructions below to make a Foldable to help you organize your notes as you study the chapter. You will see Foldable reminders in the margin of this Interactive Study Notebook to help you in taking notes.

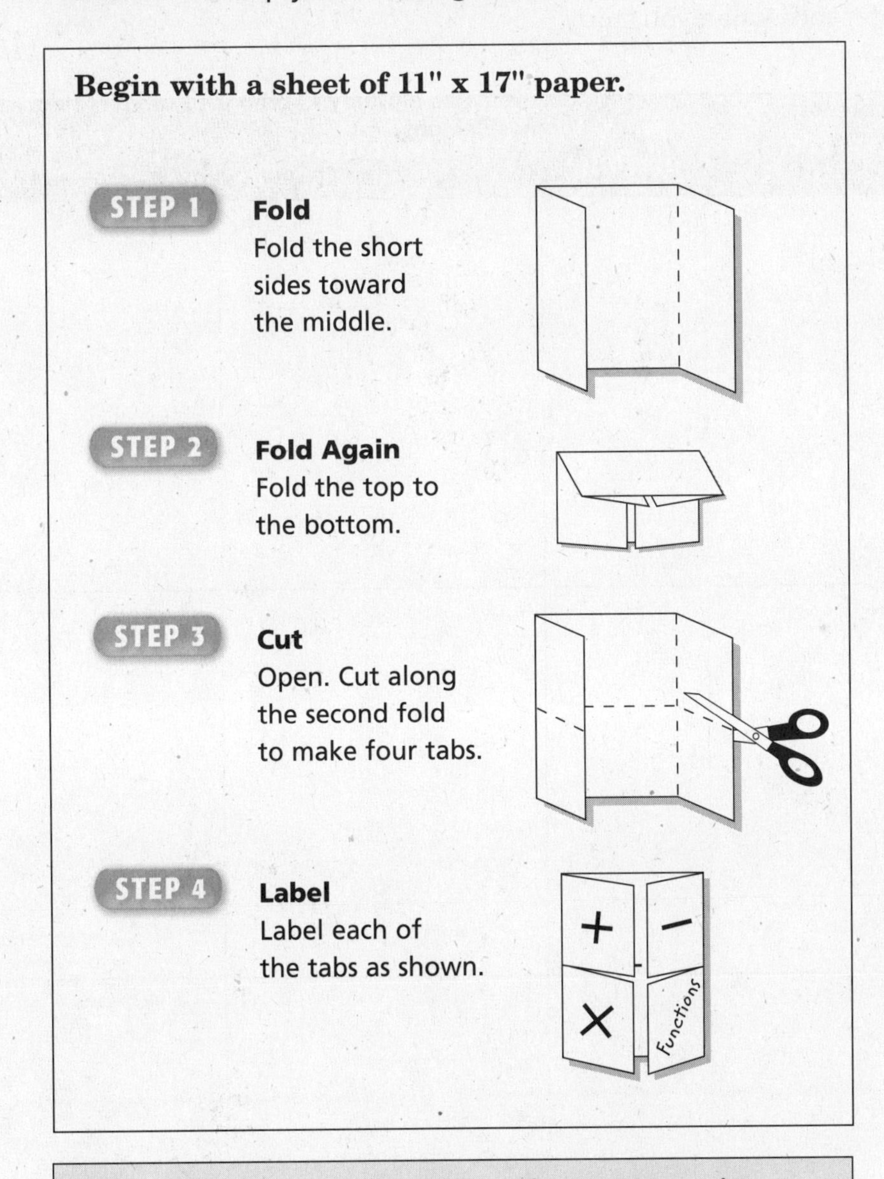

Begin with a sheet of 11" x 17" paper.

STEP 1 **Fold**
Fold the short sides toward the middle.

STEP 2 **Fold Again**
Fold the top to the bottom.

STEP 3 **Cut**
Open. Cut along the second fold to make four tabs.

STEP 4 **Label**
Label each of the tabs as shown.

NOTE-TAKING TIP: When taking notes, write clean and concise explanations. Someone who is unfamiliar with the math concepts should be able to read your explanations and learn from them.

CHAPTER 13

BUILD YOUR VOCABULARY

This is an alphabetical list of new vocabulary terms you will learn in Chapter 13. As you complete the study notes for the chapter, you will see Build Your Vocabulary reminders to complete each term's definition or description on these pages. Remember to add the textbook page number in the second column for reference when you study.

Vocabulary Term	Found on Page	Definition	Description or Example
binomial [by-NOH-mee-uhl]			
cubic function [KYOO-bihk]			
degree			

Vocabulary Term	Found on Page	Definition	Description or Example
nonlinear function			
polynomial [PAHL-uh-NOH-mee-uhl]			
quadratic function [kwah-DRAT-ink]			
trinomial [try-NOH-mee-uhl]			

13–1 Polynomials

Build Your Vocabulary (page 326–327)

An ______ that contains one or more ______ is called a **polynomial**.

A polynomial with ______ is called a **binomial**, and a polynomial with ______ is called a **trinomial**.

The **degree** of a monomial is the ______ of the ______ of its variables.

What You'll Learn

- Identify and classify polynomials.
- Find the degree of a polynomial.

Example: Classify Polynomials

1 Determine whether each expression is a polynomial. If it is, classify it as a *monomial, binomial,* or *trinomial.*

a. $\frac{-2}{x}$

The expression ______ a polynomial because $\frac{-2}{x}$ has a variable in the ______.

b. $x^2 - 12$

This ______ a polynomial because it is the difference of two ______. There are two terms, so it is a ______.

Remember It

When classifying polynomials, first write all expressions in simplest form.

Your Turn **Determine whether each expression is a polynomial. If it is, classify it as a *monomial, binomial,* or *trinomial.***

a. $x^3 + 3x^2 + 8$ ______

b. $\sqrt{x} + 5$ ______

EXAMPLE Degree of a Monomial

2 Find the degree of each monomial.

a. $-10w^4$ The variable w has degree ☐, so the degree of $-10w^4$ is ☐.

b. $8x^3y^7z$ x^3 has a degree of ☐, y^7 has a degree of ☐, and z has a degree of ☐. The degree of $8x^3y^7z$ is ☐ + ☐ + ☐ or ☐.

Your Turn **Find the degree of each monomial.**

a. $5m^3$

b. $-3ab^2c^5$

EXAMPLE Degree of a Polynomial

3 Find the degree of each polynomial.

a. $a^2b^5 - 4$

term	degree
a^2b^5	☐
4	☐

The greatest degree is ☐. So, the degree of the polynomial is ☐.

b. $2x^2y^2 + 7xy^6$

term	degree
$2x^2y^2$	☐
$7xy^6$	☐

The greatest degree is ☐. So, the degree of the polynomial is ☐.

Your Turn **Find the degree of each polynomial.**

a. $x^3y^3 + 4x^4y$

b. $-3mn^4 - 7$

HOMEWORK ASSIGNMENT

Page(s):

Exercises:

13–2 Adding Polynomials

WHAT YOU'LL LEARN

- Add polynomials.

EXAMPLE Add Polynomials

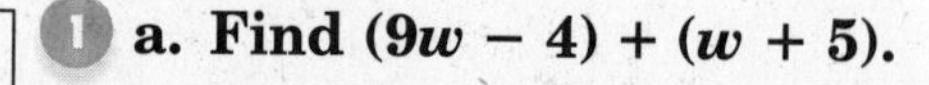

1 a. Find $(9w - 4) + (w + 5)$.

Method 1 Add vertically.

$$\begin{array}{r} 9w - 4 \\ (+) \quad w + 5 \\ \hline \end{array}$$

Align like terms.

Add.

Method 2 Add horizontally.

$(9w - 4) + (w + 5)$

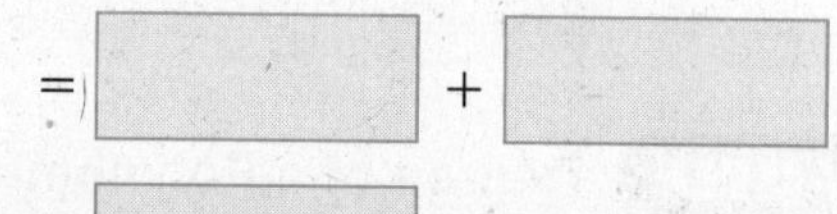

= ______ + ______ Associative and Commutative Properties

= ______

The sum is ______.

FOLDABLES™

ORGANIZE IT

Write an example of adding two polynomials with two or three terms each under the "+" tab.

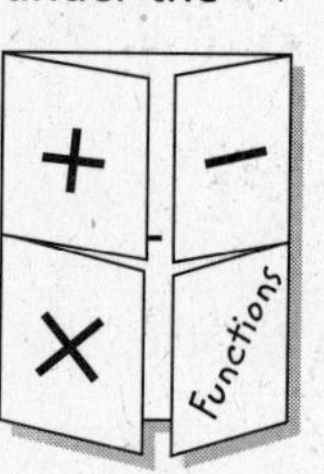

b. Find $(6x^2 - 3x + 1) + (x^2 + x - 1)$.

Method 1 Add vertically.

$$\begin{array}{r} 6x^2 - 3x + 1 \\ (+) \quad x^2 + x - 1 \\ \hline \end{array}$$

Align like terms.

Add.

Method 2 Add horizontally.

$(6x^2 - 3x + 1) + (x^2 + x - 1)$ Write the expression.

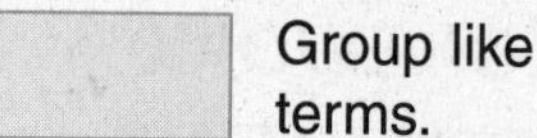

= ______ + ______ + ______ Group like terms.

= ______ Simplify.

The sum is ______.

REVIEW IT

Identify the like terms in the expression $6x + 5 + 3y - 4 - 2x$. *(Lesson 3–2)*

____

____

____

____

____

c. Find $5a^3 + (2a - 4a) + 7$.

$5a^3 + (2a - 4a) + 7$ Write the expression.

$=$ ______ Simplify.

The sum is ______.

d. Find $(4x^2 - 3y^2) + (x^2 + 4xy + y^2)$.

$$\begin{array}{rrrr} & 4x^2 & & -3y^2 \\ (+) & x^2 & +\ 4xy & +\ y^2 \\ \hline \end{array}$$

Leave a space because there is no other term like $4xy$.

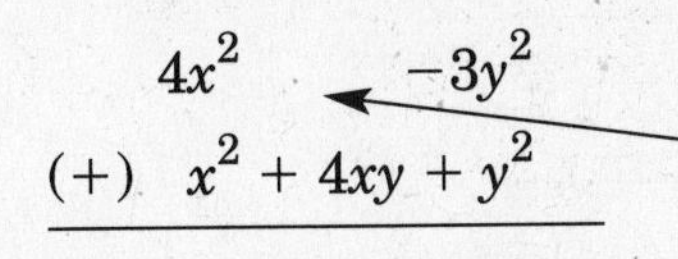

The sum is ______.

Your Turn **Find each sum.**

a. $(5b + 2) + (3b - 6)$

b. $(3m^2 - 5m + 9) + (-5m^2 + 3m - 7)$

c. $(3a^3 + 2ab - 4b^3) + (-2a^3 + 3b^3)$

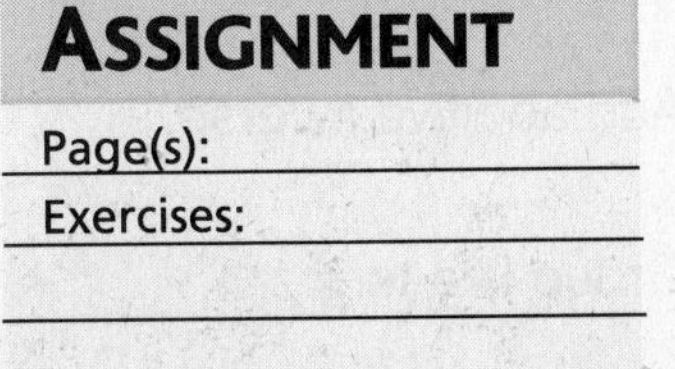

HOMEWORK ASSIGNMENT

Page(s): ______

Exercises: ______

13–3 Subtracting Polynomials

What You'll Learn

- Subtract polynomials.

EXAMPLE Subtract Polynomials

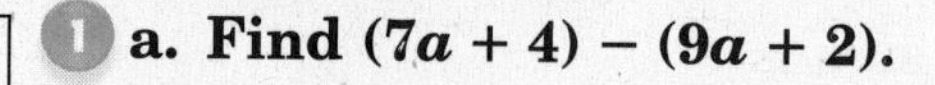

1 a. Find $(7a + 4) - (9a + 2)$.

$$\begin{array}{r} 7a + 4 \\ (-)\ 9a + 2 \\ \hline \end{array}$$

Align like terms.

Subtract.

b. Find $(8b^2 + 6) - (3b^2 + 6b + 1)$.

$$\begin{array}{r} 8b^2 \qquad\quad + 6 \\ (-)\ 3b^2 + 6b + 1 \\ \hline \end{array}$$

Align like terms.

Subtract.

Your Turn **Find each difference.**

a. $(2x + 9) - (5x - 4)$

b. $(5k^2 + 3k - 4) - (2k^2 + 1)$

Foldables

Organize It

Write an example of subtracting two polynomials with two or three terms each under the "–" tab.

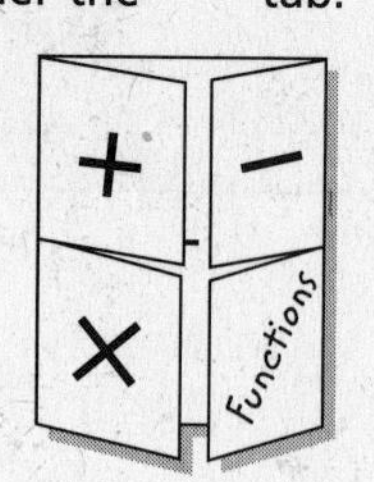

EXAMPLE Subtract Using the Additive Inverse

2 a. Find $(4x - 8) - (3x + 9)$

The additive inverse of $3x + 9$ is $(-1)(3x + 9)$ or ______.

$(4x - 8) - (3x - 9)$

$= (4x - 8) +$ ______ Add additive inverse.

$=$ ______ Group like terms.

$=$ ______ Simplify.

WRITE IT

Explain when you might use a placeholder when subtracting two polynomials.

b. Find $(7ab + 2b^2) - (3a^2 + ab + b^2)$.

The additive inverse of $3a^2 + ab + b^2$ is

$(-1)(3a^2 + ab + b^2)$ or ________.

Align the like terms and add the additive inverse.

$$\begin{array}{rl} 7ab + 2b^2 & \quad\quad 7ab + 2b^2 \\ (-)\ 3a^2 + ab + b^2 \longrightarrow & (+)\ \_\_\_\_\_\_ \\ \hline & \_\_\_\_\_\_ \end{array}$$

Your Turn **Find each difference.**

a. $(8c - 3) - (-2c + 4)$

b. $(-3xy - 4y^2) - (2x^2 - 8xy + 2y^2)$

HOMEWORK ASSIGNMENT

Page(s):

Exercises:

13–4 Multiplying a Polynomial by a Monomial

WHAT YOU'LL LEARN

- Multiply a polynomial by a monomial.

EXAMPLE Products of a Monomial and a Polynomial

1 **a. Find $-8(3x + 2)$.**

$-8(3x + 2) =$ ______ $+$ ______ Distributive Property

$=$ ______ Simplify.

b. Find $(6x - 1)(-2x)$.

$(6x - 1)(-2x) = 6x$ ______ $- 1$ ______ Distributive Property

$=$ ______ Simplify.

Your Turn **Find each product.**

a. $3(-5m - 2)$

b. $(4p - 8)(-3p)$

EXAMPLE Product of a Monomial and a Polynomial

2 **Find $4b(-a^2 + 5ab + 2b^2)$.**

$4b(-a^2 + 5ab + 2b^2)$

$=$ ______ Distributive Property

$=$ ______ Simplify.

Your Turn Find $-3x(2x^2 - 4xy + 3y^2)$.

FOLDABLES™

ORGANIZE IT

Write an example of multiplying a monomial by a polynomial with two or three terms under the "×" tab.

WRITE IT

How do you determine the degree of a polynomial?

EXAMPLE Use a Polynomial to Solve a Problem

3 FENCES **The length of a dog run is 4 feet more than three times its width. The perimeter of the dog run is 56 feet. What are the dimensions of the dog run?**

Explore You know the perimeter of the dog run. You want to find the dimensions of the dog run.

Plan Let w represent the width of the dog run. Then ______ represents the length. Write an equation.

Perimeter	equals	twice	the sum of the length and width.
P	$=$	2	$(l + w)$

Solve

$P = 2(l + w)$	Write the equation.
$56 =$ ______	$P = 56$, $\ell =$ ______
$56 =$ ______	Combine like terms.
$56 =$ ______	Distributive Property
______ $=$ ______	Subtract ______ from each side.
______ $=$ ______	Divide each side by ______.

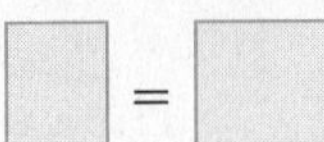

The width of the dog run is ______ and the length is ______ or ______.

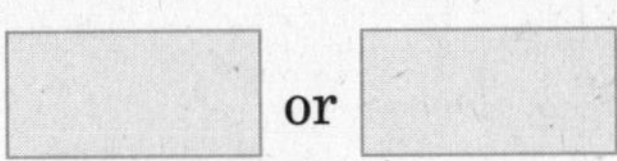

Your Turn The length of a garden is four more than twice its width. The perimeter of the garden is 44 feet. What are the dimensions of the garden?

HOMEWORK ASSIGNMENT

Page(s):

Exercises:

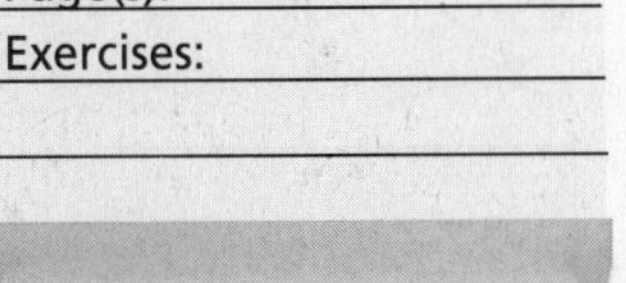

13–5 Linear and Nonlinear Functions

WHAT YOU'LL LEARN

- Determine whether a function is linear or nonlinear.

BUILD YOUR VOCABULARY (page 327)

A **nonlinear function** is a function whose graph is  a ________ line.

EXAMPLE Identify Functions Using Graphs

1 **Determine whether each graph represents a *linear* or *nonlinear* function.**

a.

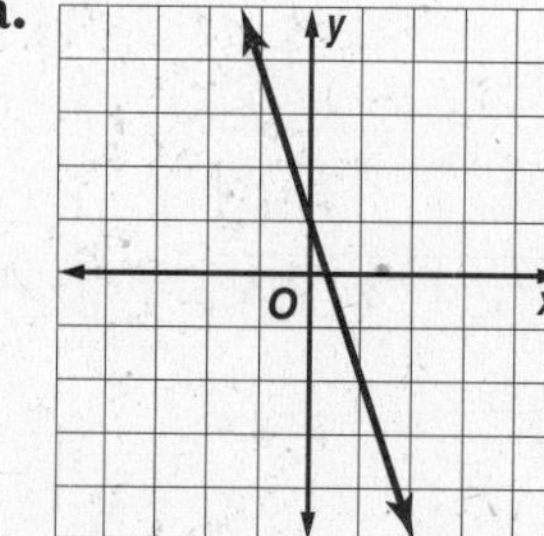

The graph is a ________ line, so it represents a ________ function.

b.

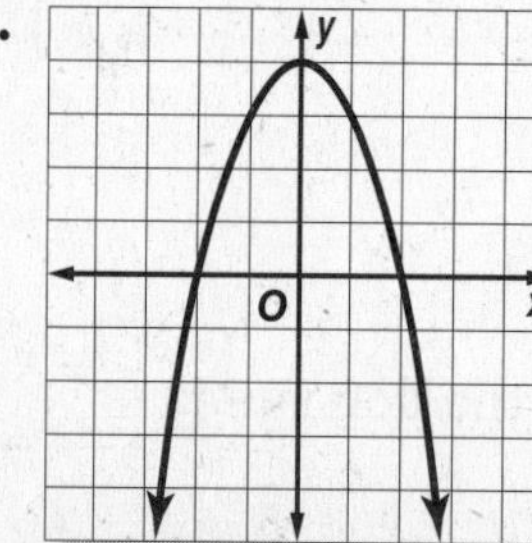

The graph is a ________, not a ________ line, so it represents a ________ function.

REVIEW IT

Why is the equation $y = 2x^2 + 3$ a function? *(Lesson 8–1)*

____

____

____

____

____

Your Turn **Determine whether each graph represents a *linear* or *nonlinear* function.**

a.

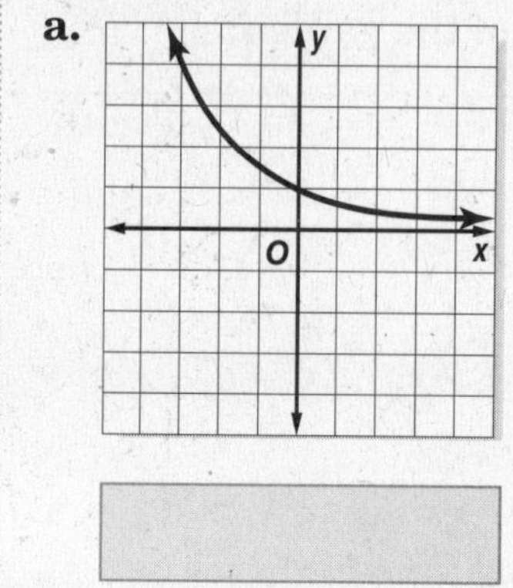

____

b.

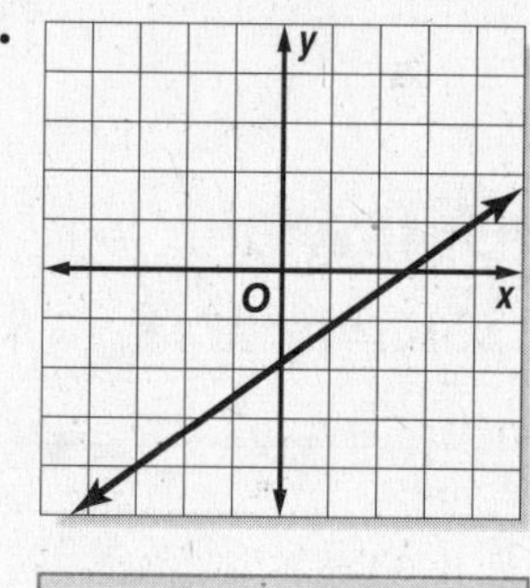

____

FOLDABLES

ORGANIZE IT

Write an example of a linear function and an example of a nonlinear function under the Functions tab.

EXAMPLE Identify Functions Using Equations

2 a. Determine whether $y = -5x - 4$ represents a *linear* or *nonlinear* function.

This equation represents a ______ function because it is written in the form ______.

b. Determine whether $y = 2x^2 + 3$ represents a *linear* or *nonlinear* function.

This equation is ______ because x is raised to the ______ and the equation cannot be written in the form ______.

Your Turn **Determine whether each equation represents a *linear* or *nonlinear* function.**

a. $y = \frac{2}{x} + 6$ ______

b. $2x + y = 4$ ______

EXAMPLE Identify Functions Using Tables

3 Determine whether each table represents a *linear* or *nonlinear* function.

a.

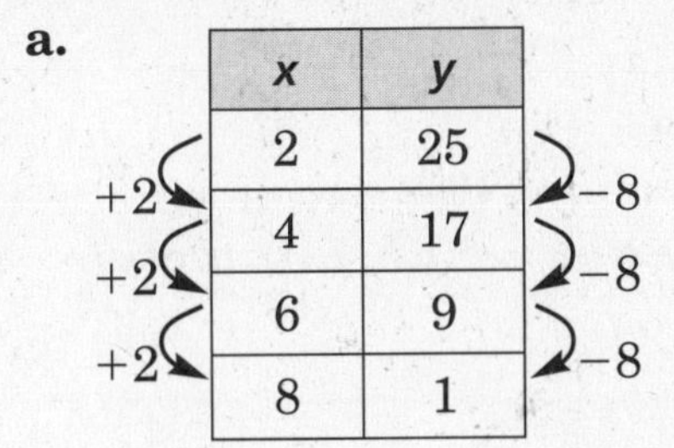

x	y
2	25
4	17
6	9
8	1

As x increases by ___, y decreases by ___. So, this is a ______ function.

b.

x	*y*
5	2
8	4
11	8
14	16

+3, +3, +3 (x); +2, +4, +8 (y)

As *x* increases by ______, *y* increases by a ______ amount each time. So, this is a ______ function.

Your Turn **Determine whether each table represents a *linear* or *nonlinear* function.**

a.

x	*y*
3	10
5	11
7	13
9	16

______

b.

x	*y*
10	4
9	7
8	10
7	13

______

HOMEWORK ASSIGNMENT

Page(s):

Exercises:

13–6 Graphing Quadratic and Cubic Functions

WHAT YOU'LL LEARN

- Graph quadratic functions
- Graph cubic functions.

EXAMPLE Graph Quadratic Functions

1 **a. Graph $y = -2x^2$.**

Make a table of values, plot the 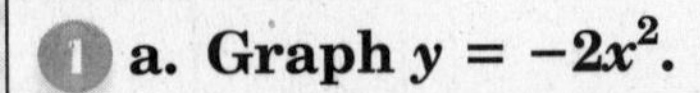, and connect the points with a ______.

x	$-2x^2$	(x, y)
−1.5	$-2(-1.5)^2 = -4.5$	
−1	$-2(-1)^2 =$	(−1, −2)
−0.5	$-2(-0.5)^2 =$	(−0.5, −0.5)
0	$-2(0)^2 =$	
0.5	$-2(0.5)^2 = (-0.5)$	
1	$-2(1)^2 =$	(1, −2)
1.5	$-2(1.5)^2 = -4.5$	

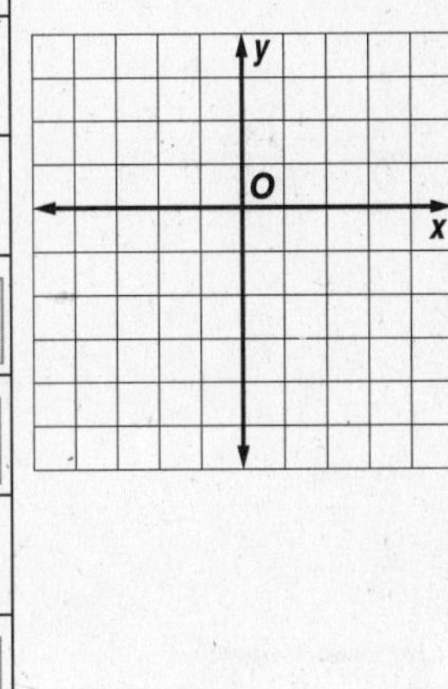

REMEMBER IT When substituting values for x in a function, consider using decimal values if necessary to find points that are closer together.

b. Graph $y = \frac{x^2}{2} + 1$.

Make a table of values, plot the ordered pairs, and connect the points with a curve.

x	$\frac{x^2}{2} + 1$	(x, y)
−2	$\frac{(-2)^2}{2} + 1 =$	
−1	$\frac{(-1)^2}{2} + 1 =$	(−1, 1.5)
0	$\frac{(0)^2}{2} + 1 = 1$	
1	$\frac{(1)^2}{2} + 1 = 1.5$	
2	$\frac{(2)^2}{2} + 1 =$	(2, 3)

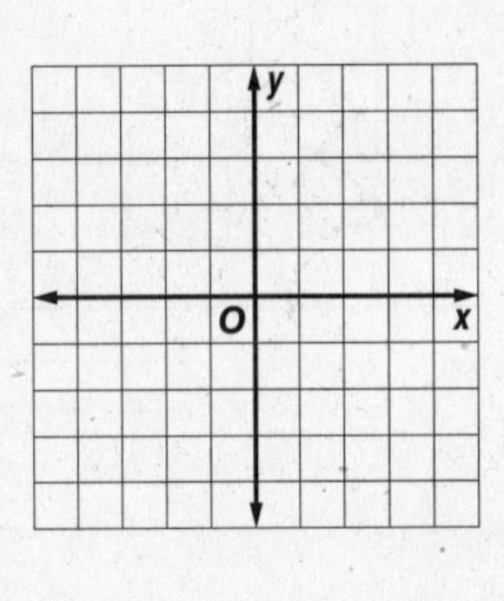

Your Turn **Graph each function.**

a. $y = -x^2$

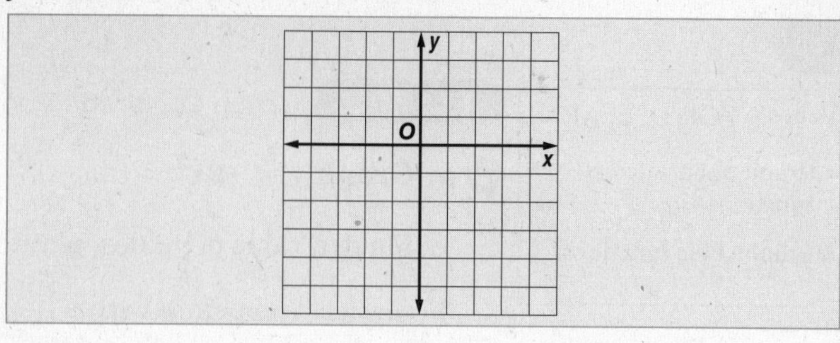

b. $y = 3x^2 - 8$

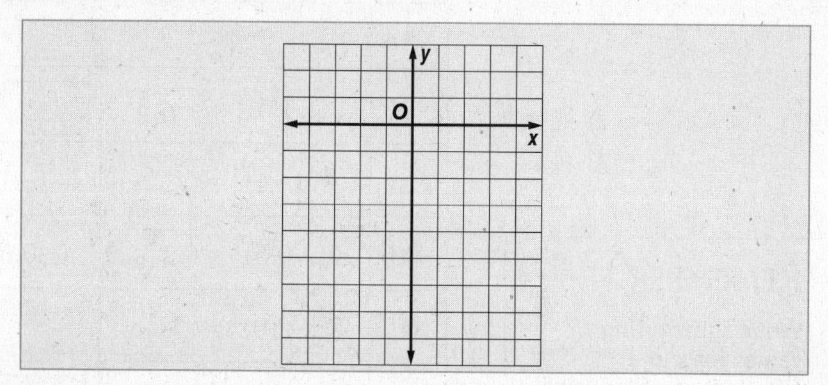

EXAMPLE Graph Cubic Functions

2 Graph $y = \frac{-x^3}{2}$.

x	$-\frac{x^3}{2}$	(x, y)
-2	$-\frac{(-2)^3}{2} = 4$	$(-2, 4)$
-1	$-\frac{(-1)^3}{2} = 0.5$	
0	$-\frac{(0)^3}{2} =$	
1	$-\frac{(1)^3}{2} = -0.5$	$(1, -0.5)$
2	$-\frac{(2)^3}{2} =$	

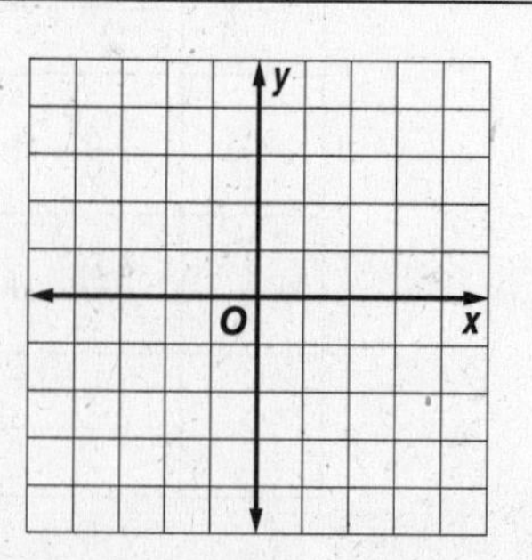

FOLDABLES™

ORGANIZE IT

Under the Functions tab, write an example of a quadratic function and an example of a cubic function. Then graph each function.

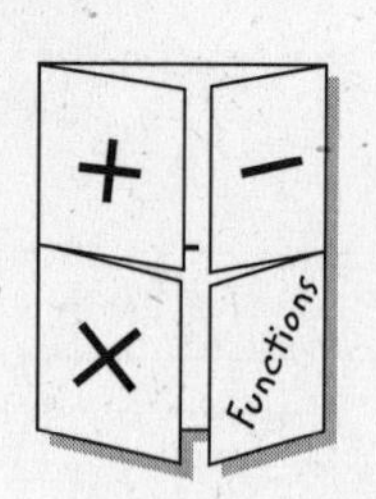

Your Turn Graph $y = x^3 - 3$.

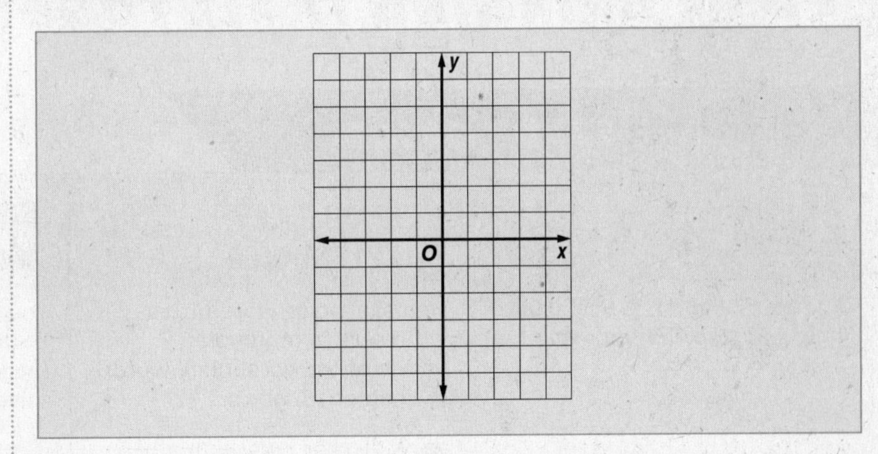

HOMEWORK ASSIGNMENT

Page(s):

Exercises:

CHAPTER 13

BRINGING IT ALL TOGETHER

STUDY GUIDE

FOLDABLES	VOCABULARY PUZZLEMAKER	BUILD YOUR VOCABULARY
Use your **Chapter 13 Foldable** to help you study for your chapter test.	To make a crossword puzzle, word search, or jumble puzzle of the vocabulary words in Chapter 13, go to: www.glencoe.com/sec/math/t_resources/free/index.php	You can use your completed **Vocabulary Builder** (pages 326–327) to help you solve the puzzle.

13-1 Polynomials

Determine whether each expression is a polynomial. If it is, classify it as a *monomial, binomial,* or *trinomial*.

1. $5m - 3$
2. $\frac{5}{c} + c^2$
3. $7 - 3y - 4y^3$

Find the degree of each polynomial.

4. pq
5. 144
6. $x^4 + x - 5$

13-2 Adding Polynomials

Find each sum.

7. $(4y - 17) + (2y + 3)$
8. $(9b^2 + 4b - 15) + (-3b^2 + 8)$

13-3 Subtracting Polynomials

Find each difference.

9. $(6x + 11y) - (10x - 2y)$
10. $\begin{array}{r} x^2 + 9xy - 12y \\ (-) \quad 2xy - y \\ \hline \end{array}$

13-4 Multiplying a Polynomial by a Monomial

Find each product.

11. $4(3q - 2)$

12. $(3y + 8)x$

13. $7a(2a^2 - 3b)$

13-5 Linear and Nonlinear Functions

Determine whether each graph, equation, or table represents a *linear* or *nonlinear* function. Explain.

14.

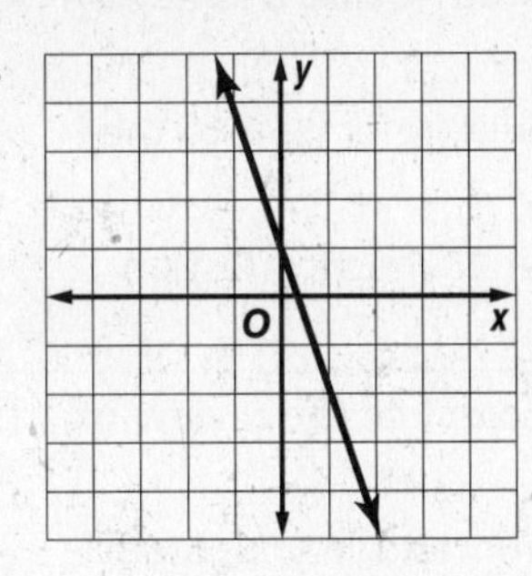

15.

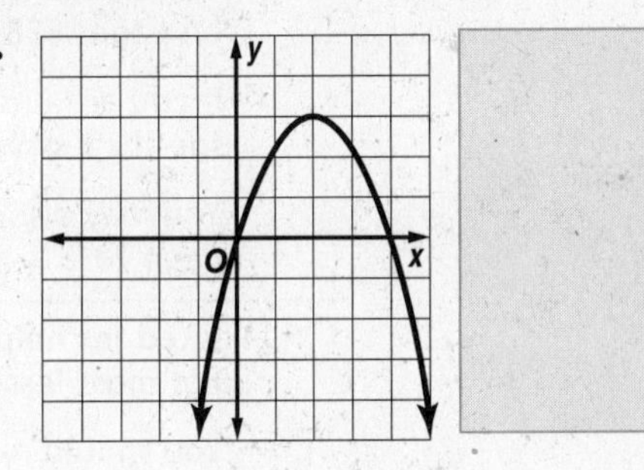

16. $y = \frac{5}{x} + 3$

17.

x	y
−3	9
−1	1
0	0
1	1

18.

x	y
−12	−3
−10	−2
−8	−1
−6	0

13-6 Graphing Quadratic and Cubic Functions

Graph each function.

19. $y = -2x^2 + 4$

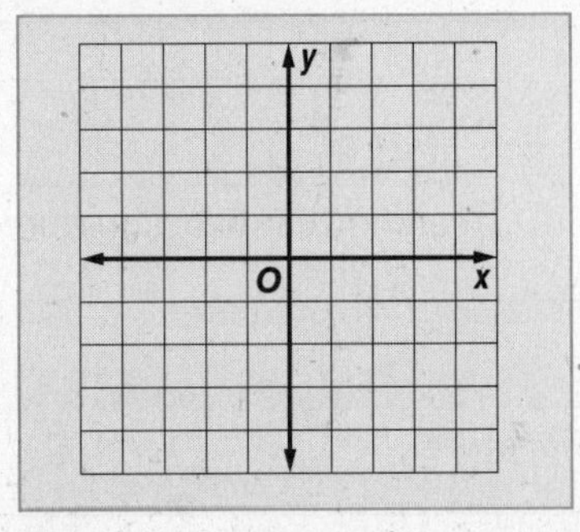

20. $y = 0.5x^3 - 2$

CHAPTER 13

Checklist

ARE YOU READY FOR THE CHAPTER TEST?

Visit **pre-alg.com** to access your textbook, more examples, self-check quizzes, and practice tests to help you study the concepts in Chapter 13.

Check the one that applies. Suggestions to help you study are given with each item.

☐ **I completed the review of all or most lessons without using my notes or asking for help.**

- You are probably ready for the Chapter Test.
- You may want take the Chapter 13 Practice Test on page 701 of your textbook as a final check.

☐ **I used my Foldable or Study Notebook to complete the review of all or most lessons.**

- You should complete the Chapter 13 Study Guide and Review on pages 698–700 of your textbook.
- If you are unsure of any concepts or skills, refer back to the specific lesson(s).
- You may also want to take the Chapter 13 Practice Test on page 701.

☐ **I asked for help from someone else to complete the review of all or most lessons.**

- You should review the examples and concepts in your Study Notebook and Chapter 13 Foldable.
- Then complete the Chapter 13 Study Guide and Review on pages 698–700 of your textbook.
- If you are unsure of any concepts or skills, refer back to the specific lesson(s).
- You may also want to take the Chapter 13 Practice Test on page 701.

Student Signature

Parent/Guardian Signature

Teacher Signature